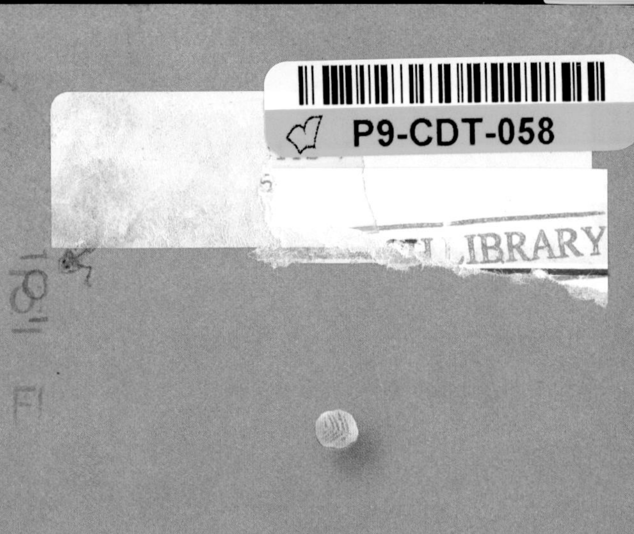

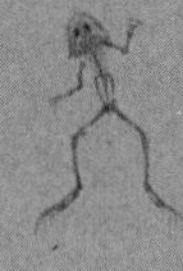

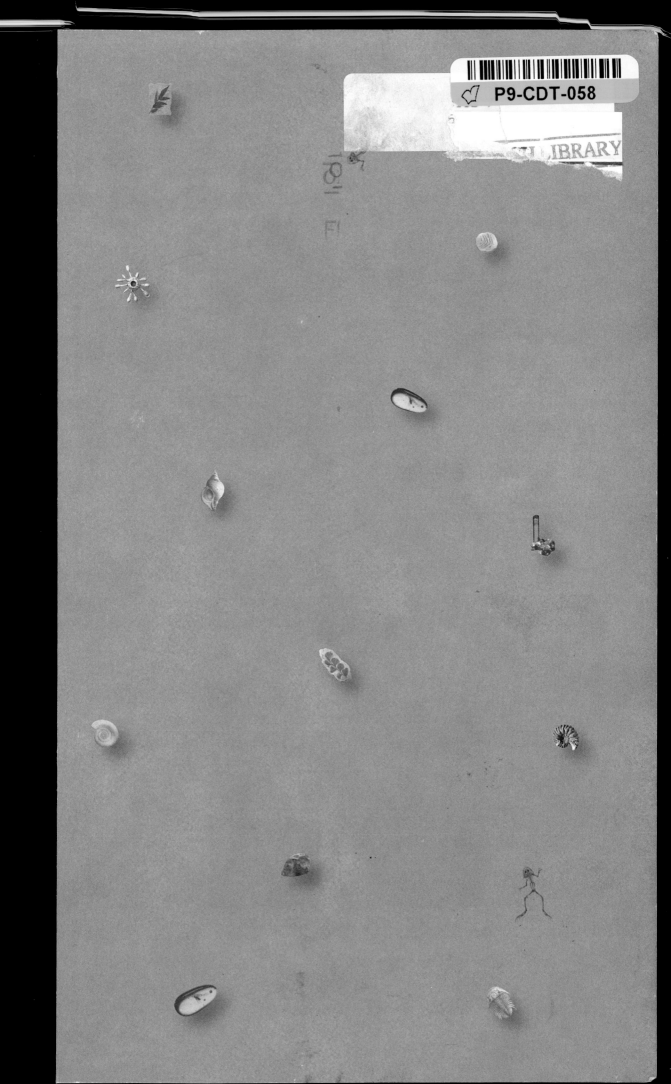

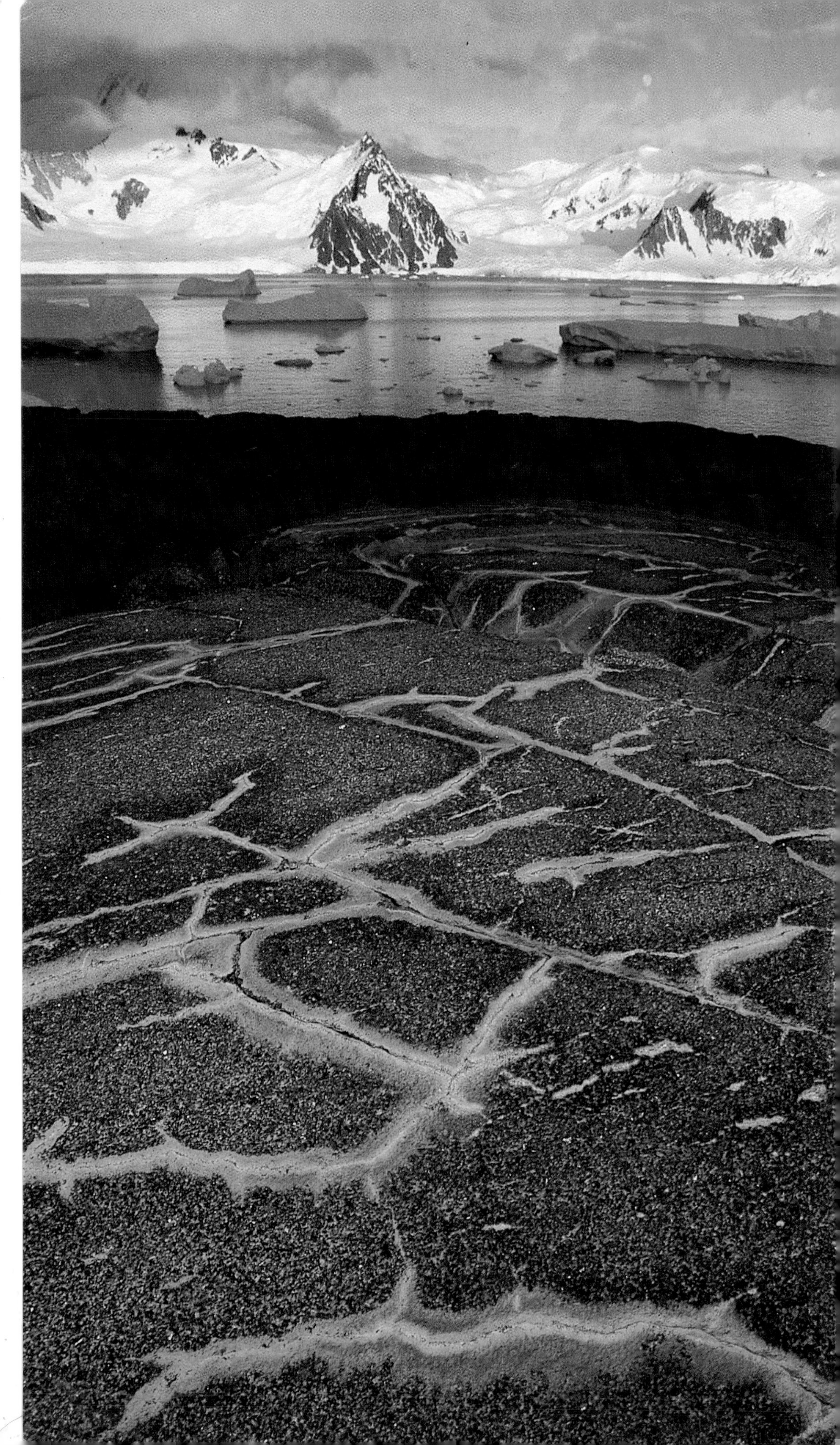

The Nature Company Guides

Rocks & Fossils

Arthur B. Busbey III, Robert R. Coenraads, David Roots, Paul Willis

Consultant Editors

David Roots and Paul Willis

The Nature Company Guides are published by Time-Life Books

Conceived and produced by Weldon Owen Pty Limited
43 Victoria Street, McMahons Point, NSW, 2060, Australia
A member of the Weldon Owen Group of Companies
Sydney • San Francisco • London

The Nature Company owes its vision to the world's great naturalists: Charles Darwin, Henry David Thoreau, John Muir, David Brower, Rachel Carson, Jacques Cousteau, and many others. Through their inspiration, we are dedicated to providing products and experiences which encourage the joyous observation, understanding, and appreciation of nature. We do not advocate, and will not allow to be sold in our stores, any products that result from the killing of wild animals for trophy purposes. Seashells, butterflies, furs, and mounted animal specimens fall into this category. Our goal is to provide you with products, insights, and experiences which kindle your own sense of wonder and help you to feel good about the world in which you live.

For a copy of The Nature Company mail-order catalog, or to learn the location of the store nearest you, please call 1-800-227-1114.

THE NATURE COMPANY
Priscilla Wrubel, Ed Strobin, Steve Manning,
Georganne Papac, Tracy Fortini

TIME LIFE CUSTOM PUBLISHING
Time-Life Books is a division of Time Life Inc.
Time-Life is a trademark of Time Warner Inc. U.S.A.

VICE PRESIDENT AND PUBLISHER: Terry Newell
EDITORIAL DIRECTOR: Donia A. Steele
DIRECTOR OF NEW PRODUCT DEVELOPMENT: Regina Hall
DIRECTOR OF SALES: Neil Levin
DIRECTOR OF CUSTOM PUBLISHING: Frances C. Mangan
DIRECTOR OF FINANCIAL OPERATIONS: J. Brian Birky

THE NATURE COMPANY GUIDES
PUBLISHER: Sheena Coupe
MANAGING EDITOR: Lynn Humphries
PROJECT EDITOR: Fiona Doig
ASSISTANT EDITOR: Greg Hassall
COPY EDITOR: Lynn Cole
EDITORIAL ASSISTANTS: Louise Bloxham, Edan Corkill, Vesna Radojcic
ART DIRECTOR: Hilda Mendham
DESIGNERS: Stephanie Cannon, Clive Collins, Lena Lowe, Mark Nichols
JACKET DESIGN: John Bull
PICTURE RESEARCH: Joanna Collard
ILLUSTRATIONS: Anne Bowman, Chris Forsey, Michael Saunders,
Mark Watson, Pictogram, Rod Westblade
PRODUCTION MANAGER: Simone Perryman
VICE PRESIDENT INTERNATIONAL SALES: Stuart Laurence
COEDITIONS DIRECTOR: Derek Barton

Library of Congress Cataloging-in-Publication Data
Rocks & Fossils/Arthur B. Busbey III ... [et al.],
p. cm. — (The Nature Company guides)
Includes index.
ISBN 0-7835-4803-6
1. Geology—Popular works. 2. Rocks—Collection and preservation.
3. Fossils—Collection and preservation.
I. Busbey, Arthur Bresnahan, 1953–. II. Series: Nature Company guide.
QE31,R66 1996 95-47661
550—dc20 CIP

Manufactured by Kyodo Printing Co. (S'pore) Pte Ltd
Printed in Singapore

A Weldon Owen Production

Observe always that everything is the result of change, and get used to thinking that there is nothing Nature loves so well as to change existing forms and to make new ones like them.

Meditations, MARCUS AURELIUS (121–180),
Roman emperor devoted to stoic philosophy

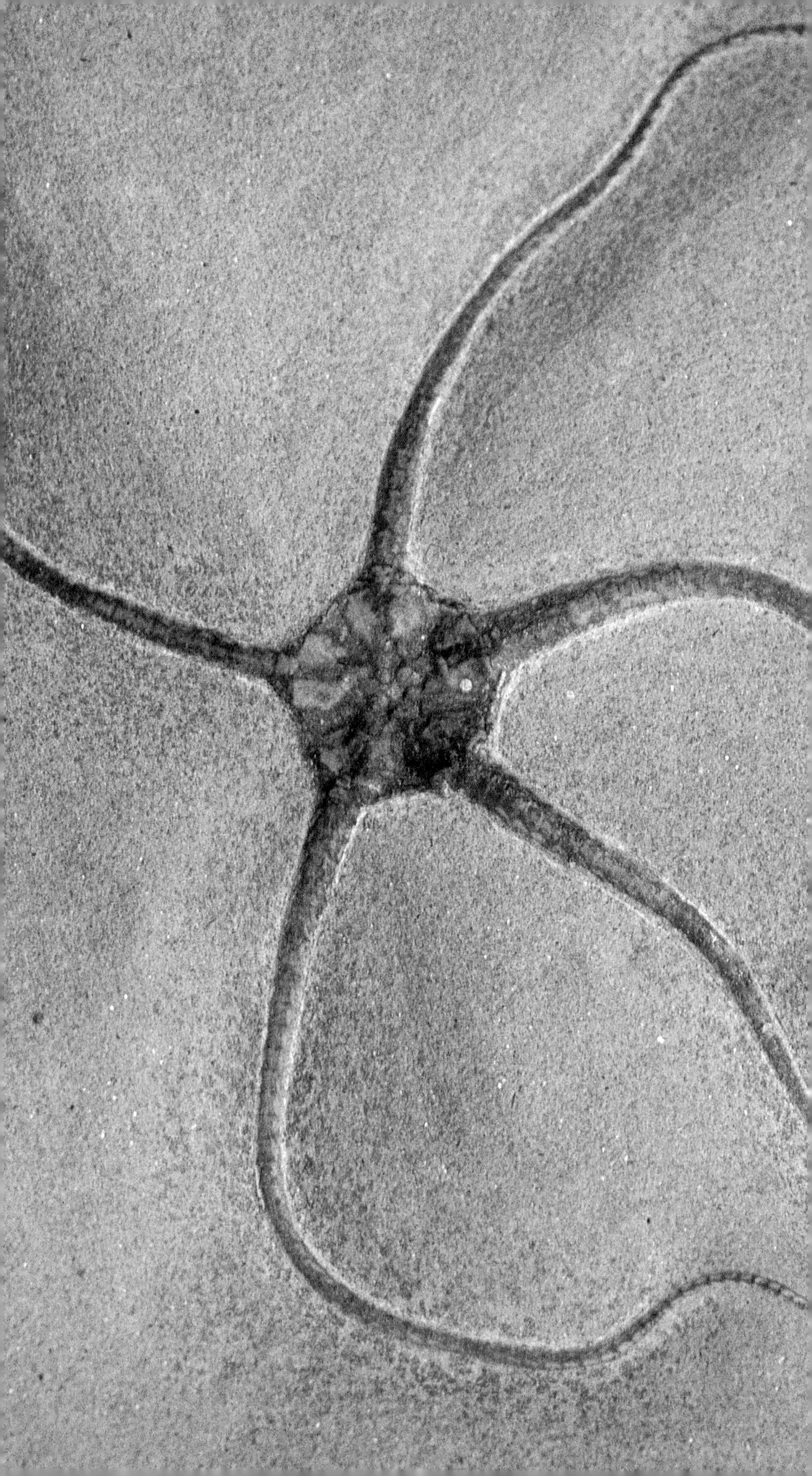

Contents

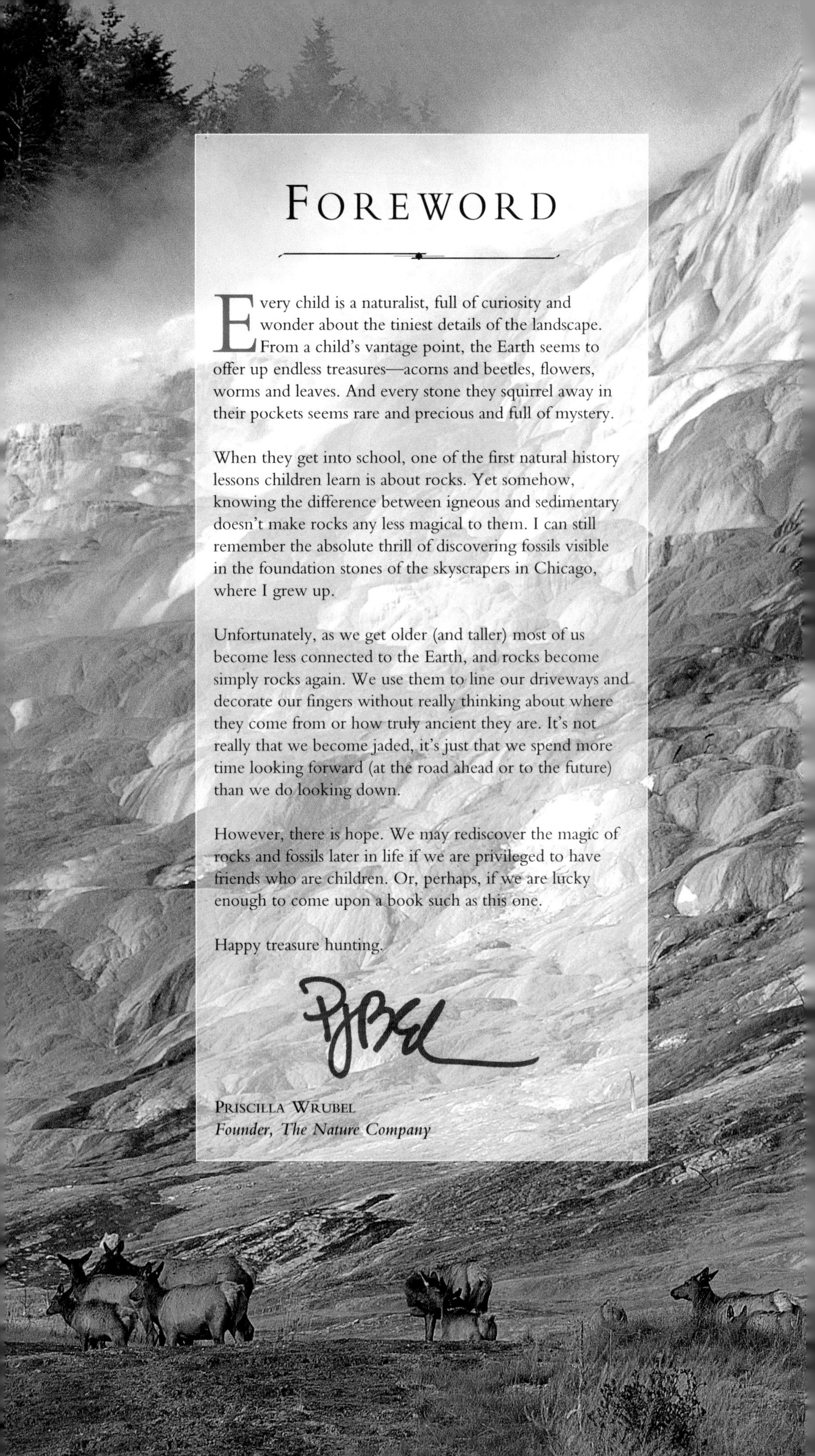

FOREWORD

Every child is a naturalist, full of curiosity and wonder about the tiniest details of the landscape. From a child's vantage point, the Earth seems to offer up endless treasures—acorns and beetles, flowers, worms and leaves. And every stone they squirrel away in their pockets seems rare and precious and full of mystery.

When they get into school, one of the first natural history lessons children learn is about rocks. Yet somehow, knowing the difference between igneous and sedimentary doesn't make rocks any less magical to them. I can still remember the absolute thrill of discovering fossils visible in the foundation stones of the skyscrapers in Chicago, where I grew up.

Unfortunately, as we get older (and taller) most of us become less connected to the Earth, and rocks become simply rocks again. We use them to line our driveways and decorate our fingers without really thinking about where they come from or how truly ancient they are. It's not really that we become jaded, it's just that we spend more time looking forward (at the road ahead or to the future) than we do looking down.

However, there is hope. We may rediscover the magic of rocks and fossils later in life if we are privileged to have friends who are children. Or, perhaps, if we are lucky enough to come upon a book such as this one.

Happy treasure hunting.

PRISCILLA WRUBEL
Founder, The Nature Company

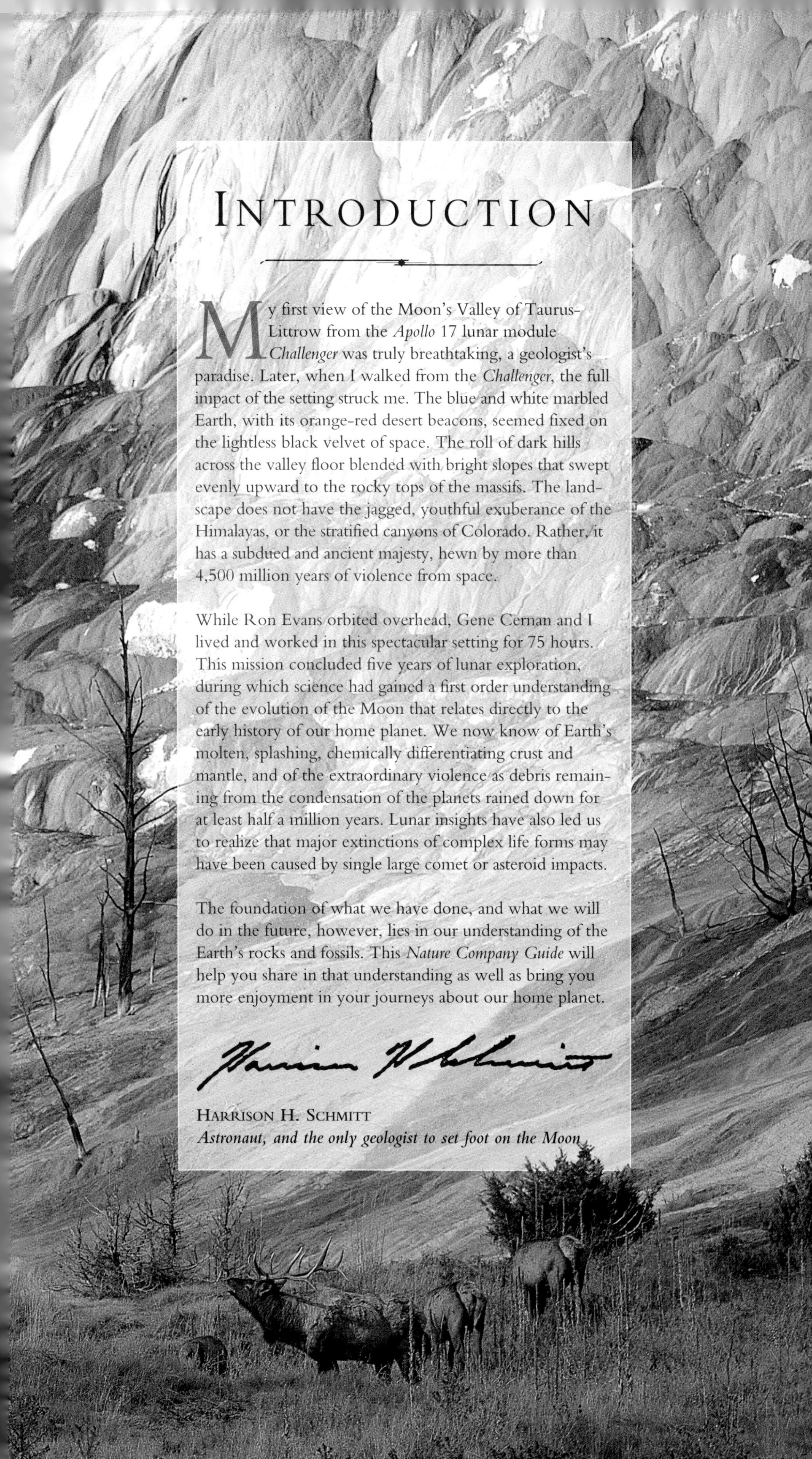

INTRODUCTION

My first view of the Moon's Valley of Taurus-Littrow from the *Apollo* 17 lunar module *Challenger* was truly breathtaking, a geologist's paradise. Later, when I walked from the *Challenger*, the full impact of the setting struck me. The blue and white marbled Earth, with its orange-red desert beacons, seemed fixed on the lightless black velvet of space. The roll of dark hills across the valley floor blended with bright slopes that swept evenly upward to the rocky tops of the massifs. The landscape does not have the jagged, youthful exuberance of the Himalayas, or the stratified canyons of Colorado. Rather, it has a subdued and ancient majesty, hewn by more than 4,500 million years of violence from space.

While Ron Evans orbited overhead, Gene Cernan and I lived and worked in this spectacular setting for 75 hours. This mission concluded five years of lunar exploration, during which science had gained a first order understanding of the evolution of the Moon that relates directly to the early history of our home planet. We now know of Earth's molten, splashing, chemically differentiating crust and mantle, and of the extraordinary violence as debris remaining from the condensation of the planets rained down for at least half a million years. Lunar insights have also led us to realize that major extinctions of complex life forms may have been caused by single large comet or asteroid impacts.

The foundation of what we have done, and what we will do in the future, however, lies in our understanding of the Earth's rocks and fossils. This *Nature Company Guide* will help you share in that understanding as well as bring you more enjoyment in your journeys about our home planet.

HARRISON H. SCHMITT
Astronaut, and the only geologist to set foot on the Moon

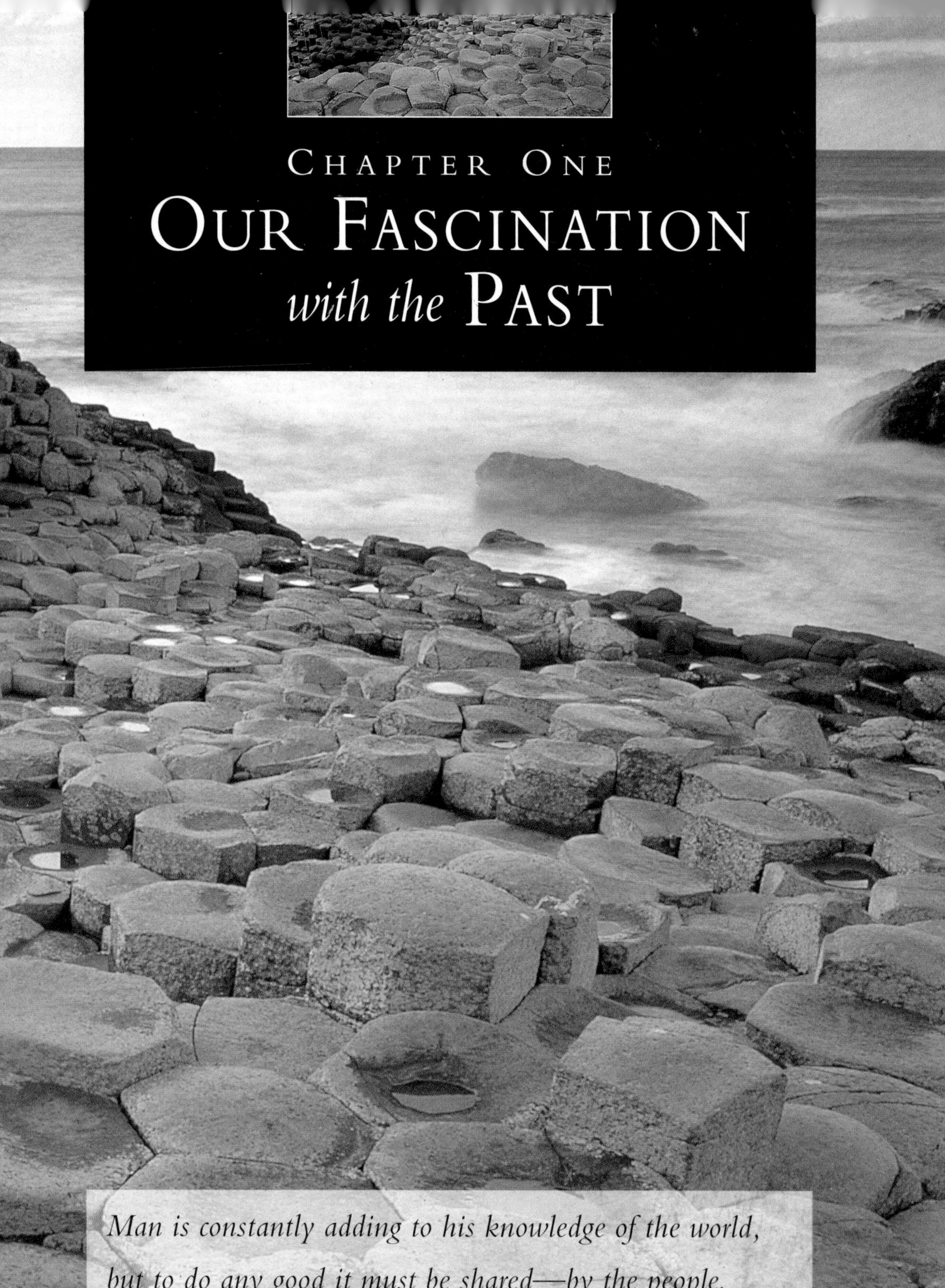

Chapter One
Our Fascination *with the* Past

Man is constantly adding to his knowledge of the world, but to do any good it must be shared—by the people.

Alistair Cooke (b. 1908),
British-born American radio and television commentator

UNRAVELING *the* PAST

Rocks are the pages of Earth's diary, and fossils are the words on the pages, recounting surprising details of the history of life.

The Earth and the rocks from which it is made have always been a benchmark for stability: we use such expressions as "solid as a rock", and "firm as the Earth beneath our feet". Rocks are the foundation upon which civilizations are built. Earth is a quarry from which we extract the minerals we need, but science views Earth and its rocks in a very different light. To a geologist, rocks are the chronicle of Earth's history. The Earth is far from stable: it is a dynamic place, constantly changing, moving, and being dramatically rearranged.

Geology not only reveals the immense history of the Earth and explains its myriad geological formations, it also gives us a new perspective on our place in Nature. From a geological viewpoint, the whole of humanity is little more than an extremely recent flash in the pan.

While geology is the science of rocks, its sister discipline, paleontology, deals with fossils, the remains of ancient organisms that have been turned to stone. While geology has demonstrated our insignificance with respect to the history of the Earth, paleontology has spelled out our peripheral importance in the history of life. Not only is humanity a recent arrival, it is only one species among many millions that share a common heritage. Humans are of the Earth and are only a small part of life.

EXQUISITE CRYSTALS *in a cave (left) delight all who see them. Beautifully worked pieces (far top left) can have great artistic value.*

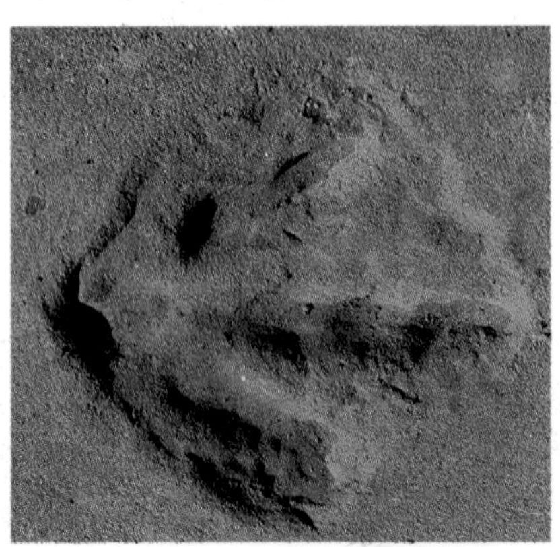

ANCIENT GIANTS *have left their mark. This dinosaur footprint (left) was found near Broome in Western Australia.*

AMATEUR ACTIVITIES

With an understanding of geology, one can reconstruct an ancient shoreline or delve into the heart of an extinct volcano. With an appreciation of paleontology, and armed with little more than a hammer and chisel, the amateur can confront the largest animals of all time. For those with minimal equipment, paleontology and geology remain the most accessible of the sciences. Indeed, many breakthrough discoveries in both disciplines are made by interested amateurs.

The great thrill of finding a rare and beautiful mineral or fossil specimen is combined with a profoundly humbling respect that comes from understanding how the material came to be formed. There is immense satisfaction in being able to read a multi-million-year-old rock formation as easily as you would read a newspaper.

There are many levels at which the hobbyist can enjoy geology and paleontology. Some people are satisfied

THE DELIGHT *of an amateur collector is to marvel at amazing landscapes, such as Mesa Arch, Utah, USA, seen above at dawn. Some hobbyists specialize in particular groups of fossils, such as the intriguing pearlescent ammonites (top right).*

simply with knowledge of the geological processes that have shaped the world we inhabit. Others collect specimens, perhaps of a particular type of mineral or group of fossils. Some people enjoy fieldwork and organize complex collecting expeditions to exotic places. Others prefer to clean specimens at home, or to cut and facet minerals into items of jewelry and other ornaments. A few are so enthused by their hobby that they become "professional amateurs" and make a career of it. The sciences of geology and paleontology are filled with practitioners who started as keen amateur collectors and later dedicated their lives to the science of their hobby.

Both geology and paleontology can be environmentally beneficial activities. Scouring geological formations in search of specimens is a most enjoyable form of exercise. The conscientious collector disturbs each area as little as possible and removes only the minimum of specimens. A deepening understanding of the history of the Earth and of life upon it comes with the pursuit of geology and paleontology as a hobby—it broadens the individual's perspective on human values and our place in Nature. If we recognize our position within the natural scheme, concern for and appreciation of the environment will follow.

THE AMATEUR'S CONTRIBUTION

The participation of hobbyists has brought rewards to both paleontology and geology. A classic instance is the find of William Walker, an English plumber and amateur collector. In 1983, while looking for fossils in a quarry in the south of England, he unearthed a fearsome-looking claw more than 12 inches (30 cm) long. When Walker showed it to staff of The British Museum (Natural History), they were rather excited—nothing like this claw had ever been seen before.

Some two tons of rock was collected from the site by staff of the museum. From it, they recovered a relatively complete skeleton of a totally new type of meat-eating dinosaur. It was named *Baryonyx walkeri* in honor and recognition of the keen hobbyist (shown below with his prize) who had found it.

DAWNING *of* CURIOSITY

Belief in Creation concepts preceded the development of inquiry into the nature of rocks and the formation of the Earth.

Early understandings of the Earth and its history were usually rooted in and intertwined with prevailing religious concepts that a god or gods had created the world in the form we know some time in the past. Variations on this theme were the ancestor myths of particular cultures. These held that certain features in the landscape had been created by specific actions of mythical ancestral beings. Native Americans, for example, hold many such creation myths, and Australian Aborigines believe that all landscape features, as well as animals and plants, were created in a "Dreamtime" by the activities of an array of totemic ancestral spirits. Beliefs such as these preceded the development of inquiry into the essential nature of rocks and the formation of Earth. The landscape was as God or the ancestors had created it and there the matter ended.

ACTS OF NATURE, *such as volcanoes (above), were seen as punishment from a deity. Ammonites (right) were named after the Egyptian god Ammon. The Rainbow Serpent (top left) appears in the creation myths of Australian Aborigines.*

Such religious ideas of Creation persisted for a very long time. In 1731, Johann Scheuchzer (1672–1733), a Swiss philosopher, published an engraving of a fossil he had found in 1726. He called it *Homo diluvii testis* (man who witnessed the biblical flood), and described it as "the bony skeleton of one of those infamous men whose sins had brought upon the world the dire misfortune of the deluge". This specimen was later shown to be an eight-million-year-old salamander.

DOWSING, *being demonstrated (left) in Paris in 1913, was used to locate mineral deposits.*

Even today, many groups believe in religious creation myths at the expense of accepting the scientific understanding of the origin and history of Earth. In such cases, the adherence to a creation myth is usually associated with restricted religious values rather than a rejection of the science-based world view.

SUPERSTITIONS

Associated with religious understandings of Earth were a number of superstitions of varying origins. Mineral deposits were not located by

AMBER *(below) was thought to cure numerous ailments, and fossils (left) were believed to have grown inside the Earth.*

To be surprised, to wonder, is to begin to understand.

The Revolt of the Masses, JOSE ORTEGA Y GASSET (1883–1955), Spanish philosopher

SEA URCHINS *(left) were thought to be magical snake eggs. The philosopher Pliny noted that they were greatly prized by the Druids.*

their geological settings but by dowsing, a pseudoscientific method by which an experienced practitioner located hidden materials by holding a divining rod that apparently responded to the attraction of the deposit. Some people still use dowsing as a way of locating underground water, but in the past, it was used for detecting an array of mineral deposits, including copper, lead, and gold. There is no scientific support for this method—any success in locating materials by dowsing is usually related to a known abundance of the material in that particular area.

Fossils also provided a long and contentious history as to their origin, significance, and use. While some of the philosophers of ancient Greece held amazingly accurate insights into the origin and formation of fossils, they were very much in the minority. For most of the past 2,500 years, fossils were shrouded in mystery. A few philosophers of medieval times thought that fossils were the remains of ancient animals and plants that had been turned to stone.

A more widespread theory was that fossils grew inside the Earth as a result of certain mystical forces flowing through rocks. Some fossils were thought to have magical powers, and others attracted names such as "devil's toenails" and "serpent stones", betraying the associations that these objects had for prescientific peoples.

DEVIL'S TOENAILS AND SERPENT STONES

A minority of prescientific philosophers believed that fossils were the remains of ancient organisms. A more popular philosophy was that fossils had grown within the Earth and were organisms that had not had a chance to "germinate" properly on the surface where they belonged.

Lay people needed to identify fossils with living or mythical creatures familiar to them. The large, curled shells of the bivalve *Gryphaea* were common fossils in many parts of Europe. These robust shells were often found on the surface, having eroded out of the rocks. At a loss to explain what they were, people called them devil's toenails because they resembled the horny cuticle from the hoof of a goat. In Yorkshire, England, the beautifully coiled shells of Jurassic ammonites were thought to be snakes that St Hilda had turned to stone. Heads were sometimes carved on the ends of them by local craftsmen.

A Scientific Viewpoint

It is only during the past 200 years or so that there has been scientific consideration of the Earth and its rocks and fossils.

The work of surveyor and geologist William Smith (1769–1839) marked the start of a period of enlightenment in England. He mapped coal mines and worked on the construction of an expanding national network of canals, needed to support the industrial revolution in the late eighteenth century. In the process, Smith noticed that the same layers of sedimentary rocks were revealed in cuttings over large areas. He also discovered that the order in which rock units were laid down never varied across their geographic extent.

These observations signaled a level of organization in the formation of rocks that had not previously been recognized, and logic decreed that the rock units at the bottom of these stacks or columns were older than those above. Smith was able to test his theories by accurately predicting the geology of places he had never been.

LAVA FLOWS *(above) and erosion (right) are ways the Earth recycles rocks. James Hutton (top left) devised his theory of recycling after observing geological processes.*

Another breakthrough in the development of geology as a modern science came with the work of a Scottish geologist, James Hutton (1726–97). Hutton contributed the theory that the history of rocks occurs in cycles. Rocks are broken down to sediment by the process of weathering. The sediment is moved by the forces of erosion and mass transportation to accumulate in a new place. The sediment consolidates to a new type of rock, which could be buried under more rock until, heated to its melting point, it flowed back to the surface as lava. Lava cools into rock that can be broken down by weathering, and the whole cycle starts over again.

COAL MINES *were mapped in England by William Smith, who noticed the similarity in the order of the rock layers.*

Combined with Smith's observations, the implications of Hutton's theory were profound. Prior to this, the Earth was thought to have had a definite beginning, being created in the form we know at a comparatively recent time. The work of Hutton and Smith raised the possibility that the Earth was far older and was continually changing and recycling itself.

The father of modern geology is the English scientist Sir Charles Lyell (1797–1875). His main contribution to the young science of geology was the introduction of the concept of "uniformitarianism".

ACCORDING TO *Charles Lyell (below), geological processes proceed at the rate they always have. Alfred Wegener (left) devised the complex theory of Continental Drift.*

This proposed that past events occurred at the same rate as they do today. The accumulations of deep units of sedimentary rock were formed by the same processes of sediment deposition, and at the same rate, as those that can be observed in the modern world. Since most geological processes were shown to be very slow, and the rock units that were formed by these processes were often very large, Lyell concluded that the Earth was extremely old. He was the first person to talk of its age in millions of years.

Today's Thinking

The application and refinement of these basic principles have generated most of the science of geology during the past 200 years. The other major theory that has so dramatically reshaped our understanding of Earth's history is the theory of Plate Tectonics. In 1912, German meteorologist Alfred Wegener (1880–1930) published *The Origin of Continents and Oceans*, in which he proposed his theory of continental drift. This held that the continents were not fixed in position but that they slowly moved across the surface of the Earth. His main evidence for this was the apparent fit of some continental margins—for example, the east coast of South America and the west coast of Africa seemed an almost perfect match. Wegener could not, however, propose a suitable mechanism by which the continents moved, so his ideas were largely ridiculed and ignored.

During the early part of the Cold War in the '50s, both the USA and Russia set up sensitive vibration detectors to monitor atomic tests conducted by the opposing side. They also logged thousands of earthquakes, and nearly all occurred in distinct lines across the Earth. It was soon clear that these lines were the edges of enormous plates that covered the Earth's surface and that earthquakes were the result of friction between the plates. Continents on the plates were slowly moving on the surface of the Earth. This is the concept of plate tectonics, which is related, but not identical, to continental drift. Wegener was vindicated, but he had died before plate tectonics was accepted as a theory. Today, most geological phenomena can be explained by these two concepts.

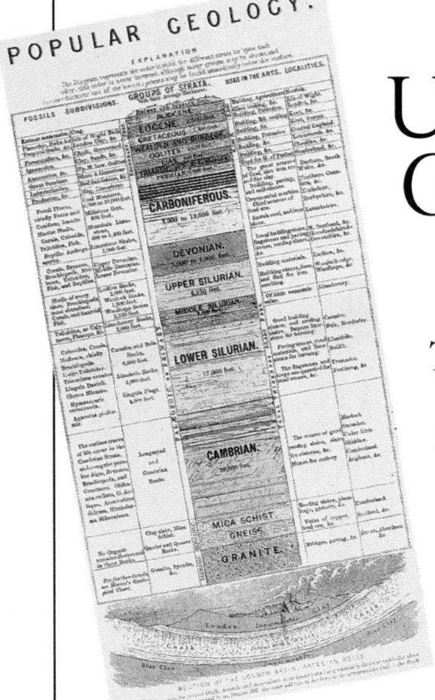

Understanding Geological Time

The immensity of geological time is staggering: it is measured in thousands of millions of years.

There are two ways of looking at geological time: as relative time, or as absolute time. The work of early geologists William Smith, James Hutton, and Charles Lyell resulted in an understanding that rocks accumulate in layers with the oldest at the bottom, and that this process occurred in the past at the same slow rate as it does today (Lyell's principle of uniformitarianism). After careful mapping of all the world's known rock units, a theoretical continuous stack, called the stratigraphic column, was constructed. From this, it is easy to determine the age of a particular rock unit relative to others.

Relative Time

To facilitate working with this cumbersome intellectual structure, the column was divided into a number of subsequences called Eras, Periods, and Epochs. These equate to the period when the subsequences were laid down and often take the name of the area where they occur. For example, the chalk deposits that form the White Cliffs of Dover were laid down in the "Cretaceous" Period, from the Greek word for chalk. So when we talk about a fossil being from the Cretaceous Period, we are assigning it a relative age.

It was apparent from Lyell's principle of uniformitarianism that the many thick deposits of the stratigraphic column must have taken a long time to accumulate. This was based on his discovery that many types of rock accumulated at imperceptibly slow rates. The Earth clearly was ancient, but no one knew just how old.

THE DEVONIAN PERIOD *was named after the rocks of Devon in southern England (below). An early rendering of a geological time chart (above), and an early depiction of a Jurassic scene (right).*

CHALK DEPOSITS, *such as those in southern England (left), gave the Cretaceous its name. This painting,* Strata at Cape Misena, Bay of Naples *(below), was done before the significance of the strata was understood.*

> *Such vast periods*
>
> *of time baffle*
>
> *the imagination.*
>
> *Life on Earth,*
> DAVID ATTENBOROUGH
> (b. 1926), British naturalist

ABSOLUTE TIME

To establish the absolute age of a rock (in millions of years), we must find and read some form of clock within the rock itself. Today, we recognize several "rock-clocks" that operate in different ways. One is the decay of particular types of isotopes (forms of an element) to other isotopes. In cases where the rate of decay from one to another is known, reading the ratio of the first (parent) isotope to the second (daughter) isotope tells how long that particular process of decay has been going on.

There are a few restrictions on this technique. The rock-clock must be set to zero when the rock is laid down, as occurs in molten rocks where the product isotope is a gas that can bubble off. After the rock solidifies, new gases formed by isotope decay are trapped inside the rock. Ideally, there should be no contamination of the rock by either parent or daughter isotope from other sources.

Another rock-clock uses a technique called fission-track dating. The common mineral zircon contains a small number of uranium atoms. Being unstable, these atoms eventually emit particles that leave minute tracks inside the crystal. The longer the zircon crystal has been around, the more tracks will be visible inside the crystal. By counting the number of tracks, we can work out how long it has been since the crystal formed.

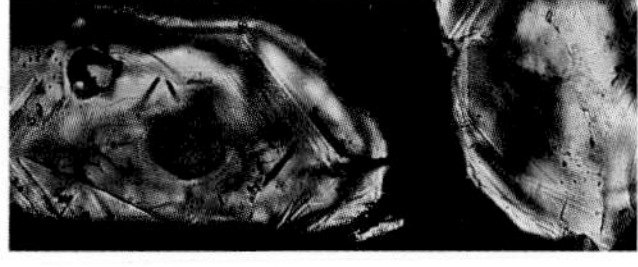

Not all rock types can be dated, but enough datable rocks occur throughout the stratigraphic column to allow most to be closely dated.

A well-known dating technique is called C14, or carbon dating. This measures the decay of one isotope of carbon to another. In this case, the original C14 accumulates only while an organism lives, and decay begins at the death of the organism. Unfortunately, carbon dating is useful for only the past 50,000 years or so. While this range is of use to archeologists (who study ancient human cultures), the periods geologists and paleontologists are interested in are far too great to be determined by this method.

TINY TRACKS *in zircon crystals (left) in a rock date it at 3,600 million years.*

THE TIMELINE

The whole of human history is no more than the blink of an eye compared to the history of the Earth.

Dating techniques and the stratigraphic column have provided us with a history of the Earth. By looking at rocks, we can date many events that took place at particular points in that history. By looking at fossils within those rocks, we can see what organisms were alive at different times. So by putting rock units in their correct sequence within the stratigraphic column, we can read the history of Earth and of life upon it.

A popular abstraction of the stratigraphic column is the geological timeline. This is a summary of time-related information from the stratigraphic column. It shows the geological periods in their correct succession, but rarely refers back to the rocks from which this information was originally derived. Other events are often superimposed on the timeline in their true, relative positions. These may include the movements of the continents, major glaciations, the first appearances of particular forms of life, and the formation of the "red beds", 2,500-million-year-old iron-rich rocks that indicate the presence of a well-oxygenated atmosphere.

SPECTACULAR EPISODES *of intense vulcanism (left) are recorded in Earth's rocks. Fossils, such as the palm (above), tell the story of ancient life forms.*

Although the division of the stratigraphic column into Eras, Periods (and also Epochs, which are not shown), is standard around the world, there are some regional variations. The best known of these is the practice in North America of dividing the Carboniferous Period into sub-periods, the Mississippian and Pennsylvanian. These sub-periods refer to the lower

MAKING A TIMELINE

The average toilet roll contains 500 sheets. Imagine a roll with 40 sheets removed. Each of the 460 remaining sheets represents 10 million of Earth's 4,600-million-year history, a basic scale model for a timeline. The length along the roll represents time, one end being the beginning of the Earth and the other representing the present. Unicellular life started on the 350th sheet back from the present end. Multicellular life appeared almost 300 sheets later (60 sheets from the present end). Dinosaurs emerged 22 sheets from the present end and disappeared 15½ sheets later. Our human species first appeared just 1/20 of a sheet from the present end of the last sheet. On this scale, our current calendar (less than 2,000 years old) would be represented by the last 1/5,000 of a sheet from the present end. Plot these points on your timeline as well as other events indicated on the timeline shown here.

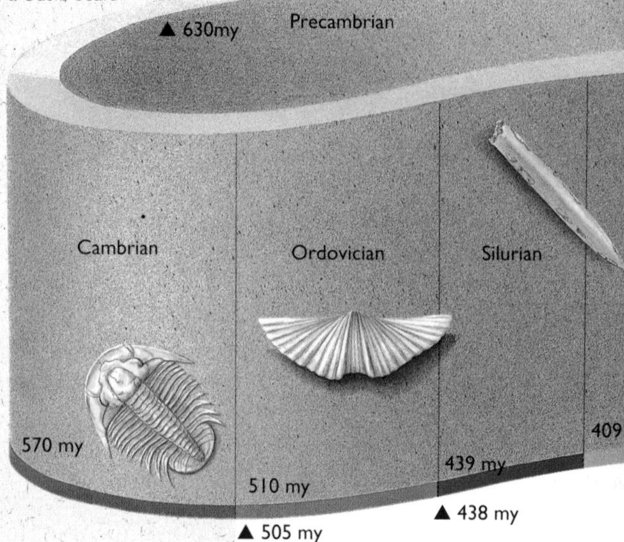

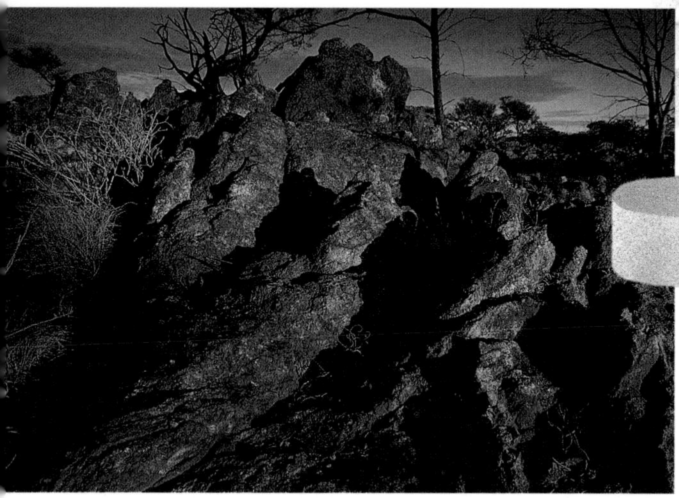

and upper Carboniferous, respectively. For consistency, this book refers only to the Carboniferous.

In the representation of a geological timeline shown here, the sequence of geological periods from the present back to the beginning of the Earth, 4,600 million years ago. It is very difficult to show the timeline in its correct proportions because so many well-documented and interesting developments have occurred in just the past 600 million years.

EARTH'S OLDEST ROCKS *(above) are the granitic foundations of the Western Gneiss Terrane in Western Australia.*

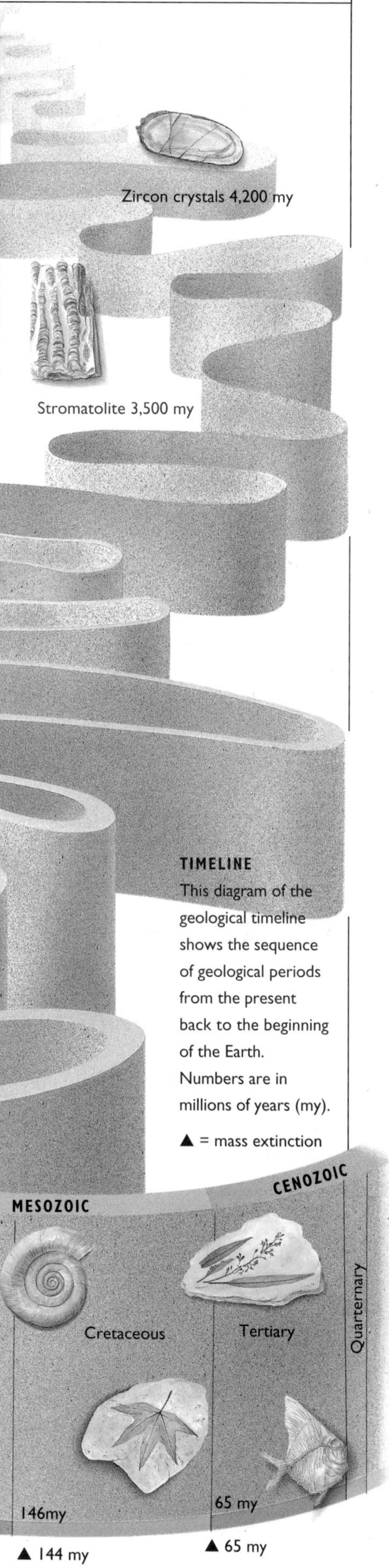

Zircon crystals 4,200 my

Stromatolite 3,500 my

TIMELINE This diagram of the geological timeline shows the sequence of geological periods from the present back to the beginning of the Earth. Numbers are in millions of years (my).

▲ = mass extinction

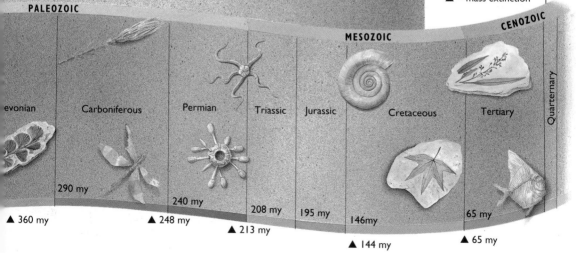

PALEOZOIC

MESOZOIC

CENOZOIC

evonian

Carboniferous

Permian

Triassic

Jurassic

Cretaceous

Tertiary

Quarternary

290 my

240 my

208 my

195 my

146my

65 my

▲ 360 my

▲ 248 my

▲ 213 my

▲ 144 my

▲ 65 my

SEPARATING FACT *from* FICTION

Rocks and fossils have always been a source of wonder, but sometimes imagination rather than reason takes over in their interpretation.

The systematic methods of the sciences of geology and paleontology ensure that hoaxes, superstitions, suppositions, and even outright fraud will eventually be explained. A number of naturally occurring phenomena are erroneously thought to be fossils. One of the most common of these false fossils, or pseudofossils, are the thin deposits of manganese oxides that form fern-like patterns, called dendrites, on rock surfaces.

More insidious are deliberate hoaxes that have been planted in the field. Although highly unusual, such an episode occurred in the 1720s in Würzburg, Germany. Johann Beringer found a huge variety of carved "fossils" representing many different types of animal and plant, as well as depictions of comets and stars. Today, we know that these were fabrications, but the maker of them is even now a matter of debate.

Some suspect it was a student prank, but a more sinister scenario suggests that Beringer's professional rivals deliberately made and planted the specimens for Beringer to unearth. It is not entirely clear if Beringer was taken in, and study of his work indicates that he was uncertain of the origin of his finds. Legend has it that Beringer finally realized the extent of the fraud when he found his own name reproduced on one of the pseudofossils. Unfortunately, he made this crucial discovery the day after his huge manuscript on the specimens had been published, and spent a considerable amount of his time and money buying up as many copies of the manuscript as he could.

"DRAGONS' TEETH" *sold by Chinese apothecaries were credited with curing a multitude of ailments.*

FRAUDS

Cases such as this are, thankfully, rather rare in geology and paleontology. One of the most infamous cases was the Piltdown Man. In the early 1900s, the fossilized remains of a half-human, half-ape were discovered in a quarry in southern England. This specimen received the name Piltdown Man and was heralded as the missing link between apes and humans.

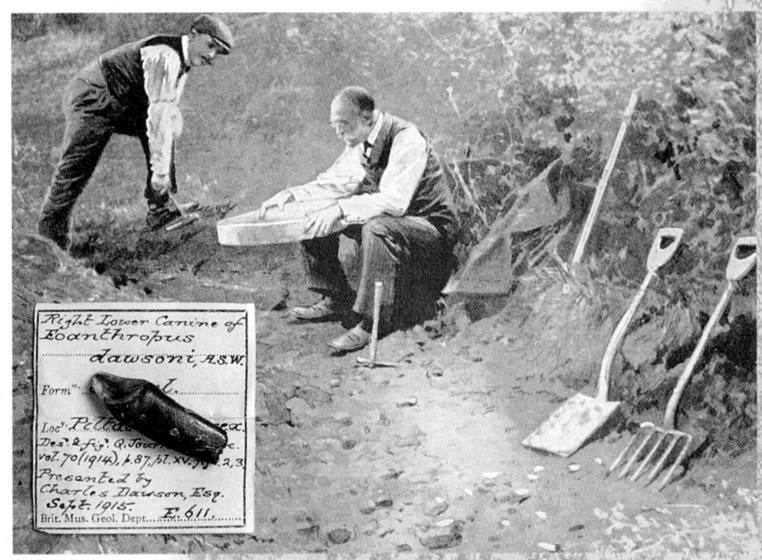

EXCAVATING PILTDOWN MAN *(right). The bones and tooth (inset) of this missing link were "found" in 1912 but were later shown to be fraudulent. Dendrites (top left and bottom right) resemble fossil ferns.*

THE HIMALAYAS *(right), said to be one source of rock and fossil specimens "found" by Dr Gupta, an eminent Indian paleontologist.*

It was not until during the 1950s that the fossils were shown to be fakes consisting of a mixture of chemically treated human and orangutan bones. The person behind this fraud is still unknown.

More recently, there was the saga surrounding Dr Gupta in India. For most of the 1960s through the late 1980s, Gupta was considered the pre-eminent authority on the geology and fossils of the Himalayas and northern India. He produced specimens of fossils and rocks, supposedly found at particular localities in the mountains, and published his results. By the mid 1980s, it had become obvious that work associated with him did not tally with that of other workers in the same region. Slowly, a vast and systematic fraud was revealed—Gupta had been buying specimens from around the world and pretending to have found them in the Himalayas. His motives are unclear and he still protests his innocence, despite all the evidence.

CRYSTALS AND THE NEW AGE

The perfect geometrical shapes of crystals have fascinated people for a long time. Before the processes of their formation were understood, crystals were thought to have magical powers and were sometimes used in religious and sacred ceremonies.

By and large, these misunderstandings have disappeared in favor of more rational explanations of their structure, but during the past decade or so, the New Age movement has revitalized interest in the supposed power of crystals. Certain types of crystal are claimed by various practitioners to have healing properties for some ailments. There is no scientific evidence to support such claims. It is no coincidence that, since the advent of this New Age of mysticism, the price of crystal specimens has increased dramatically.

A reasonable probability

is the only certainty.

Country Town Sayings,
EDGAR WATSON HOWE
(1853–1937), American journalist

MISTAKEN SUPPOSITIONS

With hindsight, many mistaken identifications made before rocks and fossils were understood are rather humorous. At one time it was thought that all fossils were the remains of animals and plants that had perished in the biblical flood and that, before this, giants roamed the Earth. Given this background, it is not surprising that a fossil measuring two feet (60 cm) in diameter and weighing 20 pounds (9 kg) should be identified as the scrotum of a giant human. English scientist Robert Plot had described the find in 1677, but it was identified by others. We now know that *Scrotum humanum* is actually the bottom part of the thigh bone of the meat-eating dinosaur *Megalosaurus*.

In China, fossilized teeth of dinosaurs and other animals, thought to be dragons' teeth, were ground down for use in traditional medicines. In fact, the cave deposits that contained the fossils of Peking Man (*Homo erectus*) were located after an amateur fossil collector noticed teeth from the deposit on sale in a Chinese apothecary's shop.

SCROTUM HUMANUM *(above) was really the base of a dinosaur thigh bone.*

THE COLLECTOR'S CODE

Rock and fossil collectors have their own code, but the basic tenets are the same for collectors of all natural objects.

COLLECTING *organic gems such as coral can damage the balance of the reef ecosystem (right). Restrict fossil collecting to choice specimens, such as the delicate crab (left).*

When collecting natural objects, there are a few rules and a certain code of etiquette that should be observed. These ensure that there is minimal disturbance to the environment and that enough specimens are left for other collectors. The code also ensures that the good reputation of collectors is guarded, which allows them to continue with their hobby without a backlash caused by the thoughtless actions of others. Individuals are part of a community of collectors and the basic rules of behavior apply to all members.

WHERE TO COLLECT

Most collecting sites are on private property, so the first rule is to get permission from the property owner. You need to be quite clear about why you are there and what you intend to do. If you wish to camp overnight and light a campfire, be sure to ask specifically for permission to do so. Explain where on the property you wish to go, how long you intend to be there, how many people are in your group, and whether you want to collect and remove specimens. Property owners are more likely to cooperate if your intentions are clear.

When seeking permission to enter an area, it's a good idea to check if there are any special conditions the owner would like you to observe. Farmers may ask that you keep out of a particular field for a host of reasons—perhaps it has just been sown, or there may be livestock in it.

Gates should always be left as you found them, either open or closed. In mines and quarries, where blasting or excavating work may be in progress, your safety and well-being depend on your being fully informed about unsafe areas.

Having received permission to enter an area, obey any special instructions to the letter. If permission is refused, never try to sneak in by another entrance. This is trespass and property owners are quite entitled to prosecute.

There are certain areas where you should never collect. National parks, nature reserves, and historic sites usually forbid the gathering of rock, fossil, or mineral samples in the area. Although such places are usually not rigorously policed, it is up to visitors to respect any "no-collecting" rules. A responsible rock and fossil collector never tries to recover a specimen in a forbidden area just because no one is looking.

SEEK PERMISSION *to search at quarries (above) and keep in mind that they are risky places. Never enter abandoned mines (right).*

While Collecting

The cardinal rule is to take only what you need for your collection. Greed by a few individuals has stripped many sites bare. More often than not, surplus specimens taken by such individuals are stored away and later thrown out. These specimens should have been left *in situ* for other collectors to find and cherish.

Although collecting rock and fossil specimens often means digging and excavating, try to minimize your impact on an area. If you have permission to dig, be sure to back-fill all holes before you leave. Never dig in areas that are susceptible to erosion, such as on steep hillsides or creek banks. Never remove trees or shrubs in your quest for specimens. Always leave campsites clean and tidy, making sure any fires are properly extinguished.

When working on cliffs and steep banks, beware of falling debris. If you are at the bottom of a cliff, wear a hard hat. If you are at the top of a cliff, be careful not to dislodge any material, and never throw anything over the edge. Disused mines are particularly dangerous areas, so do not venture inside.

IMPORTANT SITES *may be reserved solely for professional scientists (right) involved in particular research work.*

Common Sense

Collecting etiquette is really the application of common sense. Never act on someone else's property in a way that you would not want them to act on yours. Never put yourself or others in danger. Be mindful of collectors who will follow in your footsteps, and act with appropriate care. These simple considerations will ensure that there are always specimens to collect, places to collect them, and a friendly reception from the wider community.

NATIONAL PARKS, *such as Yellowstone (right) in Wyoming, USA, are areas where collecting is forbidden.*

Rocks do not recommend the land to the tiller of the soil, but they recommend it to those who reap a harvest of another sort—the artist, the poet, the walker, the student, the lover of all primitive open-air things.

The Friendly Rocks,
JOHN BURROUGHS (1837–1921), American naturalist and poet

Chapter Two
Forces Shaping *the* Earth

In Search *of a* Theory

During the twentieth century, a revolution occurred in the Earth sciences that has resulted in the textbooks being rewritten.

Until the 1950s, almost nothing was known of the ocean basins, comprising some 70 percent of the Earth's surface. Some thought that the continents progressively collapsed inwards, creating oceans, and that the Earth was shrinking like a drying orange, with mountains forming as the surface wrinkled. Deep ocean trenches were known from the Pacific Ocean, but ocean floors were thought to be mostly flat and evenly covered with sediment.

The differences between types of volcanoes were puzzling, but there were simple models to explain them. Steep volcanoes were located next to trenches, but the reason for this was a great mystery, and it was not understood how or why they formed. The occurrences of granite and other igneous rocks were mapped, but why they occurred where they did was another puzzle.

Knowledge of the internal structure of the Earth was based on information gathered from the rate that earthquake noise traveled through rocks of various compositions. Work with sound waves, seismic studies, revealed that the Earth had concentric layers that mostly increased in density and rigidity with depth. It was known that gravity pulled dense material to the center, displacing less dense material outward.

The change to our present understanding of the Earth really began in 1912, when climatologist Alfred Wegener drew maps showing how the scratches left by the Permian glaciation would form a clear radial pattern out from the South Pole if the several continents on which they occurred had been joined in a tight polar cluster during the glaciation. Although now widely dispersed, and with the patterns pointing in odd directions, when reassembled as they once were, the scratch marks fit together perfectly. Such an alignment brought Africa and South America together into a close fit, closing the Atlantic Ocean. Wegener argued that

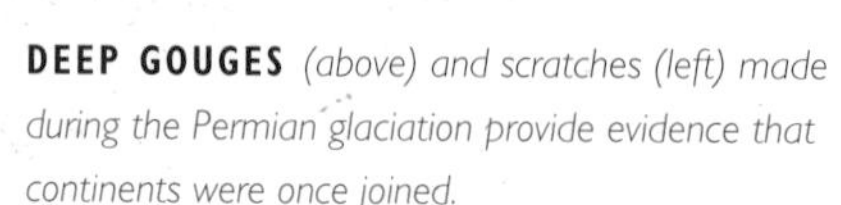

DEEP GOUGES *(above) and scratches (left) made during the Permian glaciation provide evidence that continents were once joined.*

CHANGES IN UNDERSTANDING *An early depiction of subterranean lakes and rivers (left). Later theories tried to explain volcanoes (above), but Wegener's theory (right) introduced the concept of plate tectonics.*

the Atlantic coasts of Africa and South America were too similar in shape to have been produced independently, and showed how closing the Atlantic Ocean would make sense of many such puzzling biological, geological, and climatological observations.

Wegener's theory was hotly debated, but foundered on the arguments of the most notable physicists of the day. They wanted him to propose a mechanism that caused the drifting, something he could not do. They were also sure that the Earth was younger than Wegener's theory implied, believing that had it been older, it would have cooled to a solid long before.

Breakthrough

In 1913, geologist–physicist Arthur Holmes (1890–1965) argued that the slowly released energy from newly discovered radioactive elements could have kept the Earth hot. Holmes also explained how the breakdown of the small quantities of these naturally occurring radioactive isotopes in igneous rocks could be used to determine when the rocks had solidified. He also suggested that Earth's hot mantle convected sluggishly like soup in a pot, setting up a motion that made continents collide and rift.

Although the theory of continental drift was unpopular for several decades, a dramatic about-face occurred in the 1960s, when our knowledge of the ocean floors exploded with new technology. Echo sounders recorded ocean depths and mapped the huge mid-ocean ridges. Oceanic magnetic patterns were found to be symmetrical about mid-ocean ridges. This simple pattern, absent from continents, suggested that the ocean floor, actively forming at mid-ocean ridges, was younger than the continents.

With this flood of information, scientists have built a model of Earth, continuously convecting, covered by thin, rigid surface plates driven by convection to rift, overlap, and fold. The plates grow at mid-ocean ridges, sink at trenches, and fold at mountain ranges, creating oceans and moving continents.

HYDROTHERMAL VENTS *deep in the ocean (right) mark the mid-ocean ridge. The symmetry and young age of the oceans are quite recent discoveries.*

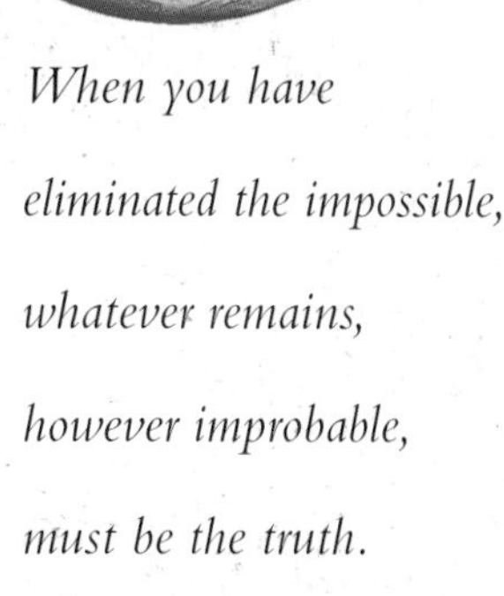

When you have eliminated the impossible, whatever remains, however improbable, must be the truth.

The Sign of Four,
Sir Arthur Conan Doyle (1859–1930), English writer

Ancient Lands

All of the continents in the world today are fragments of the supercontinent Pangea, which began to break up some 200 million years ago.

The surface of the Earth appears to us to be stable but it is actually moving continually. We now know that the Earth has had a long and complicated history. With sufficient time, the slow movements of its surface plates are enough to rip apart old continents and put together new ones.

By 250 million years ago, all of the landmasses on Earth had collided to form one supercontinent, Pangea. It was a long, narrow continent stretching from Pole to Pole. The single ocean that lapped both its shores was a larger version of the Pacific, and covered more than 75 percent of the Earth's surface. Modern mountains, such as the Andes, Himalayas, and Rockies, did not yet exist.

BEECH LEAVES *(left). Ancient relatives of these trees once forested Gondwana. Mount Washington, (right) in the North American Appalachians, which were once connected to the Caledonian Mountains in Europe.*

The present Appalachian Mountains, in eastern USA, are the remnants of a giant mountain range that formed when earlier continents collided during the formation of Pangea.

If we could reassemble the continents as they were before the Atlantic Ocean opened more than 150 million years ago, the continental shelves of North and South America would fit precisely together with those of Africa and Europe. The Appalachian Mountains of North America and the Caledonian Mountains of Europe would fit together to form a single, continuous mountain range. That is why today these two separate mountain ranges contain the same rock layers in the same sequence, with the same fold patterns, and with identical fossils of the same age.

Fossils provide further strong evidence of continental connections. For example, fossils of the small, amphibious, freshwater reptile *Mesosaurus*, are found in both Africa and South America. Reassembling the continents brings these areas together, and the 270-million-year-old rocks in which the fossils are found match precisely. The only way that the living *Mesosaurus*

EVIDENCE *of Gondwana's final split (below) at Cape Raoul, in Tasmania, Australia. Rocks on the northern tip of Antarctica match these rocks exactly.*

30 mya

present

THE SHIFTING CONTINENTS

About 200 million years ago, the supercontinent Pangea began to split into Gondwana and Laurasia. By 65 million years ago, these continents had begun to split apart.

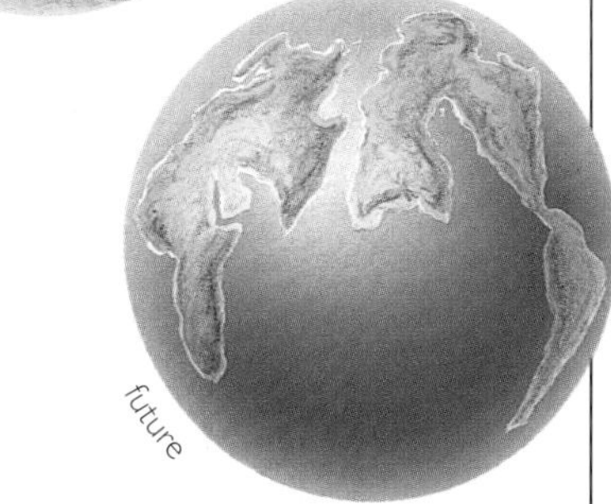

FUTURE EARTH?

How the Earth's continents might appear in 65 million years.

could have occurred on both continents would be if they were once joined.

The Breakup Begins

Pangea began to break up about 200 million years ago. The evolution of life had begun long before, so as the supercontinent cracked and the continental fragments rifted apart, they carried with them a common basic stock of the species living at the time. Isolated on separate continental fragments, these life forms evolved in different ways to produce flora and fauna unique to each fragment.

The rifting process first split Pangea into two smaller supercontinents. The southern part (Gondwana) contained all the modern Southern Hemisphere continents as well as India. The northern sector (Laurasia) was composed of the present Northern Hemisphere continents. These two supercontinents then began the ongoing breakup that has resulted in the continents of today. The rifting of Gondwana and Laurasia into smaller continents formed the Atlantic, Indian, and Southern oceans. These ocean basins have increased in size at the expense of the Pacific Ocean, which has shrunk steadily since Pangea's breakup.

Detective Work

Scientists have deduced this story from Alfred Wegener's mappings of glacial scrapings, the fossil record, the present-day distribution of species, and from marine, continental, and satellite geophysics. Further evidence lies in the identical matching of bedrock geology of continental shelf areas that were once joined. The most precise evidence currently available, however, comes from marine geophysics and from the precision of the modern satellite-dependent Global Positioning System (GPS). The GPS can track the exact position of all continents on a daily basis. We can measure their movements so precisely that we know the exact distances and directions they are traveling. In general, the continents are separating at a rate of up to 3 inches (7.5 cm) per year—about the same rate that your fingernails grow.

The Future

With this knowledge, we can speculate how the Earth may look in the distant future. One likely scenario is that the Atlantic Ocean will keep opening, and the Pacific Ocean closing. Australia will continue moving north until it collides with South East Asia and Japan, squeezing all of these into a mountain range. At about this time, the Asian continent will rift in two through China, spreading east-west. As a result, the western half of the Atlantic Ocean will begin to close. The rifting of China will speed the collision between North America and the Australia–East Asia mass. As a result, Alaska and Siberia will overlap to become the highest mountains in the world.

DISTANT RELATIVES

Mesosaurus *(above left) occurred on two separate continents. Australian waratahs (left) and South African proteas (right) share a Gondwanan ancestry.*

The Dynamic Earth

Inside the "solid" Earth, huge amounts of energy, in the form of heat, continuously move the continents and reshape the oceans.

Although the Earth appears to be solid, it has evolved, during the 4,600 million years of its life, from a totally molten planet to one with a rigid skin. Today, we know the outer skin, or crust, as either continents or ocean floor.

Originally, the whole Earth was composed of magma without a surface crust. Gravity separated the magma, bringing the lighter elements (mostly silica) to the surface where, being too buoyant to sink, they collected and cooled into a floating rock scum. With further cooling, this surface material solidified to form the early continents and the exposed molten magma between the continents also solidified into a thin skin of basalt (the most common rock to solidify from magma). This situation is still apparent today: the light continents ride higher than the more dense basalt.

CONVECTION CURRENTS *crack the surface of molten lava, mimicking the action beneath the Earth's crust.*

Gases also rose to the surface and formed the early atmosphere. Among these gaseous elements were oxygen and hydrogen, combined as water. The water remained in a gaseous state, however, until the surface of the Earth was cool enough for water to condense and stabilize as a liquid. The water pooled in the lowest points to form the first oceans. Invariably, the oceans formed on the basalt, while the higher continents shed water and became dry land.

Earth's Engines

The semi-molten interior has always circulated in what are known as convection cells, where the hottest magma rises, spreads out, releases heat at the surface, cools, and descends again. The heat is slow to escape through continents, which results in convection cells rising and rifting, cracking the continents, opening oceans, and creating mountains and trenches elsewhere.

To picture a single convection cell at work, think of a pot of soup on a burner. Heated from below, a hot plume rises in the center, and the hottest soup spreads out across the top, then subsides around the edges. Many such convection cells interact inside the Earth.

Where hot rising fluid at the center of each cell (an "up plume"), reaches the surface and spreads out, it pushes the surface scum outwards, so the area is called a divergent zone.

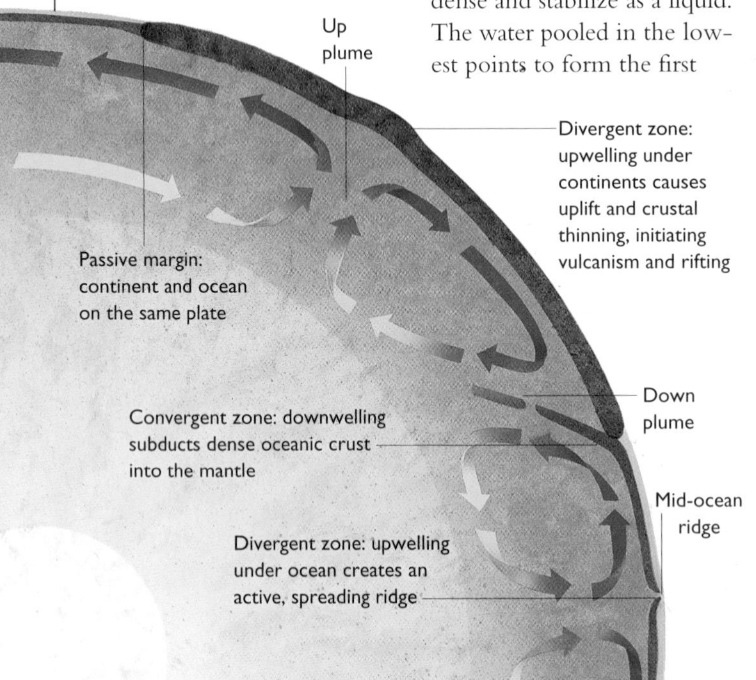
Up plume
Divergent zone: upwelling under continents causes uplift and crustal thinning, initiating vulcanism and rifting
Passive margin: continent and ocean on the same plate
Down plume
Convergent zone: downwelling subducts dense oceanic crust into the mantle
Mid-ocean ridge
Divergent zone: upwelling under ocean creates an active, spreading ridge

CONVECTION CELLS *form beneath both oceanic and continental crust (left). The rising and falling heat they generate causes continents to crack, oceans to spread, and mountains to form.*

SHIFTING GROUND *The movement of the Earth's plates shapes the land. The Andes (above) rise over a subduction zone. The San Andreas Fault (left) is responsible for many major earthquakes.*

Where two convection cells meet, the cooler material sinks back (a "down plume"), leaving a collection of buoyant material on the surface. Where two cells touch in this way, the line of contact is called a convergent zone.

Large rigid areas of the Earth's crust are called plates, and these are separated from each other by distinct boundaries. Where two plates meet, they actively collide, separate, or just pass by. These activities generate vibrations that we feel as earthquakes. Earthquakes are easy to locate and plotting them identifies all plate boundaries.

The location of plate boundaries is largely controlled by the convergent and divergent flows in the mantle beneath. Divergent flows pull continents apart (a phenomenon known as rifting), and basalt fills such continental rifts. As they open further, more basalt forms along the rift, creating new ocean floor.

At convergent zones, plates come together and different phenomena can happen, depending on the type of plate involved. The dense basaltic oceanic plates are sucked down into the convergent zone (subduction) but the lighter, silica-rich continental plates are too buoyant to be subducted. Instead, they fold and thicken with the force of the collision. These differences produce either oceanic trenches or mountain ranges.

A region where two plates slide by each other is known as a transform boundary. While such boundaries are prone to earthquakes, continental crust is neither created nor destroyed there. A major transform boundary is the San Andreas Fault, which cuts through the west coast of the USA and represents the boundary between the Pacific and North American plates.

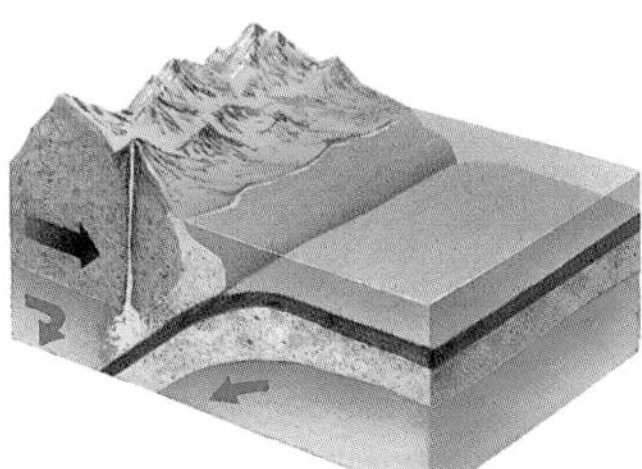

THE ANDES *formed at a convergent zone where an oceanic plate was subducted under a continental plate.*

THE HIMALAYAS *formed at a convergent zone where two continental plates collided and squeezed together.*

THE SAN ANDREAS FAULT *marks the transform boundary between the Pacific and North American plates.*

THE EAST AFRICAN RIFT VALLEY *represents a continental divergent zone. Africa will split and a new ocean will form.*

THE MID-ATLANTIC RIDGE *is a mid-ocean divergent zone where oceanic crust is formed continually.*

PLATE TECTONICS *in* ACTION

All geological and geographical phenomena can be explained with respect to plate tectonics.

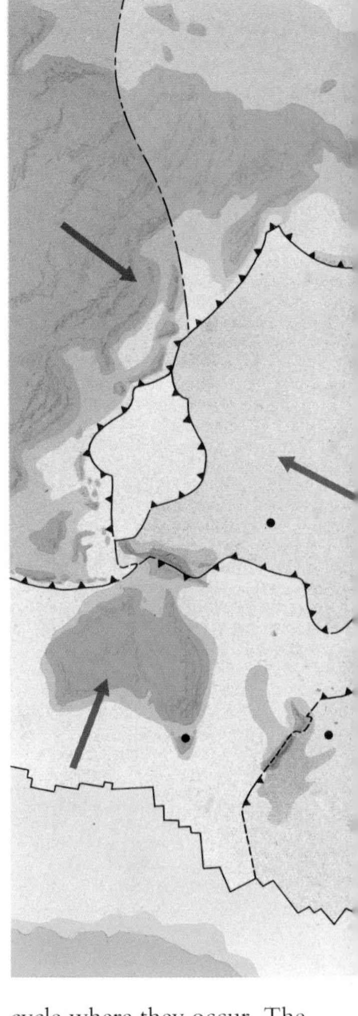

The formation and movement of the Earth's plates is a continuous process. Cooling, convecting mantle pulls down oceanic plates and causes continents to collide, forming supercontinents. These land masses prevent heat escaping from below and the underlying magma begins to heat up and expand. Eventually, the underlying magma begins to rise, reversing its previous direction. When this happens, the supercontinent rifts apart, sending smaller continents in opposite directions. They eventually collide with other continents to form new supercontinents where oceans once existed, starting a new cycle.

ROWS OF VOLCANOES *mark plate edges like dot-to-dot puzzles. A series (above) of volcanoes in Iceland. Snow-covered volcanoes (below) in Chile.*

ACTIVE AND PASSIVE MARGINS

All geological and geographical features have a tectonic "address", a certain environment within the tectonic cycle where they occur. The basic nature of coastlines is a good example. Those on the side of a continent that face a new mid-ocean ridge are "passive margins". They tend to be flat and have wide continental shelves. These coasts usually have deltas from large rivers flowing into them and

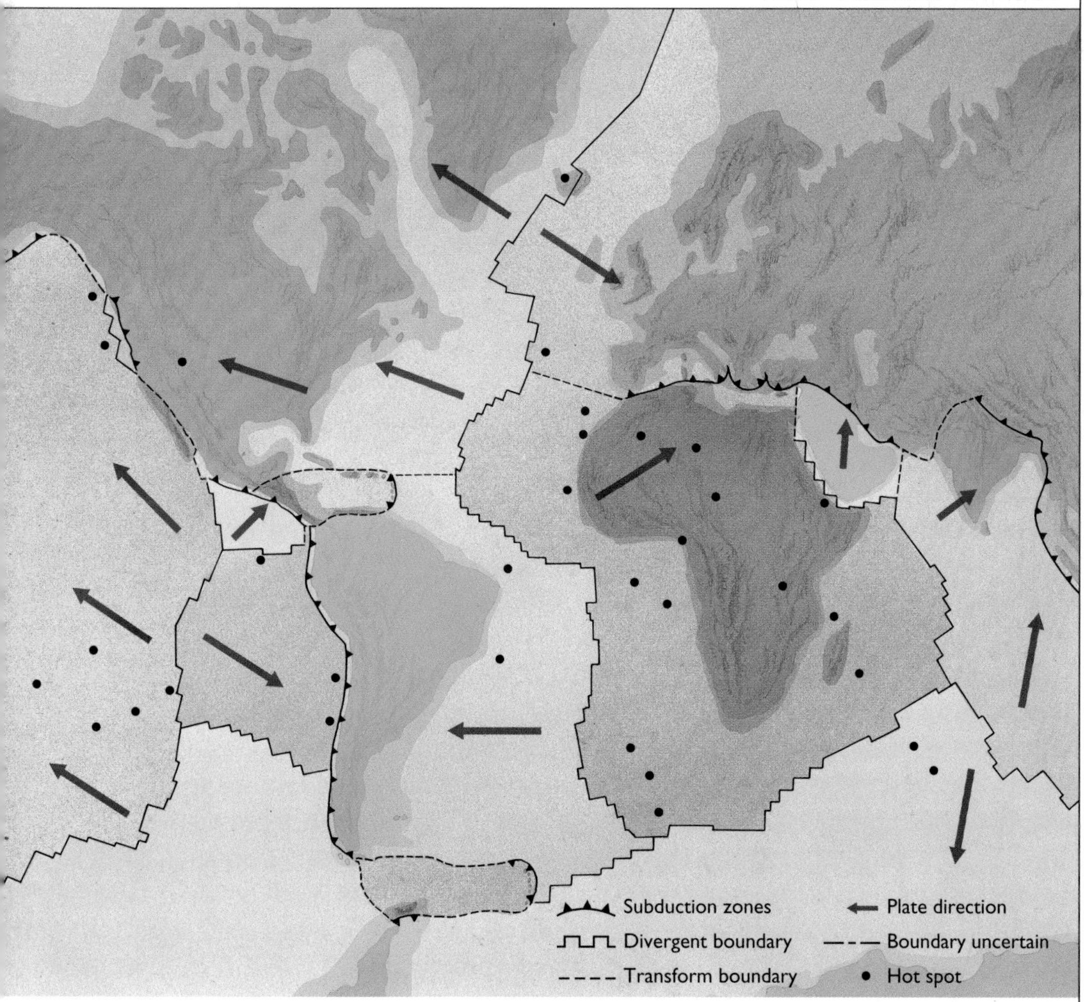

Subduction zones
Divergent boundary
Transform boundary
Plate direction
Boundary uncertain
Hot spot

THE WORLD AS WE KNOW IT *(above), showing the tectonic plates, continental shelves, and hot spots.*

so they have extensive sedimentary deposits that are not folded. Metamorphic and igneous rocks are rare. Mineral deposits, such as salt, coal, and iron, are usually associated with these flat-lying sedimentary beds. There is no vulcanism associated with passive margins and earthquakes are rare. The coasts surrounding the Atlantic Ocean are all typical passive margins.

Continental margins that face subduction zones are completely different, and are referred to as "active margins". These are mountainous, with thin or non-existent continental shelves. Rivers on these active coastlines tend to be short and turbulent and carry little sediment. As a result, most of the sedimentary rocks that occur here are from sediments that have been scrapped from the downgoing ocean plate at the subduction zone. These rocks are folded, faulted, and fractured into convoluted beds.

Metamorphic rocks are common on active margins, brought up from deep within the crust and exposed at the surface as a result of mountain building. The extensive vulcanism in the mountains produced by the rising magma means that igneous rocks are also common. Minerals associated with metamorphic or igneous rocks, such as gold, silver, lead, and copper, are most likely to be found on active margins.

Volcanoes are common in mountains along active margins and earthquakes are frequent and often violent. The west coast of North and South America is typical.

PASSIVE MARGIN *(right) in Argentina, showing the typically flat topography.*

BIRTH *of* OCEANS

Oceans are born when, pulled by the moving magma beneath, continents crack. A new basalt ocean floor forms progressively, filling the rift.

THE NORTHERN TIP *of the Red Sea (above), part of the great African rift. Now widening, this embyonic ocean is separating Africa from the Saudi Peninsula. The Galapagos Vent (far left) warms surrounding water, creating an isolated ecosystem.*

Major oceans begin their lives in the middle of continents. Breakup (rifting) occurs when a heat-generated divergent zone develops under a continent creating a bulge in the crust. The greatest stress on the crust is directly above the upwelling magma, and eventually, the bulge will crack three ways—the three splits being at 120 degrees to each other and growing from the central point outward. The East African Rift Valley, the Red Sea, and the Gulf of Aden provide a classic example of a three-split rifting system.

The first feature to form during a rifting episode is a rift valley, such as the rift valley of eastern Africa. Rift valleys are typically long and steep-sided with active volcanoes along their length. Both surface erosion and erosion by the upwelling magma beneath thins the crust at a rift valley by as much as 80 percent. This thin crust very slowly pulls apart, splitting the old continent.

Each of the three rift arms extends outward and, eventually, one arm will reach an ocean. Since the rift floor is below sea level, ocean water enters and floods the rift. The hot, spreading basalt floor of the rift causes rapid evaporation of the sea water, which

PERFECT SYMMETRY

Ocean floors are built from vertical basalt dykes, or rather half-dykes. These form when hot rock enters the vertical cracks that develop across rifting plates, and later mid-ocean ridges. Basalt lava rises to fill the cracks. As these dykes chill to the crack sides, the center holds heat longer and this weaker core is pulled apart by continuing plate separation, leaving half the dyke on each side. The next intrusion enters between the halves. Continued spreading pulls every following dyke in half. This produces the perfect symmetry of mid-ocean ridges between their flanking continents.

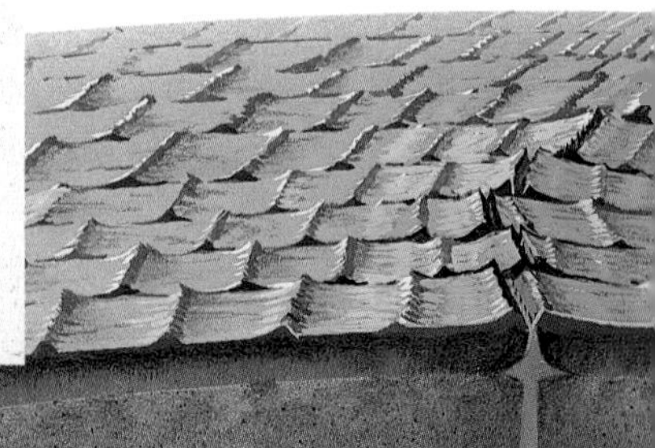

OLDUVAI GORGE *(above left) in the East African Rift Valley. Volcanoes often occur in this rift, as seen here in Virunga National Park, Zaire (left). A mid-ocean ridge runs through Iceland (above).*

concentrates the salt. With rising salinity, the salt is precipitated. This, in turn, stimulates rapid breeding of microorganisms that produce carbon-rich muds. These muds accumulate in the rift, creating an excellent environment for the later production of oil and gas. Rifting continues until a second rift arm reaches an ocean and creates a through passage for water, preventing more salt from being deposited.

Spreading continues at what has now become a mid-ocean ridge, and the continents move progressively farther apart as the new ocean opens. As the new continental margins cool, they sink a little into the underlying magma. Cool continental edges float lower in the mantle, sinking below the ocean surface and creating continental shelves. This explains the presence of salt deposits and oil fields on such continental shelves—originally, they were all rift valley deposits.

Marks on the Ocean Floor

Across the ocean floor, beneath the sediment, are long marks that show the paths that plates have taken as they separated. These marks run between opposed continental shelf edges and are called fracture zones. Fracture zones restrict the possible movements of plates and can be used to unravel plate movement and reassemble past supercontinents.

Once a plate begins to move, it rarely changes direction, unless its movement is interrupted by continents colliding. When pulled apart, basalt breaks at right angles to the direction of pull, as does a toffee bar.

So mid-ocean ridges are composed of many short, straight segments at 90 degrees to the opening direction of the separating plates. These segments are connected by short "transform faults", which record the opening shape of the rift and are the cause of the fracture zones on the ocean floor.

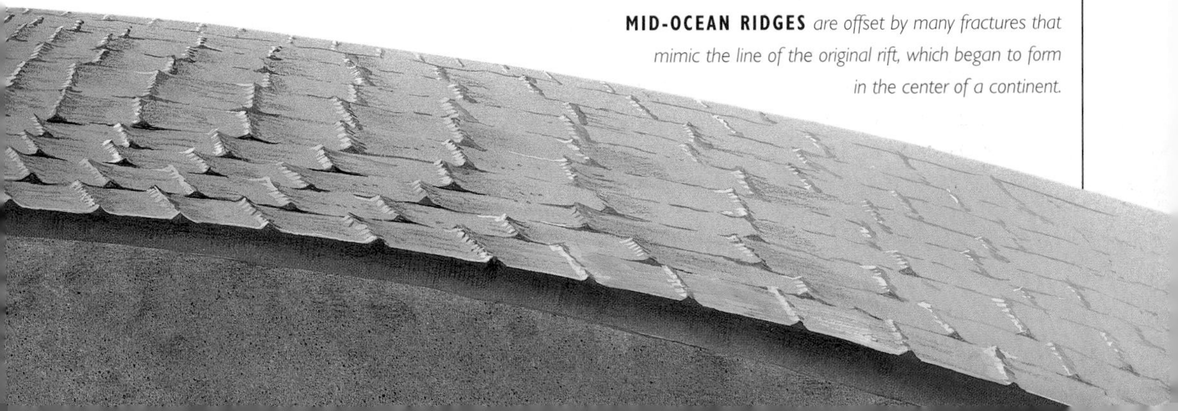

MID-OCEAN RIDGES *are offset by many fractures that mimic the line of the original rift, which began to form in the center of a continent.*

COLLISIONS *and* SUBDUCTIONS

For millions of years, the continents have moved slowly about the surface of the Earth, their progress producing and altering complex geographic features.

The Earth has been about the same size for most of its life and its surface area has remained relatively constant. This being so, crust being consumed in some areas must be balanced by what is being formed at about the same rate elsewhere. Convergent zones where plates collide are consumers of crust.

In convergent zones, plates come together. These plate collisions produce different features, depending on the type of crust involved. The basaltic crust of oceanic plates is dense enough to be sucked down into the convergent zone, a process called subduction. Eventually, the sinking oceanic crust will melt into the magma again. Subduction zones create oceanic trenches, the deepest parts of the ocean, which can extend to a depth of seven miles (11 km).

SATELLITE VIEW *of the eruption of Sakura-jima, Japan, taken in 1984. This volcano has erupted often since AD 708.*

The silica-rich continental crust is too light to be sucked into a convergent zone, so continents above convergent zones tend to buckle up into mountain ranges—think of a rug being pushed against a wall, where it buckles and folds up into mini-mountain ranges. This is precisely the way the world's major mountain ranges were formed.

When an oceanic plate collides with a continental plate, the oceanic plate is sucked under. At the convergent zone, magma that contains both quartz and basalt collects. The resultant liquid rock rises through the overlying continental crust to erupt volcanically. This type of volcano is always present on the continent side of the subduction zone, and the resulting active volcanoes can line the plate edge.

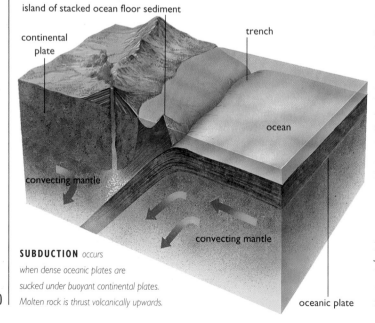

island of stacked ocean floor sediment
continental plate
trench
ocean
convecting mantle
convecting mantle
oceanic plate

SUBDUCTION *occurs when dense oceanic plates are sucked under buoyant continental plates. Molten rock is thrust volcanically upwards.*

RIM OF FIRE

An almost complete ring of such volcanoes exists parallel to the subduction zones around the Pacific margin, from the southern tip of South America, up the west coasts of both South and North America, across the Aleutian Islands, through Japan and the western Pacific to New Zealand. The Rim of Fire is a name often used for the active circum-Pacific

VAST MOUNTAIN RANGES, *such as the Andes (above), and active volcanoes, such as Mt Bromo (right), Indonesia, mark the edges of subduction zones.*

volcanoes that mark where subduction now occurs. The earliest volcanoes are still at the continental edges, while younger ones have moved out into the Pacific, forming island arcs.

In the western Pacific, it appears that the underlying convergent zone is moving into the Pacific, creating new subduction zones as it goes. Island arcs form above these new subduction zones and, to compensate for the consumption of crust, rifts appear to the west that eventually form small ocean basins.

Japan is a classic case of this moving subduction zone. Japan is composed mostly of a rifted fragment of continental eastern Asia, and Japanese volcanoes mark the location of an underlying subduction zone. The rift is now the Sea of Japan. Similar rifting phenomena have occurred around the western margins of the Pacific Ocean creating island arcs such as the Marianas.

A similar process of island arc formation probably took place in the eastern Pacific. The opening of the Atlantic Ocean pushed the Americas over island arcs that formed off their Pacific coasts, pushing the islands into the Andes Mountains to the south, and adding a complex set of terranes to the Pacific coast of Canada and the USA. Olympic National Park, Washington, for example, is composed of very rough, stacked ocean floor sediments that include a marine volcano. These all attest to North America having been pushed over the Pacific Ocean floor as the Atlantic Ocean expanded.

VOLCANOES

Molten rock erupting through Earth's crust forms three types of volcano, varying in shape and the chemistry of their lavas.

In 1943, Mr Mimatsu, a postmaster on the island of Hokkaido, Japan, found a volcano starting to form in a local potato field. Steam, then plumes of ash and lava, burst forth and a volcano was born. In 1995, it was 1,320 feet (403 m) and still growing.

Mimatsu, who had once been a geologist's assistant, knew he was seeing a momentous event. With all the geologists away at war, he wanted to record proceedings, but how? Watching through the mesh on the office window, he noticed that it was squared like a sheet of graph paper, so month by month, he graphed the volcano's growth from his desk. When the geologists eventually read his reports, they were staggered by their precision. Mimatsu was an amateur, but his method is now famous.

His volcano, Showa Shinsan (meaning "New Volcano"), is above a long-active subduction zone and the lava that rises there is andesite, which can be considered a basalt with a high silica content.

RED-HOT LAVA *flows from Kilauea (left), in Hawaii. The 1883 eruption of Krakatau (top left), Indonesia, resulted in a tsunami that killed nearly 40,000.*

FIERY ERUPTIONS

At subducting plate boundaries, volcanoes puncture the overriding plate in very narrow, continuous lines. For example, the Banda Island arc of Indonesia and the Antilles Island arc in the Caribbean, have many active volcanic islands, each in a band stretching for thousands of miles.

Above subduction zones, earthquakes abound, and very steep andesite volcanoes repeatedly form, explode, and rebuild. They are closely spaced in bands that precisely parallel the trench, yet are uniformly offset to one side. Rising majestically above their surroundings, they steepen upward in a conical shape that has great beauty—until the next eruption.

Volcanoes near trenches occur where wet sediment sitting on the descending plate mixes with molten rock. This releases gases and fluids into the molten rock above the subducting plate making it more fluid. The fluids and gases escape through the overlying plate, and are followed by molten rock, which reaches the surface to form volcanoes.

The eruptions that occur above subduction zones are spectacular, violent, and noisy. For example, the Krakatau eruption in Indonesia in 1883 blew away a 2,620-foot (798-m) volcano, leaving an under-sea caldera 985 feet (300 m) deep.

Seawater flooding into this hole and dropping onto red-hot rock, caused repeated violent, noisy explosions.

ASTRIDE *the mid-ocean ridge, Icelanders live with the threat of constant eruptions (left). Showa Shinsan volcano (above), observed by Mimatsu.*

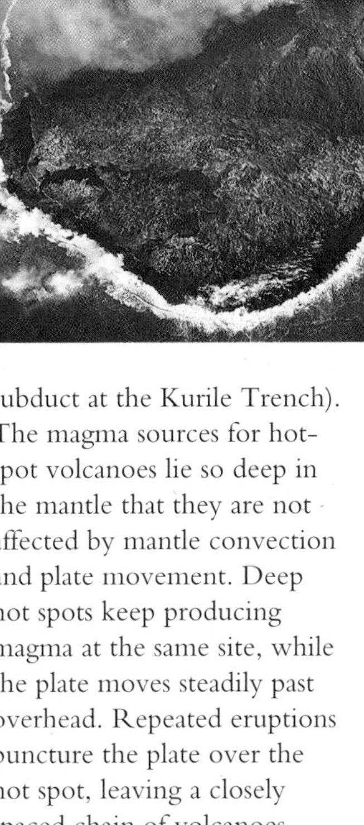

THREE TYPES *of volcano: Anak Krakatau, Indonesia, (far left) and Mt Redoubt, Alaska, (below) both sit above subduction zones; Kilauea (top), Hawaii, is a hot-spot volcano; and Surtsey Island (left), near Iceland, is a mid-ocean ridge volcano.*

The terrifying din was heard 2,250 miles (3,650 km) away, a tenth of the way around the world. The volcano is now rebuilding, as Anak Krakatau (Child of Krakatau), toward another spectacular eruption.

Hot-Spot Volcanoes

Straight lines of hot-spot volcanoes can occur mid plate. They are the exception to the tectonic rule that nothing much happens in the middle of plates. Being easy-flowing, low-silica basalt, these volcanoes form wide and very low-angled cones. In contrast to island arcs, where all the volcanoes may be active, only the few youngest in the hot-spot chains are active.

The Hawaiian Islands are at the active end of the Hawaiian–Emperor sea mount hot-spot chain. This is a chain of hot-spot volcanoes, mostly submerged, that stretches in an unbroken line for 3,000 miles (5,000 km) from Hawaii to Kamchatka, north of Japan (where the oldest of them subduct at the Kurile Trench). The magma sources for hot-spot volcanoes lie so deep in the mantle that they are not affected by mantle convection and plate movement. Deep hot spots keep producing magma at the same site, while the plate moves steadily past overhead. Repeated eruptions puncture the plate over the hot spot, leaving a closely spaced chain of volcanoes. The age of rocks in each volcano can be determined, so a volcanic track across the plate allows the speed and direction of the plate to be calculated.

Mid-Ocean Ridge

The third type of volcano is part of the world-encircling mid-ocean ridge (MOR), visible in Iceland. The MOR is really a single, extremely long, active, linear volcano, connecting all spreading plate boundaries through all oceans. Along its length, small, separate volcanoes occur.

The MOR exudes low-silica, highly fluid basalt and, having produced the entire ocean floor, must rate as the largest volcano of all and the largest single structure on the face of the Earth.

EARTHQUAKES

More than a million earthquakes, ranging from tiny tremors to cataclysmic disasters, are recorded on Earth every year.

Earthquakes are very common events. They occur inevitably when stress that has built up in the Earth's crust is released suddenly as the crust breaks. Evidence of such crustal fractures can sometimes be seen after earthquakes as vertical- or sideways-thrown surface faults.

EARTHQUAKE ZONES

Virtually all the world's major earthquakes occur at tectonic plate boundaries, particularly those that are converging or sliding past one another. The San Andreas Fault, along the west coast of the USA, is a famous example of two sliding plates, and it can be traced for miles, offsetting roads and river channels. At such a fault, the two plates may move at a slow, steady rate and the resulting movement of the Earth will be imperceptible. However, if the plates lock fast, stresses build up and when the plates finally shift, the massive release of energy can have devastating effects.

The shock waves, known as seismic waves, from such a release can be highly destructive in built-up, populated areas. Initial damage, such as collapsed buildings and dams, is sometimes surpassed by subsequent fires and explosions resulting from broken electricity and gas lines.

Even earthquakes that occur on the sea floor can be destructive. Movement of the sea floor causes waves in the ocean, known as tsunamis, that move out concentrically from the source. Barely perceptible in the deep ocean, these waves rise in shallow coastal waters, flooding shores and causing major damage.

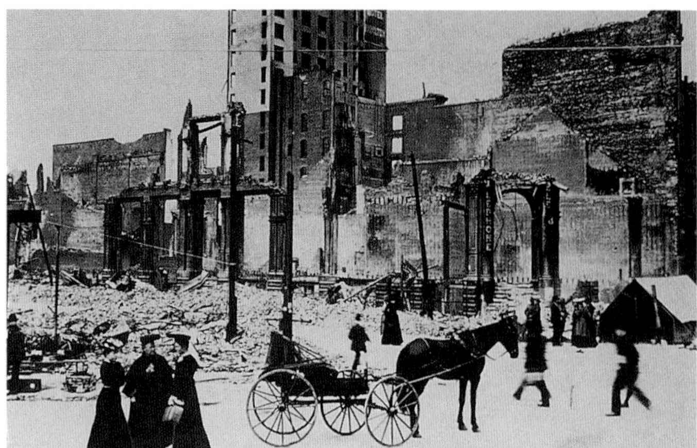

PAST AND PRESENT *San Francisco in 1906 (top) after being struck by an earthquake measuring 8.3 on the Richter scale. A cartoon (top left) from that year. Kobe, in Japan (above), following the 1995 quake, which registered 7.2 on the Richter scale.*

VALUABLE LESSONS

As terrible as they are, earthquakes have given us a greater understanding of the Earth's interior, and of the workings of plate tectonics. Scientists are now able to calculate the Earth's structure and composition from the velocity of seismic waves. These waves are reflected and refracted as they pass through layers of different densities. The absence of seismic reverberations on the opposite side of the planet from an earthquake is known as a "shadow zone", and from this, scientists have been able to deduce the existence of a liquid outer core.

DEVASTATION *in California, USA, following powerful quakes in 1989 (left) and 1994 (far left).*

SHAKY GROUND *A railway line in Kobe (above) after the 1995 quake. An ancient Chinese seismometer (below): a tremor would cause a ball to fall into a frog's mouth.*

The concentration of earthquakes and volcanoes in narrow belts around the globe lends great support to the theory of plate tectonics. Such a belt, known as the "Rim of Fire", runs around the perimeter of the Pacific Ocean, marking the region where the Pacific plate is being dragged beneath adjacent plates. Exact plotting of the earthquake foci define the precise plane of the plate as it dips away from the surface near the oceanic trenches. Earthquakes do not generally occur beyond a depth of about 430 miles (700 km), because at this point, the subducting slab becomes hot and plastic and all stress is relaxed.

> *What happens to us is irrelevant to the world's geology, but what happens to the world's geology is not irrelevant to us.*
>
> HUGH MACDIARMID (1892–1978), Scottish poet

MEASURING EARTHQUAKES

The energy released by earthquakes is recorded on seismometers. Consisting of a freely suspended mass in a frame that is firmly attached to the bedrock, a seismometer measures the relative movement of the frame, which shakes with the Earth, with respect to the inertial mass, which remains relatively still.

The size of an earthquake at its source is known as its Richter magnitude. On this scale, a magnitude of one can be detected only by instruments, while a magnitude of eight is recorded only during major quakes (the San Francisco earthquake of 1906 had a magnitude of 8.3). Each increase of one on the scale represents a sixtyfold increase in energy released.

Rocks *and* Minerals

Rocks, composed of one or more minerals, are the thin, cool skin of the Earth, keeping us insulated from the inferno within.

Rocks are the solid materials of the outer Earth. Minerals are naturally occurring inorganic crystalline solids, natural assemblies of one or more known chemical elements. Although there are thousands of different minerals, just a handful of these make up virtually all of the Earth's rocks. The most common rock-forming minerals include quartz, olivine, feldspar, amphibole (hornblende), pyroxene (augite), and mica (muscovite and biotite). Feldspar alone makes up more than 50 percent of rocks. If you can recognize these common rock-forming minerals, you will be able to identify almost any rocks you might find (see pp. 70–75).

There is a geological history inherent in each rock. Basalt is the most abundant rock, covering more than 70 percent of the Earth's surface, most of this under the ocean. By comparison, all other rocks are rare. We name rocks to define their aggregate minerals, appearance, mode of formation, and particular plate tectonic address. One scheme for naming rocks separates them into three classes, each indicating a physical mode of formation. Igneous rocks form when magma, molten earth material, cools and crystallizes. Sedimentary rocks are cemented fragments of rocks, minerals, or biological material. Metamorphic rocks were once one of the former two, but have been changed by heat and/or pressure.

Rocks can form in less obvious ways. Some precipitate from water, for example, chert, phosphorite, and some limestones. Some settle in water from volcanic dust clouds, for example, bedded tuff, an igneous–sedimentary rock. Some form when coral reefs, built by microorganisms, consolidate to limestone.

LIMESTONE CAVES *(left), cut by groundwater, are later decorated as lime-rich water evaporates to create intricate calcite formations. White veins form as quartz fills cracks in sandstone (below).*

Spreading Plate Margins

There is a direct relationship between specific rock locations and plate tectonics.

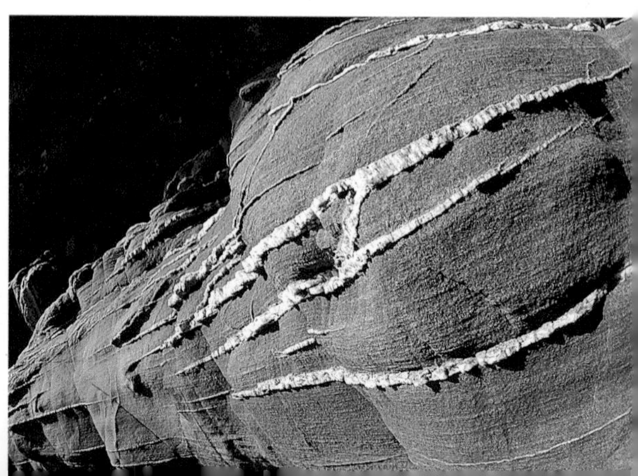

IGNEOUS ROCKS *Ignimbrite tuff (left) from the 1912 eruption of Katmai, Valley of 10,000 Smokes, Alaska, USA. Granite (far left) showing feldspar (pink), quartz (white), and mica (black).*

Before the theory of plate tectonics evolved, every rock seemed of equal importance. We now see greater significance in the few rocks that characterize certain locations.

Basalt is the primary melt from the upper mantle, and the only rock produced at spreading ridges. A rift valley opens across a rifting continent, fills with basalt then, later, the sea enters to flood the widening gulf.

The basalt basement of oceans is uniform in composition but the plate thickens with time to about 50 miles (80 km). Olivine, a magnesium iron silicate, and similar minerals increase in proportion with depth, until the basalt changes to peridotite. When found with chert and marine sediment, this rock indicates past ocean floor, whereas basalt is too common to be diagnostic.

Convergent Plate Margins

The igneous rocks andesite, ignimbrite, and granite, all high in quartz, characterize subduction zones, where all three collect as magmas. Andesite outpourings above subduction zones can cause collapse depressions, or calderas, that might fill with hot, foaming clouds of andesite-composition ash, consolidating to welded tuff, or ignimbrite. Granite collects at depth until it is buoyant enough to rise, sometimes up the feeder pipes of volcanoes to reach the surface.

When continents collide, rivers are born that strip sediments from the uplifted areas and deposit them at continental edges, possibly thousands of miles distant. So this sediment characterizes the collision, but not the site of the collision.

Metamorphic rocks also characterize subduction zones. The subduction process folds, squeezes, and heats both igneous and sedimentary rocks as they plunge to hot depths. This produces twin bands of metamorphic rocks, paralleling the subduction zone.

Away from Margins

Many rocks are formed far from all the activity at plate margins. Mid-plate processes are slower, extending for longer periods than is possible at plate margins, so the end products are fully weathered. Plate margins, in contrast, usually produce incompletely weathered rocks.

Most mid-plate volcanic rocks form from hot-spot basaltic volcanoes, both in ocean basins and on continents. Mid-plate sedimentary rocks, mostly marine, are formed from particles from a variety of sources. Ocean-floor sediments are carried far from the continental margin, where they originated as eroding rocks, or the debris of living organisms. Recognizing the indicator rocks of past tectonic locations allows the ancient history of most locations to be unravelled.

SPECTACULAR MINERALS *include aragonite (above), and amazonite embedded with smoky quartz crystal (right).*

ROCKS *from* MELTS

Igneous rocks solidify from molten magmas, and their composition depends on where they collected and how quickly they surfaced.

Watching white-hot molten basalt pour slowly from a lava tube directly into the ocean engenders an almost primeval feeling. Seeing a cloud of ash particles surging violently from the base of an explosive volcanic eruption makes a very different and terrifying impression. In contrast, the slow rise of crystalline granite plutons 60 miles (100 km) through the Earth's crust seems pretty tame, but when magma cools in any of these ways, igneous rocks form.

Of the hundreds of named igneous rocks, we shall look at only three—granite, andesite, and basalt. Each has a different composition, according to where its magma collected. The type of igneous rock, determined by its mineral composition, can be gauged from its relative darkness. Being mostly quartz and feldspar, granites are light in color. They form from magma high in silica. Andesite, containing

feldspar, hornblende, quartz, and micas, is darker and forms from magma of moderate silica content. Basalts, which rarely have any quartz, contain feldspar, micas, and hornblende, and are darker still.

Most igneous rocks have well-developed crystal structures, although a microscope may be needed to see them. The grain size of any igneous rock is increased by slow cooling and low viscosity, which allow elements to migrate through a melt and reach sites where crystals are growing. When basalt magma cools rapidly on the Earth's surface, it is fine-grained; when it cools at depth, its crystals will be larger—this form is called dolerite (or diabase). Even deeper cooling, taking millions of years, produces a coarser form called gabbro, still with the same chemistry.

THE FLUID *nature of magma is apparent when it surfaces as lava (left). Cooling basaltic lava forms columns (above) as it shrinks. Andesite (top left).*

FORMATION

Magma is a fluid, or rather a mixture of fluids. Earth's upper mantle is partly molten, with minerals floating in the melt. Many minerals join this magma soup from rocks that have been pushed deep and partly melted. Within a melt, minerals have great mobility, so buoyant minerals rise and dense minerals sink. Think of adding marbles and marshmallows to boiling water. The marbles roll to the center of the pot, but do not rise. The marshmallows float to the outside of the pot but do not sink. We now have a marshmallow magma, and a marble magma, each in distinctly located reservoirs, held in position by convection. Similar mechanisms operate in the Earth's mantle.

Silica-rich melts, such as granite and andesite, form above subducting slabs, but rarely elsewhere. The silica content of over-slab volcanoes is higher near continents than farther out to sea. Silica is melted from beneath the

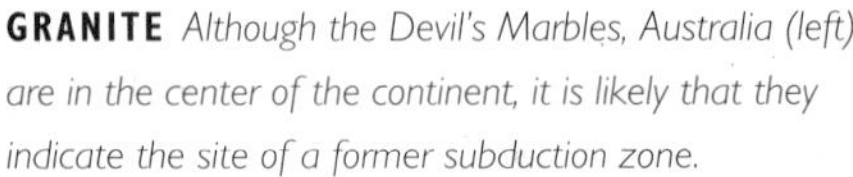

GRANITE *Although the Devil's Marbles, Australia (left) are in the center of the continent, it is likely that they indicate the site of a former subduction zone.*

WHEN ANDESITIC *volcanoes, such as Ruapehu, New Zealand (background above) erupt, they are sudden and violent, flinging out clouds of ash (left).*

continent and carried to the top of the slab by convection. In this way, some silica, water, and carbon dioxide may be added to magma at the subduction zone. Water and carbon dioxide flux the semi-solid mantle, producing a very mobile, silica-rich magma. Being buoyant, this magma does not convect down with the convection flow, but collects above the slab and sometimes rises through the overlying crust.

Andesite forms in this way during active subduction, reaching the surface to build explosive volcanoes. Later, after subduction has ceased or moved on, slower-collecting, slower-rising, and slower-cooling granite will follow, at much the same locations.

Basalt, on the other hand, forms from partial melting of the mantle. A rise in the temperature triggers the melting, and the heated basalt rises to the surface. A localized area of hot magma, a "hot spot", deep under a plate will release hot basaltic magma upwards, producing isolated basaltic volcanoes. Hot-spot volcanoes occur in a series with other extinct volcanoes produced as the plate moves over the hot spot (see p. 187).

Andesite indicates past subduction, as does granite. Granite takes longer to rise and typically arrives well after the start of andesitic vulcanism. All other igneous rocks form from melts when magmas mix, or lose some components, or when gravity separates minerals in a melt by density, allowing different compositions to be present in a single magma chamber. Some volcanoes change composition through time, and such changes are evidence that, deep within the mantle, geological processes are at work.

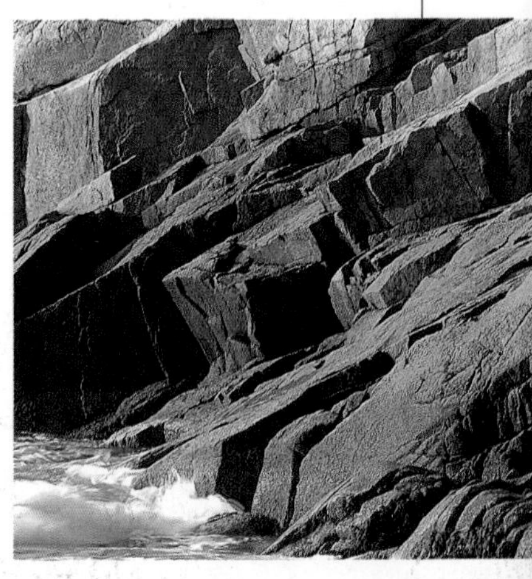

EXFOLIATING GRANITE *(right), a common weathering pattern caused by expansion and erosion of surface layers.*

Rocks *from* Particles

Sedimentary rocks hold information about past environments, telling of the origin, travels, and final deposition of their particles.

The Grand Canyon, Arizona, USA, provides geologists with a nearly continuous sedimentary record of Earth's history. Those 5,000 feet (1,500 m) of bedded horizontal layers, all stacked in sequence with the oldest at the bottom, rest upon even older folded rocks at river level.

Many layers of these sedimentary rocks contain fossil life forms, and these increase in complexity from the floor of the canyon to its rim, revealing the progression of life through time. The layers of conglomerate, shale, sandstone, limestone, ash, and lava indicate past environmental changes as deserts were replaced by rivers, then lakes and inland seas. Erosional breaks interrupt the sequence, but even so, this canyon shows an unusually detailed slice of Earth's story.

Cements

Sedimentary rocks are formed by the cementing of loose particles produced by the breakdown of other rocks. Most sediments settle in water in horizontal layers with flat upper and lower surfaces. Cementing them are the common materials, quartz, calcite, and iron oxide. These lock in the record of the past.

THE LAYERING *of this lakeshore sandstone (right) has formed intricate patterns. Conglomerate (above) is comprised of particles of various sizes.*

SEDIMENTS *are deposited at river deltas, such as the Mississippi River Delta (below), Louisiana, USA. Calcite (right) cements particles.*

The cements that hold sedimentary rocks together are precipitated among the grains. Quartz cements most sediments; particles of limey composition are usually cemented by calcite; while iron oxide, because of its small particle size, finds its way into many rocks with the groundwater and is easily identified by its dark red-brown color.

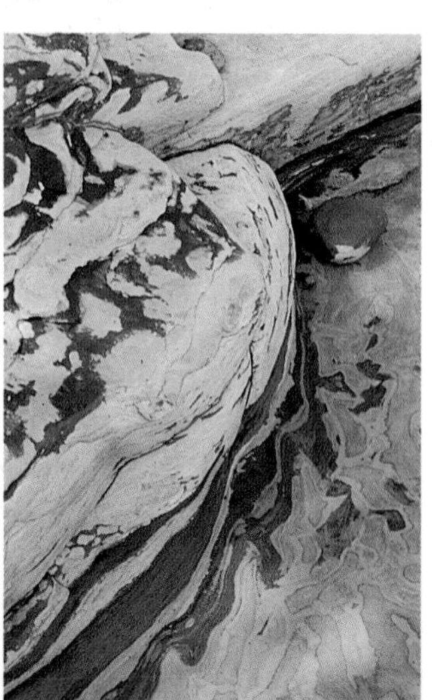

Sediments collect in slow-moving streams, or in swamps and lakes. The sediment may have traveled vast distances, along the Mississippi, for example, or have fallen from a nearby volcano. Where stream gradients are gentle, river channels meander and floods cover the land with mud and sand. Some sedimentary deposits collect in steep alluvial fans when infrequent flash floods discharge from rocky gorges into desert lake beds, such as in Death Valley, California, USA.

Most sediments, however, collect in oceans. They are

THE GRAND CANYON, *Arizona, USA, (left) reveals a nearly complete sedimentary record of 360 million years of Earth's history. The chalk cliffs (below), typical of south-eastern England, were formed from organic debris.*

The direct...evidence for this history lies in the archives of the Earth, the sedimentary rocks.

Life on Earth,
DAVID ATTENBOROUGH
(b. 1926), British naturalist

deposited on continental shelves, then flood down submarine canyons, slump into the deep ocean, and travel for hundreds of miles as turbid flows. Turbid flows are a mix of water and muddy sand with twice the density of water and great gravitational power. They hug the ocean floor, spreading out in low-angle alluvial fans and depositing muddy sand over large areas. Turbid flows first drop coarse sand, then finer grades, in order of reducing size.

One slump produces a number of cascading flows, so the deposits are rhythmically layered. The resulting sedimentary rocks, turbidites, are a major component of all rocks deposited in ocean basins. These are squeezed when plates collide, becoming the predominant continental sedimentary rock.

Particle Size

A simple approach to the study of sedimentary rocks is to consider three grades: mudstone, sandstone, and conglomerate. Mudstone has no visible grains, and such fine sediments settle gradually on quiet flood plains, in lakes, or deep ocean environments. Sandstone has visible grains like coarse sugar, up to almost ⅒ inch (2 mm). To move particles of this size, the water must have some force, such as in a river. For larger particles, which form conglomerate pebbles, greater force, such as a high flood, is needed.

Descriptive words can be added to grain size to indicate grain composition, color, cement, and other visible features, for example, "black fossil-bearing mudstone with pyrite grains". Sediments may have internal patterns that reveal their depositional environments, such as the desert sand dune cross-beds that typify the Navajo Sandstone of Zion National Park in the USA.

Many sediments are carried by wind. For example, volcanic ash blows into nearby lakes and the ocean. Some falls into rivers and travels to the sea as mud. Airborne ash drops over vast areas, and will become a broadly recognizable future sediment layer.

Some sedimentary rocks are formed from organic debris, for example, coal. Such matter, collected by flood waters, is dumped in basins and covered by massive riverborne sand waves that squeeze out air and prevent rotting.

Changed Rocks

Heat and pressure can change rocks and begin a new episode in their life story—the variety of metamorphic rocks is exceptional.

Rocks that are exposed to heat and pressure, can change in character, a process known as metamorphism. There are two different types of metamorphic rocks, formed in different environments.

When continents collide, rocks can be metamorphosed over an entire region, which is known as regional metamorphism. Localized, or contact, metamorphism occurs when hot igneous rock rises to the surface and heats surrounding rocks, causing them to be recrystallized. These changes alter the physical environment in which the original minerals formed, forcing new minerals to form. The effects of metamorphism range from simple compaction to a total remake of the rock.

Marble is perhaps the best known metamorphic rock, and the most famous marble is the Carrara marble from Italy. This lustrous white stone has been quarried for 2,000 years, and was the choice of Bernini and Michelangelo for sculpting. Marble is limestone that has been recrystallized, a process that pushes impurities outside the growing new crystals. This results in large, clean grains surrounded by dark shadings of impurities, which creates the beautiful marbled effect. When large grains grow, they often obliterate earlier textures, such as fossil shapes, from the limestone.

GLACIAL EROSION *has sculpted this exquisitely foliated schist (above). Strong regional squeezing crystallizes minerals such as pyrite (top left) and garnet (red, far right).*

Regional Metamorphism

When plates collide, the squeezing and burial of rocks produces high-pressure minerals, whereas heating rocks forms minerals stable at high temperatures. Subducting crust is subjected first to high pressure, then to high temperature. This produces a gradation from high-pressure to high-temperature minerals, which can be recognized in the field. Colliding plates cause folds that are regionally aligned, and squeeze all the minerals in the same

THIS TIGHTLY FOLDED *metamorphic structure features visible quartz veins.*

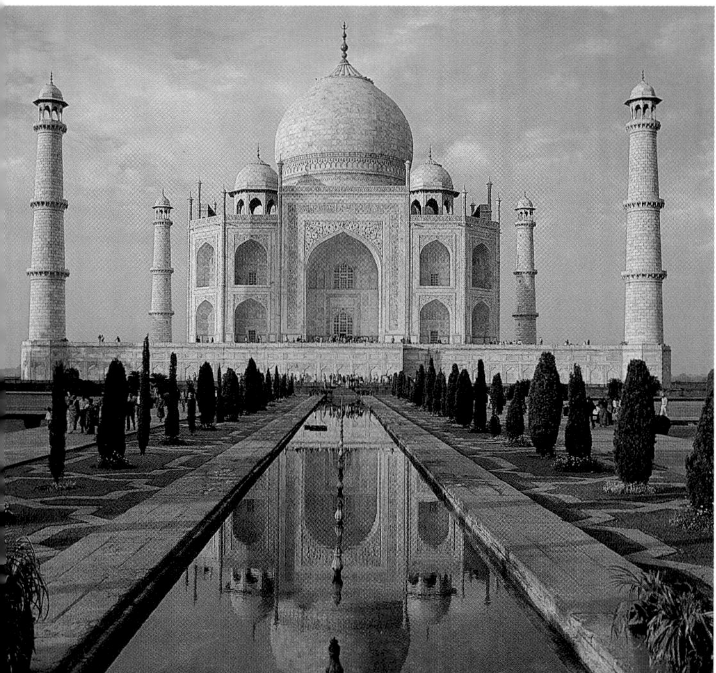

THE TAJ MAHAL *(left), in India, was constructed from marble.*

direction. There is no such alignment where contact metamorphism occurs.

A regional squeeze generates many mica minerals, all precisely aligned. Regional pressure assists such thin, flat, disc-shaped minerals to grow. They do so with the flat sides perpendicular to the squeeze. Mica splits very easily into flat sheets, and with all mica crystals identically aligned, the whole rock splits into sheets or slabs. Slate, formed from fine-grained rocks that have been mildly heated and strongly squeezed over geological time, is a good example.

If the squeezing continues, slate changes to phyllite, a rock with more strongly developed micas, but with some higher-pressure minerals, such as garnet and pyrite, starting to crystallize. The minerals in metamorphic rocks can either be randomly spread throughout the rock, or foliated, meaning that the minerals are arranged in rough planes, producing a striped appearance. Foliation occurs at temperatures approaching melting point, when elements have greater mobility.

Even higher pressures and temperatures produce schist, a coarse-grained, much more foliated rock than phyllite, with larger minerals forming and segregating. Amphibolite is similar, but from a parent rock of basaltic composition rather than from a more quartz-rich origin. Schist is distinguished by the growth of the mineral garnet in a body of micas. Garnet shows how strong, round crystals can grow as temperature rises. Gneiss forms after schist and is a strongly foliated, coarse, metamorphic rock that resembles granite in its minerals. Beyond gneiss, rocks become molten and igneous.

Contact Metamorphism

Where rock is metamorphosed by contact with igneous bodies, chemical changes are possible. Elements may be added or subtracted by fluids escaping from granite plutons (rising bodies or "blobs" of hot molten rock).

Temperature decreases outward from the granite pluton, so the degree of contact metamorphism lessens with distance from the intrusion. Some of the minerals that characterize metamorphic grades are (in order of increasing temperature): chlorite, biotite/andalusite, garnet, kyanite, and sillimanite. The high-pressure minerals are (in order of increasing pressure): mica, garnet, kyanite, and sillimanite.

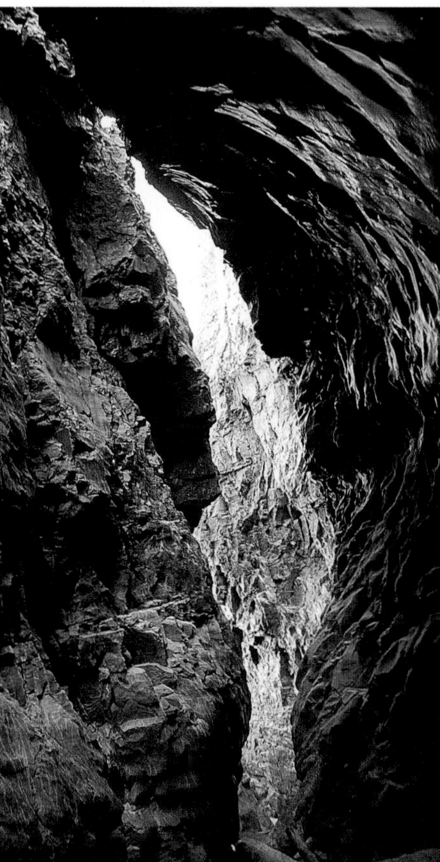

QUARTZITE *in this gorge in the MacDonnell Ranges, Australia, is 1,000 million years old.*

THE RECYCLING *of* ROCKS

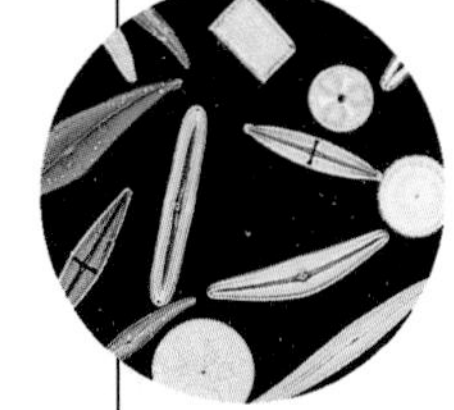

The Earth recycles rocks by natural processes that, even as they are finishing in one place, are starting again elsewhere.

All of Earth's minerals and rocks follow a cycle of continuous breakdown and rebirth, a process driven by plate tectonics. On the one hand, minerals are being crystallized from magma, sediments are being cemented into rocks, and rocks are being reformed by high heat and pressure. On the other hand, as physical changes occur, particles are being eroded from rocks, dissolving in water, and being moved by rivers, ice, wind, and gravity. As continents collide, plates sweep sediment into fold belts and bury them so deeply that they melt, to begin the cycle again. All minerals are driven toward a stable form, but they alter as their environments change.

HALF DOME *(above), Yosemite, California, USA, a vertical cliff of granite. When the skeletons of diatoms (above left) are cemented with dissolved quartz, they form chert.*

Standing on the edge of Half Dome in Yosemite, California, USA, looking all the way down the vertical, granite cliff to the valley, the deep gorge appears to have been cut by the small Merced River and Tenaya Creek, but how? The answer lies in the changing environments affecting the valley. The Yosemite granite began as deep, hot magma under high pressure, and crystallized progressively as it rose and cooled. Part-way up, it became solid, with all minerals stable at that level and tightly interlocked. As the rise continued, both temperature and pressure fell, upsetting stability and starting the process of breakdown.

Falling temperature shrinks minerals, while falling pressure expands them, so the dimensions of each mineral alter as the granite rises. Cooling produces a network

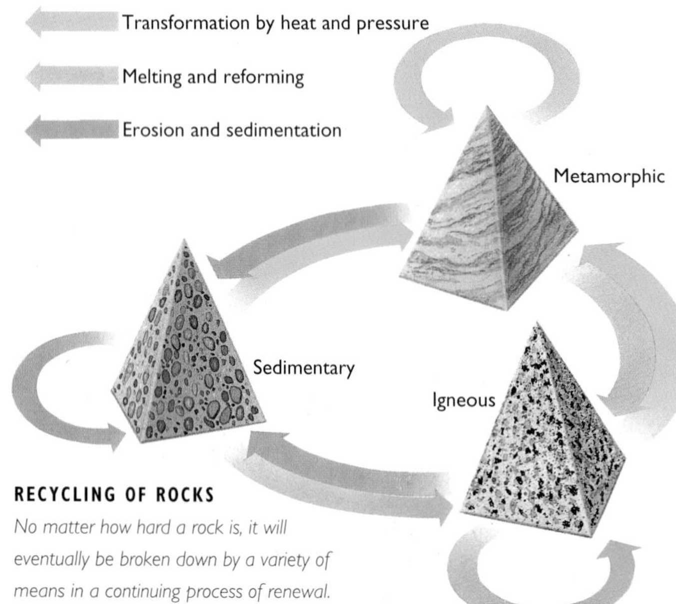

RECYCLING OF ROCKS
No matter how hard a rock is, it will eventually be broken down by a variety of means in a continuing process of renewal.

CLEARLY DEFINED *layers (left) of sandstone and diabase, Antarctica. Clay particles collect on flood plains (below).*

Nothing is wasted,

nothing is in vain:

The seas roll over

but the rocks remain.

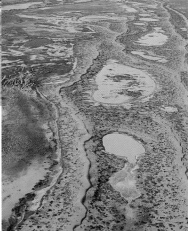

Tough at the Top,
A. P. HERBERT (1890–1971),
English writer and poet

of shrinkage cracks throughout the granite. Cracks develop among adjacent grains as the minerals adjust to changing stresses, and this allows water and oxygen to penetrate deep into the granite. Stresses fracture feldspar grains along internal cleavage planes, speeding the removal of soluble elements.

The once-solid granite gradually changes to a collection of loosely attached grains, ready to be eroded by water, wind, and ice. With the breakdown of the easily soluble components of feldspar, particles of clay are formed. Rivers move clay and quartz to the sea, where it collects in deep marginal basins, waiting to be folded and melted, continuing the cycle. About 40,000 years ago, the onset of glaciation carved out a U-shaped valley at Yosemite, truncating ridges, moving debris, and creating Half Dome's vertical cliffs. As the glacier melted, the valley floor filled with debris, leaving the river little to do but rework and sift clay from quartz sand.

WEAK POINTS

Easily soluble minerals are the weak points of rocks. As soon as one mineral is penetrated by groundwater, others come under attack. Every crack speeds rock breakdown—faces erode slowly, edges faster, and corners fastest, leading to the roughly spherical blocks that characterize all granite areas.

Many other processes speed breakdown. Limestone is directly dissolved by rainwater, assisted by soluble gases and plant acids, which greatly accelerate the process. All minerals are soluble eventually, even quartz. In the swamps of brackish Macquarie Harbour, Tasmania, Australia, brown water, rich in organic acids from button grass, dissolves glass bottles by about 1⁄50 inch (0.5 mm) every 50 years. In geological terms, this is a phenomenal rate for dissolving the most common sedimentary sand grain, quartz, from which glass is made.

As a result of the vast total volume of dissolved quartz that rivers bring to the ocean, quartz is spontaneously precipitated as light showers of super-fine crystals that settle as an ooze on the deep ocean floor. This acts to cement the decaying silica skeletons of ocean-dwelling diatoms (microorganisms that extract dissolved quartz from sea water) to form the rock chert.

Soluble minerals may remain in solution as part of the saltiness of the ocean, or form salt lakes in areas of restricted drainage. Clay mineral particles are very small and are easily transported in suspension. They settle only in quiet conditions, so collect in lakes, oceans, and on flood plains—until the next plate collision, when the cycle continues.

SOLITARY PINNACLE *left by erosion at Hells Half Acre, Wyoming, USA.*

How Soils Form

We take soil very much for granted, but we must heed the lessons of the past so that we do not waste its precious nutrients.

GLACIERS *are the mechanism by which new soil is created (left). Pitcher plants (top left) supplement poor soil by consuming insects.*

Soil is a mixture of weathered fragments of rocks and biological material. It mostly forms from the local bedrock and debris moved in from elsewhere, sometimes from quite distant areas. The value of soil depends on the nutrients it contains, and this includes both the breakdown products of plants, and elements contained in rocks that plants extract. Primary rocks, such as basalt and granite, bring a variety of nutritious elements up from the Earth's mantle. If sandstone contains little but quartz, it is of no value to most plants. Being a recycled rock, made from weathered grains of other rocks, the nutrients once present have been removed by chemical and mechanical means and by plants. When crops are grown in quartz-rich soils, they must usually be supported by the addition of fertilizer.

Glaciers bulldoze all past soils and mix them with newly ground till, the debris of glacial scraping. They then cover the scraped land with thick deposits of new till. This is, perhaps, the best of all soils, because plants have direct access to the nutrients in the finely milled rock "flour". Glaciers continuously deposit large quantities of till as they melt. Extending glaciers push up large mounds of till, while retreating glaciers dump hummocky sheets of it over the land. The till is then reworked by the strong winds that are a feature of glacial areas, and the more consistent winds of the open plains.

North v. South

There is a fundamental difference between the soils of the northern and southern hemispheres, arising from their different geographic locations during the most recent ice age. Between 40,000 and 10,000 years ago, atmospheric cooling created major Arctic and Antarctic ice caps. The massive sheet glaciers that formed at the poles spread away from these centers, gouging and scratching the rocks beneath, grinding some to dust, and moving the debris across the land.

All the northern continents, connected as Laurasia around the North Pole, were

HAYSTACK ROCKS

As one travels across the endless northern prairies and through New England, North America, large "haystacks" appear in the distance. These are solitary monolithic rocks. No other rocks may be visible in any direction. Close inspection shows that they sit on soil, but no amount of digging reveals any bedrock. They are glacier-transported haystack rocks, or erratics, the source of which may be 1,000 miles (1,600 km) closer to the North Pole than where they stand. Most haystack rocks bear no resemblance to local bedrock, but if you follow the direction of past glacial travel back toward its source, the origin can be found. They were carried and dumped amid the debris of glacial scraping by a melting glacier in the most recent glaciation. Prairie winds then totally exposed them.

RICH SOILS *are the heritage of northern continents (above), while those of the south are poor, except in volcanic areas, such as the Andes foothills (right).*

covered in ice in the most recent glaciation. The glaciers moved south over Europe, Asia, and North America, spreading till as they melted and retreated. Antipodean glaciers ran into the ocean around the perimeter of Antarctica and no other southern continents were affected. The result is now reflected in the agricultural wealth northern hemisphere countries enjoy, all top-dressed with freshly milled rock flour. Southern hemisphere continents (Australia, Africa, and South America), largely unaffected by that glaciation, have to rely on older, poorer soils.

The real difference lies in the available nutrients. Northern-hemisphere soils consist of a deep layer of rock flour. Southern-hemisphere soils, often very shallow, have little crystalline rock mineral nutrient and each crop of plants waits on the death and decay of the previous generation. When crops are harvested from southern hemisphere soils, nutrients for subsequent plant generations are totally removed, and cannot be replaced from the local barren soils. For example, wheat crops from Australia and trees used for wood chips in the Amazon Basin of South America both grow in quartz-sand soils. Harvesting from these areas guarantees future deserts in both locations. Although clear division is made here between northern and southern hemispheres, the real division lies between areas that have and have not undergone a recent glaciation.

In southern hemisphere countries, the soil that was present at the time of the assembly of Gondwana is still in use. Rain during 250 million years has leached soluble minerals from this soil, leaving only the most insoluble residues: quartz, iron oxide, and aluminum oxide. These are the typical components of desert sands, and a long, slow succession of plants is needed to return such areas to forest.

DEFORESTATION *in Madagascar has caused severe soil erosion (right).*

Happy the man whose lot is to know
The secrets of the Earth.
EURIPIDES (480–405)
Greek tragic playwright

Chapter Three
Unlocking *the* Secrets *of* Rocks

How Geology Affects Us

All plants and animals, including ourselves, are composed of the Earth's elements, and have evolved to survive the conditions that exist here.

The Earth and all its elements are stardust, produced in a cosmic process that began between 10,000 and 20,000 million years ago. We are dependent upon the Sun's rays for energy, with Earth's atmosphere providing a protective blanket against harmful solar radiation. Energy from the Sun is also stored on the Earth's surface and in its upper layers, in the form of wood, coal, petroleum, and natural gas.

Life forms have evolved to survive the specific conditions that exist on Earth. Any changes to these conditions, such as atmospheric composition, barometric pressure, the temperature on the surface, or the strength of the gravitational field, would produce devastating results. Even relatively slight changes can be detrimental, as demonstrated by the difficulties that we face when exposed to depths, high altitudes, or extreme temperatures.

In the same way as our bodies have evolved in response to naturally occurring conditions, so too have our cultures and customs. The length of a day and a year, and our seasonal schedule of planting and harvesting crops, are determined by the Earth's rotation around both its own axis and the Sun.

The Earth's Elements

Elements are made up of atoms, which are in turn made up of protons, neutrons, and electrons. The elements occurring on Earth are listed in the Periodic Table. There are 109 elements, listed in order of the number of protons contained in their nuclei (their atomic weight), although only 94 of them actually occur naturally. Eight of these elements make up virtually all matter, with the others occurring only in tiny amounts.

The first and lightest element on the table is hydrogen, whose atom contains only one proton and one electron. It is so light that it cannot exist as a free gas in

ELEMENTS *are arranged on the Periodic Table, a portion of which is shown above with the native element gold. Humans and the Earth itself consist mainly of only a few elements (below). Hydrogen, the lightest element, was used in airships, but its flammability led to disasters such as the loss of the* Hindenburg *(left) in 1937.*

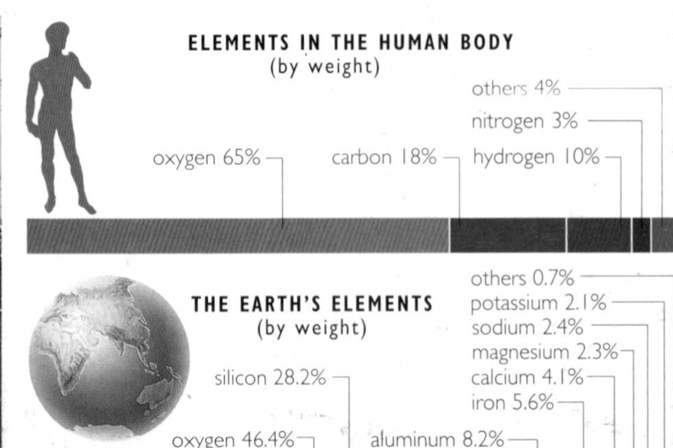

our atmosphere, escaping from the Earth's gravitational pull into space. At the bottom end of the table, the atoms are so large and complex that they are likely to break down into lighter, more stable "daughter products", releasing vast quantities of nuclear energy in doing so. Unstable elements, such as uranium, comprise about a third of the table but are very rare.

Minerals are naturally occurring elements or chemical compounds, and have distinct, measurable physical and chemical properties. A few elements, such as sulfur, carbon, silver, gold, and copper, can occur alone in their natural state, and are known as native elements or elemental minerals. Others exist only in various combinations. For example, sodium chloride, or common salt, is made up of sodium and chlorine and occurs naturally as the mineral halite. On the basis of chemical composition, the main mineral groups are the silicates, carbonates, oxides, sulfides, halides, sulfates, and phosphates.

There are now more than 3,600 mineral species known to science and the number is steadily increasing. Rocks, in turn, are composed of minerals in various combinations.

The Human Element

Surprisingly, human beings are made up of very few of the simplest elements: carbon, hydrogen, oxygen, and nitrogen, with some others present in tiny amounts. We also rely on a steady intake of elements, such as calcium, iron, sodium, potassium, and phosphorus, with our food to maintain our physiological well-being.

A single morning in our busy lives will see us encountering about a fifth of the elements on the Periodic Table. The house in which we live, the food, utensils, and power we use to make our food, and the clothes and jewelry we wear, are all derived from the same building blocks. By the time we've entered the world of modern technology, virtually all of the table's stable elements will have been encountered.

THE PERIODIC TABLE

The Periodic Table was formulated in the mid nineteenth century by a Russian chemist, Dimitry Ivanovich Mendeleyev (1834–1907), who grouped the 62 elements known at the time in order of increasing atomic weight. Element 101, a recent addition to the Periodic Table, is known as mendelevium, in recognition of his contribution.

The most common element in the universe is hydrogen, accounting for 90 percent of all known matter. The rarest of the naturally occurring elements is astatine, with only 0.0056 of an ounce (0.16 gram) present in the Earth's crust. The lightest solid element is lithium and the heaviest is osmium.

New elements are being added to the table all the time. In addition to the 94 naturally occurring elements, a further 15 heavier elements have been artificially developed. In 1982, physicists working in Darmstadt, Germany, announced that they had created an isotope of element number 109, unnilenium, from the decay of a single atom.

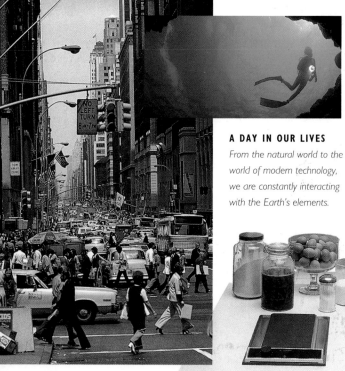

A DAY IN OUR LIVES

From the natural world to the world of modern technology, we are constantly interacting with the Earth's elements.

Starting *a* Rock Collection

To observe, study, and collect rocks allows one to marvel at the natural beauty and complexity of our planet.

Rocks and minerals are found all around us, whether we live in a crowded city or the remote countryside. They are used to build almost everything we use, and are always present beneath our feet.

Get to know the rocks around you. If you live in a city, a wealth of rocks can be found making up the buildings themselves. Note any information you can find about them. How are the rocks being used? Are they igneous, metamorphic, or sedimentary? Note their colors and textures and, of course, their location.

For rocks in natural formations, note any characteristic structures such as columns, spires, or stacked boulders. These usually typify particular rock types. How far do you have to go before the rocks around you change? If there is a cover of soil on the rocks, note its thickness, color, and texture, and whether the vegetation it supports changes as the rocks do. Try to identify the minerals that make up the rocks using the simple tests described on pages 70–75, and note what elements make up these minerals. Persevere, bearing in mind that identification skills will come with practice.

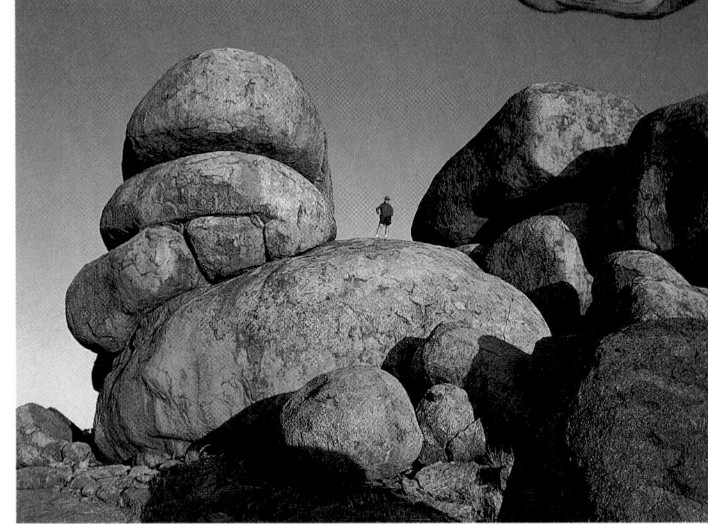

START BY IDENTIFYING *the rocks around you. These boulders in Australia (above) are granite, an igneous rock. Your collection might include carved stones, such as this turquoise medallion (top left), or cheaper ornamental stones, such as agate (top right).*

The Rock Collection

One of the best ways to learn about geology is to start a collection. You could begin with simple carved and polished items, such as statues, book ends, or slabs, and get to know the rocks and minerals used. Inexpensive ornamental stones, such as agate, onyx, limestone, marble, obsidian, and hematite, make a good starting point. If you begin by gathering rocks in the field, don't worry about trying to find the "perfect" specimen at first. Any sample of a particular rock or mineral is all that is necessary for you to become familiar with its natural features.

Eventually, most collectors are forced by the overwhelming variety of material to specialize in one particular

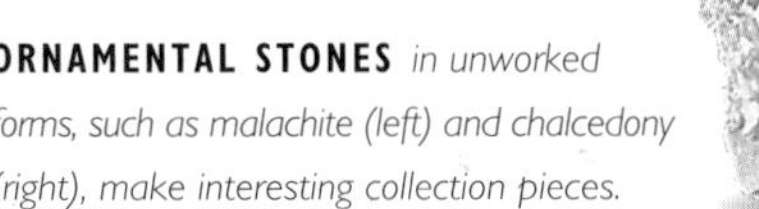

ORNAMENTAL STONES *in unworked forms, such as malachite (left) and chalcedony (right), make interesting collection pieces.*

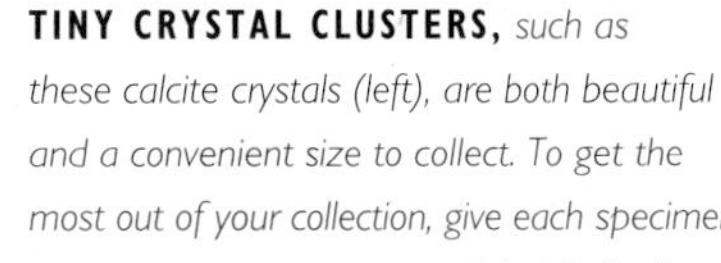

TINY CRYSTAL CLUSTERS, *such as these calcite crystals (left), are both beautiful and a convenient size to collect. To get the most out of your collection, give each specimen a label (below) and then add a description.*

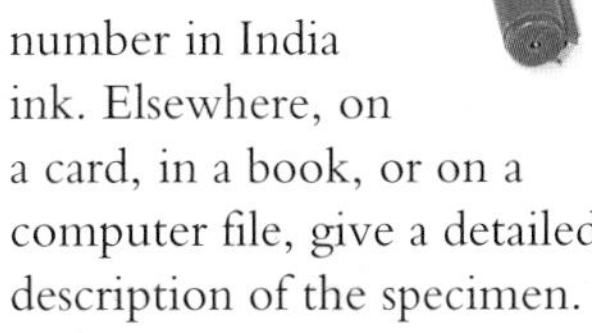

branch, for example, crystal specimens, gem minerals, ore minerals, building stones, or even particular mineral groups. You may wish to display each in a variety of forms: a fine worked piece, a cut stone, the rough material, and, if possible, a natural crystal.

Growing in popularity are miniature rock and mineral collections known as micro mounts. Tiny specimens, such as crystal clusters, are fixed in small transparent boxes that can be stored in larger trays. Ideal for apartment dwellers, such collections are enjoyed with the aid of a hand lens or microscope and are generally more beautiful and complete than collections of conventional-sized specimens for two main reasons. First, crystal clusters in cavities tend to be closer to perfect the smaller they are, and second, the rarity of some of the material in large sizes means that they are prohibitively expensive.

Any cabinet or set of drawers can be used to house a collection. Those with glass sides or tops are preferable as the exhibits can be enjoyed from all angles and less dust will settle on them.

Labeling the Collection

Labeling your collection can be just as much fun as acquiring it. Assign each specimen a catalog number by numbering the box in which it sits. If the specimens are handled frequently, you may wish to number them directly, using an inconspicuous dot of white enamel paint, and writing a number in India ink. Elsewhere, on a card, in a book, or on a computer file, give a detailed description of the specimen.

If it is a rock, list its name and constituent minerals, as well as where and when it was found or bought. If it is a mineral, list its name; chemical formula; physical properties such as hardness, specific gravity, luster, streak, and crystal system; and uses of the mineral or any elements that may be extracted from it. Labels summarizing each specimen's information can be placed in the cabinet, giving the collection a professional appearance and allowing visitors to enjoy it fully.

FINE SPECIMENS, *such as these topaz crystals growing on white quartz (left), are impressive, but may be expensive. Instead, you might collect examples of the same mineral in a variety of forms, such as turquoise (above). You can use any cabinet to house a collection (right).*

Getting Prepared

Knowing which equipment is required in the field and being well prepared will increase the safety and enjoyment of your collecting trips.

The basic gear you need for mineral and rock collecting is simple, although you can add to it over time. For a start, a geological map and a compass will help point you in the right direction, while a notebook and pencil are essential for recording details of locality, geology, and sample descriptions. You should also carry old newspaper and tape to wrap and protect each specimen, an indelible marking pen to number each package, and a backpack or a cardboard box in which to carry your specimens once wrapped.

For breaking rocks to see unweathered surfaces, or to trim up samples, a standard geological hammer is most suitable. The flat end is used for breaking rocks, while the pointed end is used for digging or gouging in soft material. Some collectors prefer a short-handled mallet with a heavier head to make breaking safer and easier. Sturdy gloves and eye protection should be worn at all times, as there will invariably be sharp chips of rock flying about.

Chisels are handy to trim down heavy specimens in the field and are relatively light to carry. A flat chisel allows you to break selectively along bedding planes or cracks and fractures. Remember, only trim a good sample if you are sure that you will not damage it. Otherwise, have it cut up later with a rock saw.

For serious rock breaking, such as when searching for mineral-filled cavities in hard basalt, a sledge hammer is recommended, along with overalls, hard hat, goggles, face protection, leggings, and gloves, to protect yourself against rock chips. When working below cliffs or quarry faces, always wear a hard hat.

Other useful items include a hand lens with ten times magnification to see finer detail, a magnet for detecting magnetic minerals, a porcelain streak plate, a Mohs' hardness set or equivalent standards (see p. 71), and a well-sealed plastic bottle with very dilute hydrochloric acid for testing carbonates, chalk, and limestones. (See pp. 70–75 for details on mineral testing.)

Metal detectors can be useful when looking for gold nuggets, as well as antiquities, such as coins, medals, rings, or bullets. Based on the principle that metals conduct electricity, they will, unfortunately, also detect cans, bottle caps, wire, and nails, but experienced operators with modern

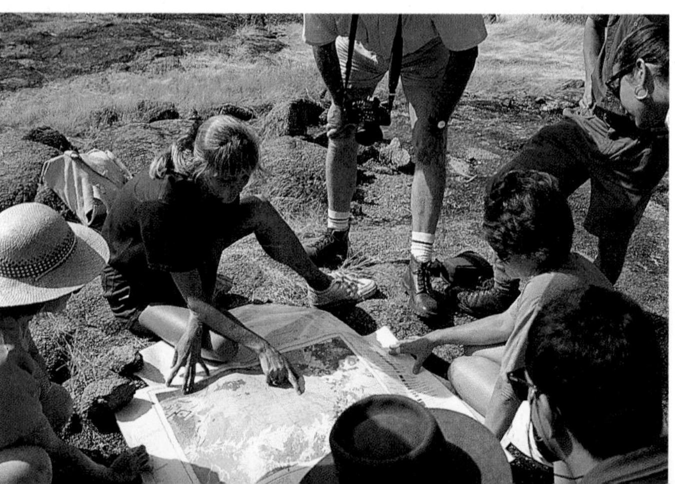

THE RIGHT EQUIPMENT *for a field trip includes a geological map (right) and clothing to protect you from flying chips of rock (top and above right).*

HOW *to* READ *the* LANDSCAPE

An understanding of the forces that shape the Earth makes geological exploration easier and more rewarding.

Minerals and gemstones are not distributed evenly throughout the Earth's crust. Rather, they tend to be concentrated within particular rock types related to specific geological events that have occurred deep beneath the ground. Often, the surface features of the landscape offer clues as to what lies beneath. However, in other cases, the geological signs are hidden far below the surface and may be glimpsed only by drilling and through the use of advanced geophysical techniques.

TECTONIC FORCES

The massive tectonic forces that move the continents and shape the ground beneath our feet transform and redistribute the Earth's rocks and minerals. A basic understanding of these processes and the effects they have on the local geology provides the key to successful geological exploration.

Mountain ranges form when continental plates collide, squeezing, buckling, and lifting the once-horizontal sedimentary strata. If the squeezing is intense enough, the sedimentary rocks will change under pressure into metamorphic rocks. These often contain a host of interesting and precious minerals, such as garnet, ruby, mica, sapphire, kyanite, andalusite, and staurolite.

The zones where ancient continents once collided are often marked by long serpentine belts and highly deformed marine rocks, such as chert, jasper, and limestone. These rocks are remnants of the once-great oceans that were crushed between the continental plates. A range of interesting minerals is associated with such rocks, including chrysotile and antigorite, chromite (the ore of chromium), pyrolusite (the ore of manganese), nephrite jade, asbestos, talc, rhodonite, and cryptocrystalline quartz.

RISING MAGMA

Beneath the Earth's thin crust, lies a hot, plastic interior. As crustal plates shift and collide, magma is generated and works its way toward the surface, sometimes erupting violently to form volcanoes. Often, however, it cools before reaching the surface, forming granite intrusions through the underlying strata. Element-rich fluids, known as pegmatites, penetrate cracks in the rocks

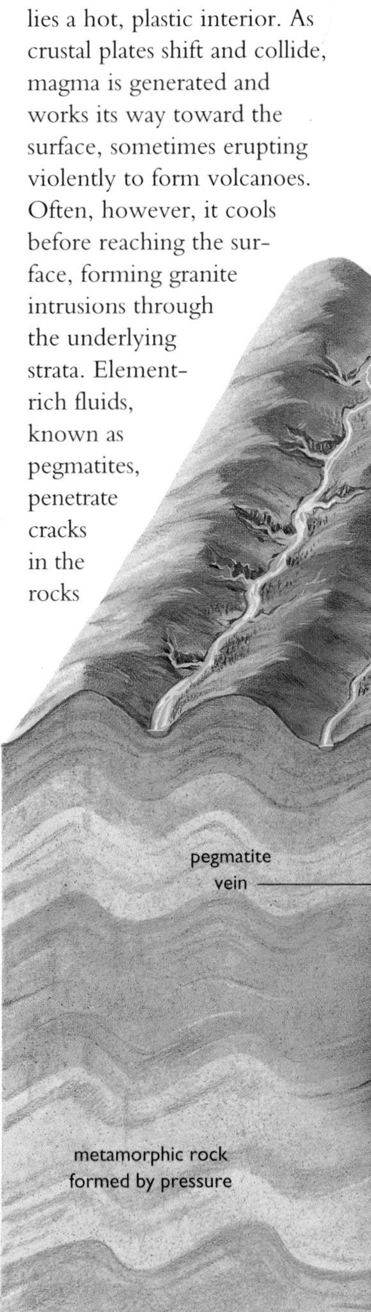

MINERALS AND GEMSTONES *are associated with particular geological features, many of which are shown in this illustration (right). Active volcanoes, such as Kilauea Iki, in Hawaii (left), may carry interesting minerals, such as peridot (top) to the Earth's surface.*

COASTAL CLIFFS, *such as this one (above) in Dorset, England, provide plently of broken rocks. Panning for gold in a river in northern Madagascar (below).*

In the field, creek beds, road cuttings, and quarries are some of the best locations to see fresh rock exposures, but always ask permission before entering quarries and keep clear of unstable areas and falling rocks. Mine dumps are a good source of fresh material and often contain minerals that miners have left behind. In Australia, collectors walking over opal mine dumps are often surprised by a good find, especially after a fall of rain, when the opal pieces, washed clean of dirt and dust, glint in the sunlight.

Coastal cliffs provide excellent opportunities for collecting as there are always piles of fallen, broken rocks at their base. Mountainous areas are also good, with large sections of rockface often free from soil cover. In areas of natural beauty, do not damage the cliff faces. Collect only from those rocks that are already broken.

Searching in creeks and rivers is a good way to find durable, heavy minerals. Check areas where they are likely to have accumulated, known as trap sites. When water slows down, its carrying capacity decreases and heavy minerals are likely to collect there, so be on the lookout for features that might slow the water, such as potholes, crevices, large boulders, rock bars running across the direction of flow, or even a sudden widening of the creek or river.

For good results, persevere with trap sites. Always excavate right back to bedrock and process the material found there, as it is against this surface that heavy material, especially gold, will accumulate. Constantly check fine material in the pan for traces of gemstones. If fine gemstones are present, you are certain to come across larger material sooner or later. Heavy storms will recharge good trap sites so there should always be something to find.

The World is the geologist's great puzzle box.

LOUIS AGASSIZ (1807–73), Swiss naturalist

HOW TO USE A COMPASS

Take a compass, hold it in front of you, and turn a full circle. There are 360 degrees in a full turn. The magnetic needle will always point north (take your magnet and other metal objects out of your pockets for this exercise or it might point to those instead).

To follow a particular direction (for example, 40 degrees north east), turn the rotating ring on the compass until 40 degrees lines up with the center line. Holding the compass before you, turn your body until the north end of the magnetic needle lines up with the north direction (N) on the dial. You can now walk in the selected direction by keeping the north needle on the N mark as you walk. This simple exercise is the basis of navigation.

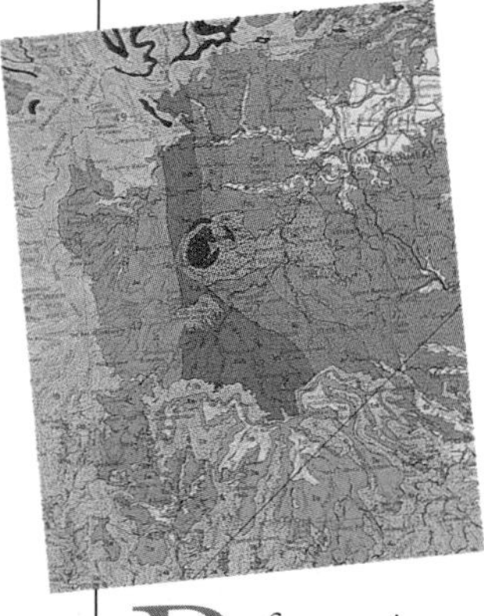

IN *the* FIELD

The shapes of mountains, the patterns of rivers, and the types of soil and vegetation on the ground, all hold clues to the geology beneath.

Before setting out to collect rocks and minerals, do a little research about the area you are exploring, and the rocks and minerals you might find there. Your collecting will be more productive and enjoyable.

READING THE SIGNS *The color of the Red Mountains (right), in Colorado, USA, indicates the presence of iron. Rivers flowing from volcanoes, such as in the Andes (below), are often a good source of gemstones. On this geological map (left), an old volcanic neck (pink) is shown surrounded by its lavas (yellow).*

GEOLOGICAL MAPS

Your rock collecting will be assisted by basic navigation skills and an understanding of geological maps. These are the same as any other maps, except they also show where different rock types occur. With such a map, you can take yourself to the most likely areas to find particular rocks, minerals, or fossils.

For every rock type, a geological map has a corresponding color and/or symbol. The key gives pertinent details about the rock, such as its appearance, age, and, often, its fossil or mineral content.

Learn to recognize the different rock types on the map and collect samples of each to compare. You will soon come to associate certain minerals and gemstones with particular rock types. This is the basis of geological exploration.

WHERE TO LOOK

Landform types, soils, and vegetation are like beacons guiding you to particular rock types. Streams and rivers often mimic the patterns in the underlying geological strata, so look at the pattern of the rivers on a standard map. Do any of the rivers head radially out from an area, or converge into a depression? This arrangement may reveal the presence of old volcanoes, often a source of gemstones. Look at the soils for further evidence. They may be a different color from other soils in the area, perhaps rich red or brown, indicating the presence of iron-rich volcanic rocks. These soils will, in turn, support a different type of vegetation or landuse pattern. The more familiar you become with these features, the easier rock collecting will be.

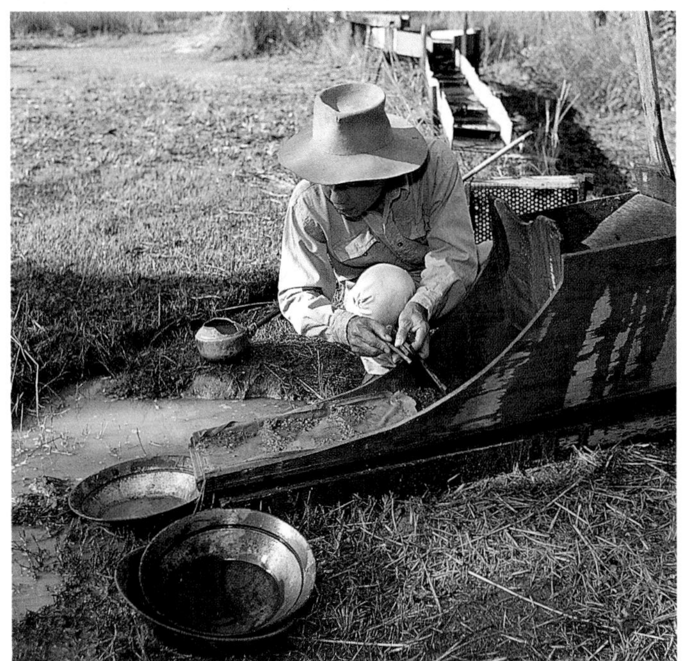

GOLD COLLECTING *with a "cradle" (left). Gravels and sediments are washed across an old carpet, which will trap any fine gold particles.*

machines can discriminate among these. Other equipment, generally beyond the scope of the amateur collector, includes ultraviolet lamps for detecting luminescent minerals and scintillometers for radioactive minerals.

Collecting in Creeks and Rivers

Creeks and rivers are good places to look for heavy minerals, such as gold, sapphire, ruby, zircon, spinel, topaz, tourmaline, chrysoberyl, and cassiterite. In order to reach the bedrock, you will need a pick, a spade, and a trowel to move small rocks and sediment. A crowbar may also come in handy for moving larger rocks, and hand-held suction pumps will help you reach material in narrow crevices, where such minerals tend to concentrate.

For fossicking on a larger scale, more sophisticated motorized equipment can be used to process more material, more efficiently. Dredges or suction pumps draw gravels from the river bed into a hand-operated pipe, and discharge them over gravity-separation machinery, known as riffle boxes or jigs. These operate like huge sieves, separating and concentrating the heavy minerals. They are generally emptied only at the end of the day, making for an exciting culmination to a day's collecting.

Precautions

Whenever you are rock collecting, dress sensibly—good, sturdy footwear and clothes that will protect you against cold, heat, rain, or sunburn. Be aware of the dangers particular to your area, whether they be tropical diseases, dangerous wildlife, unstable ground, or extreme weather conditions. Always carry maps, let others know where you are going, and get permission before collecting on private or public land. Never take unnecessary risks and be particularly careful near abandoned mine sites and quarries. Underground mine shafts are often unstable and should never be entered.

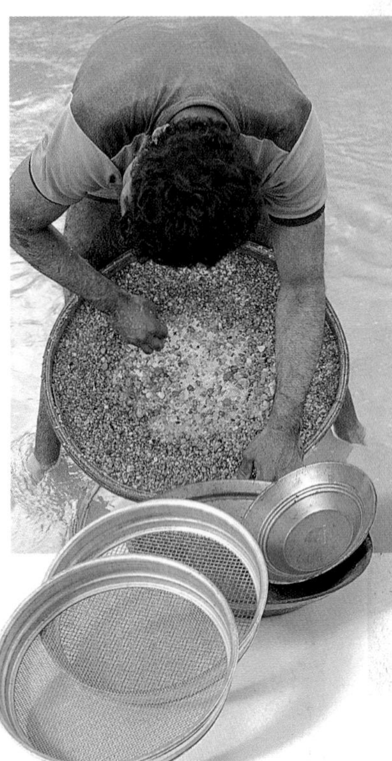

PANNING FOR GOLD AND GEMS

To isolate heavy minerals, first sieve the gravels into various sizes. Stack a coarse sieve, with holes of about ⅕ inch (5 mm), on a fine sieve, with holes of about ⅒ inch (2.5 mm), and place a gold pan underneath. Heap all of the sample material into the top sieve and work both sieves and the pan in a circular motion in water. This will concentrate the heavy minerals toward the center of the sieves and throw the lighter material toward the edges, while allowing the finest material to pass into the pan. After several minutes of sieving, flip the contents of each sieve onto a hessian bag and pick out any gemstones with tweezers. If gold or small gems are present, they will be concentrated in the bottom of the pan.

THE PATTERN *of drainage of the Green River, in Utah, USA, is starkly illustrated in this aerial view (right).*

surrounding the intrusion and produce large and beautiful crystals of minerals such as topaz, beryl (emerald), tourmaline, cassiterite (the ore of tin), quartz, feldspar, gold, and mica.

Around the edges of these intrusions, heat from the granite transforms the surrounding rocks to produce new minerals such as calcite, garnet, wollastonite, and lapis lazuli. Erosion will eventually expose these granitic bodies as circular outcrops that can be located using a geological map.

When magma reaches the surface, it may form conical shields of basalt. The rising magma cuts through and covers up all earlier geological structures and often carries minerals, such as sapphire, ruby, zircon, spinel, and peridot, to the surface from deep within the Earth. After weathering, these minerals will be released from the basalts and end up in the rivers that drain from the volcano, from where they can be collected with sieves and pans.

Small, but explosive, volcanoes are formed when silica-poor lavas erupt from deep within the mantle. These eruptions, sometimes reaching the surface in a matter of days, can carry with them one of the most sought after of all gems—the diamond.

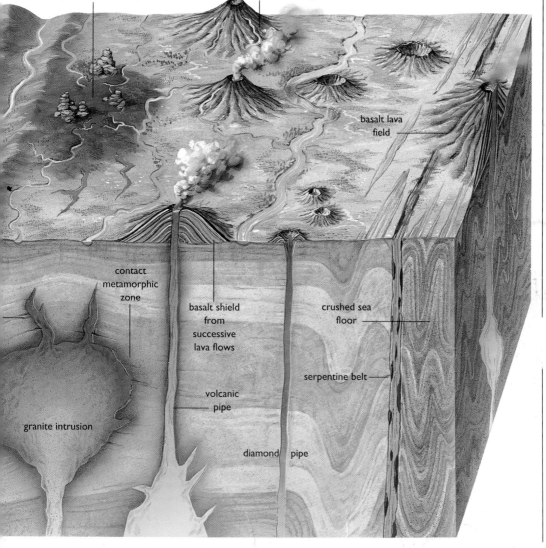

exposed granite body
radial drainage of rivers
basalt lava field
contact metamorphic zone
basalt shield from successive lava flows
crushed sea floor
serpentine belt
volcanic pipe
granite intrusion
diamond pipe

MINERAL IDENTIFICATION

Any collector, amateur or professional, can identify most minerals by using a few simple tests.

Identifying minerals is easy and satisfying. The tests you need to carry out are based on fundamental mineral properties, and most involve no equipment at all. Others involve inexpensive equipment that you can use in the field or at home. Often, one or two tests will be enough to establish a sample's identity, although occasionally, you may need to take it to a professional for further testing.

The properties that define any mineral are as follows: hardness, solubility in acid, magnetism, specific gravity, color, streak, transparency, luster, sheen, fluorescence and phosphorescence, crystal system, habit, cleavage, and fracture. These properties depend on the arrangement of a mineral's atoms, rather than merely its chemical composition. For example, diamond and graphite are identical in chemical composition, being pure carbon, but are as different in their properties as any two minerals could be.

MINERAL *identification can usually be done in the field (top). Brazilianite and mica crystals growing together (right) are easily distinguished.*

Diamond is the hardest known substance, whereas graphite is one of the softest; diamonds are transparent and shine brilliantly, whereas graphite is opaque, dull, and black. These differences are all due to the structure of the carbon atoms within the minerals. Those in diamond have a cubic arrangement whereas the atoms in graphite have a hexagonal arrangement.

With practice, you will carry out most identification tests automatically. As you pick up a mineral sample for the first time you will notice its shape and color, and whether or not you can see through it. Perhaps it looks oily or resinous, or reflects light in a certain way. With each property you identify, you greatly reduce the range of minerals that your sample might be. If this all seems a bit daunting, remember that the common rock-forming minerals are few, and with a little practice, you should easily learn to recognize them.

DIAMOND AND GRAPHITE *(above) are chemically identical but have very different physical properties.*

IDENTIFICATION TESTS

Take your mineral specimen, or the rock containing the minerals that you wish to identify (provided they are big enough to see with a hand lens), and work your way through the following tests. By comparing your results with a list of standard mineral properties, you should be able to identify your specimen.

HARDNESS

A mineral's hardness is revealed by trying to scratch it with another mineral or an object of known hardness. A scale, grading from 1 (talc) to 10 (diamond), was established by the German mineralogist, Friedrich Mohs (1773-1839). Minerals with higher Mohs' numbers will scratch those lower on the scale. For convenience, a number of everyday objects of known hardness can be used (see table).

A word of warning—never use this test on good specimens as it leaves a permanent scratch. Also, don't mistake hardness for toughness. Many a good diamond has been destroyed in the mistaken belief that diamond, being the hardest known mineral, is able to

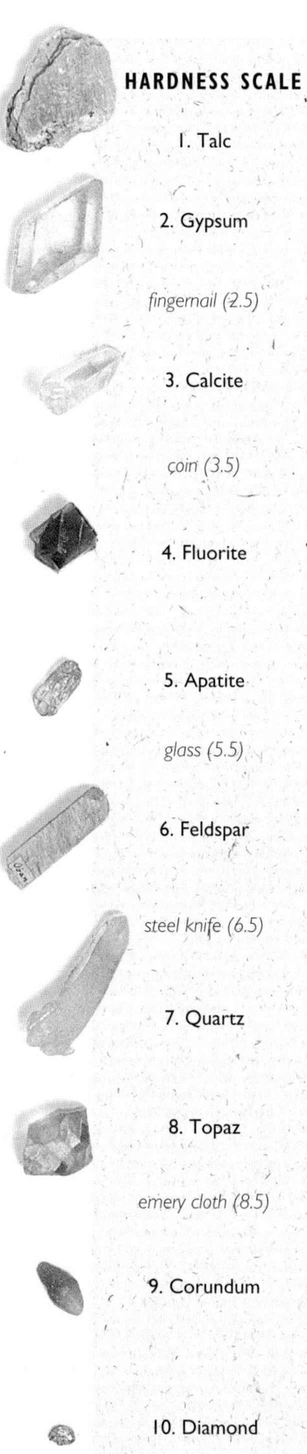

HARDNESS *Mohs' scale of hardness, from softest to hardest, combined with everyday objects you can use in the field. If an unknown mineral will scratch feldspar (6) but can be scratched by quartz (7), it has a hardness between 6 and 7.*

resist a good hammer blow. Instead, it will break into smaller fragments with a perfect octahedral cleavage.

The Acid Test

This is an easy field test for the presence of carbonate minerals, such as calcite, and carbonate rocks, such as limestone, marble, and chalk. Carefully place a drop of dilute hydrochloric acid on a fresh surface of the rock or mineral to be tested. Acid reacts with any carbonate to produce carbon dioxide gas so, if bubbles form in the acid, carbonate is present in the sample. Again, do not try this on good samples.

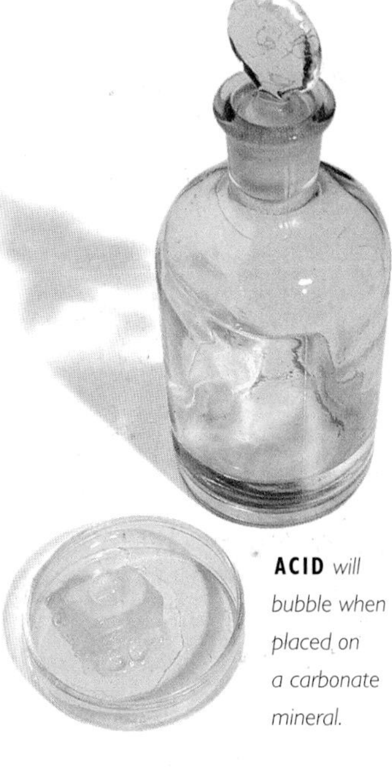

ACID *will bubble when placed on a carbonate mineral.*

Magnetism

A simple test for the presence of iron in a sample is to use a magnet. Minerals with high iron content, such as pyrrhotite, are attracted by a hand magnet. Weak magnetic samples will cause a compass needle to rotate.

Specific Gravity

Specific gravity, or density, is the relative weight of a mineral compared to the weight of an equal volume of water. Unless you are experienced, you will need to do this test at home. However, it is well worth doing, as a mineral's specific gravity is a very reliable means of identification. If your specimen contains more than one mineral, this test is not applicable.

For this test you need a balance that can weigh the mineral normally in air and again suspended in water. The mineral's weight in air is divided by the difference between the two weights to obtain its specific gravity.

With a little experience, you will be able to estimate specific gravity by hand, distinguishing between light minerals (specific gravity 1-2), such as sulfur and graphite; medium minerals (2-3), such as gypsum and quartz; medium heavy minerals (3-4), such as fluorite and beryl; heavy minerals (4-6), such as corundum, and most metal oxides and sulfides; and very heavy minerals (more than 6), such as cassiterite. Native gold and platinum are the heaviest minerals, with specific gravities of about 19.

LODESTONE *(left) was once used for navigation because of its magnetic properties. Identically sized samples of rhodochrosite and calcite reveal their different specific gravities when placed on a scale (below). Calcite is clearly lighter.*

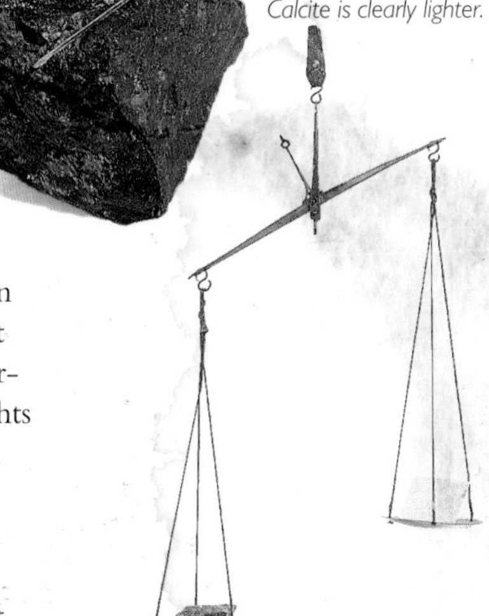

MINERAL IDENTIFICATION *by* OPTICAL EFFECTS

Perhaps the first thing you notice when you find a mineral is its appearance in the light. Color, sheen, luster, and transparency all offer clues to its identity.

A mineral's color may not always be a diagnostic characteristic. This depends on whether the color is caused by elements essential to the mineral's composition (idiochromatic minerals), or by non-essential trace elements (allochromatic minerals). Idiochromatic minerals include rhodonite and rhodochrosite (pink, colored by manganese), malachite (green, colored by copper), olivine (green, colored by magnesium), and sulfur (yellow).

The color of allochromatic minerals (particularly non-metallic ones) is extremely variable, and will not identify them. For example, the minerals corundum, apatite, tourmaline, beryl, and fluorite are colorless and completely transparent when pure, but may be found in a variety of colors, depending on the trace elements present when the crystals were growing.

In these cases, other optical effects, such as streak, transparency, luster, sheen, and luminescence can be used to identify them.

STREAK

When a mineral is rubbed against an unglazed, white, porcelain streak plate (the back of a ceramic tile will do), it leaves a distinctive streak, or line of powdered mineral. The color of this streak remains constant regardless of trace elements. For example, the allochromatic minerals corundum, apatite, tourmaline, beryl, and fluorite all leave a white streak. Streak is very distinctive for the metallic minerals. For example, brassy yellow pyrite and chalcopyrite leave a greenish-black streak, black hematite produces a cherry-red streak, black pyrolusite produces a black streak, black wolframite leaves a brown streak, and black cassiterite leaves an almost white streak.

STREAK *is constant, regardless of a mineral's color. From left to right are cassiterite (whitish streak), chalcopyrite (greenish-black), and malachite (green).*

TRANSPARENCY

A mineral's transparency refers to the ease with which light can pass through it. There are three categories of transparency: "Transparent" minerals, such as quartz, topaz, and beryl, allow objects to be seen clearly through them. "Translucent" minerals, such as opal, jade, and chalcedony, allow some light to pass through, but an object would not be visible through them. "Opaque" minerals, such as malachite and hematite, allow no light to pass through.

Minerals found in rivers tend to have abraded, frosted surfaces and these will need to be polished for the minerals to reveal their true transparency.

[illegible] fluorite, [illegible]ty of colors [illegible]tify malachite [illegible]ays green.

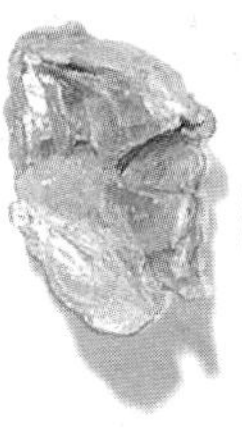
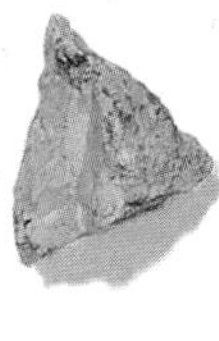

MINERAL LUSTERS *include (left to right) dull (kaolinite), silky (asbestos), metallic (galena), vitreous (calcite), pearly (talc), and waxy (turquoise).*

LUSTER

This refers to the appearance of a mineral's surface and is dependent on the way that light falling on it is absorbed or reflected. Lusters vary from "dull" (like kaolinite) to "adamantine" (diamond-like). Other distinctive lusters include "vitreous", or glassy (like quartz and beryl), "metallic" (like many of the metals, metal oxides and sulfides), "waxy" or "greasy", as if covered with a thin layer of oil (like turquoise), "resinous" (like amber), "pearly" (like pearl, moonstone, and talc), and "silky", as shown by minerals with a finely fibrous structure (such as asbestos and "satin spar" gypsum).

SHEEN

Sheen is produced by the reflection of light from within a stone, and is, therefore, caused by the internal structure, or inclusions within the mineral. Sheen must not be confused with color, nor luster, which is purely a surface effect. For example, the body color of a precious opal may be white, grey, black, or colorless, but the magnificent rainbow effect known as "play of color" is a sheen caused by the refraction of light from the ordered arrays of tiny, silica spheres that make up the opal structure. Sheen effects are best seen in cut and polished stones, but are discernible in rough material.

There are a number of terms used to describe sheen. "Iridescence" is the rainbow effect seen on the inside of some shells and on the polished surfaces of labradorite feldspar. "Adularescence", otherwise known as "schiller", is the sheen of silvery-blue light seen in semi-transparent moonstone feldspar and some opal. "Chatoyancy", or cat's-eye effect, is the reflection of light from parallel fibers or oriented needle-like inclusions within the mineral. This appears as a single, bright line when the mineral is cabochon cut. It is seen in quartz that contains parallel asbestos fibers (known as cat's-eye quartz), silicified crocidolite asbestos (tiger's-eye) and cat's-eye chrysoberyl. "Asterism" is chatoyancy occurring in several directions. This is the star-like effect for which star sapphires and rubies are famous.

LUMINESCENCE

Not all optical effects are visible to the naked eye. Certain minerals, such as scheelite, fluorite, calcite, zircon, common opal, and diamond, may be optically excited when exposed to ultraviolet light. This is known as fluorescence, and as phosphorescence if the glow continues after the light has been switched off. In the darkness, these minerals glow in stunning violets, greens and reds, quite different from the colors they display in natural light. Collectors often make special cabinets to enjoy the full beauty of these minerals.

COMMON SHEEN EFFECTS *include asterism (star ruby, left), blue-green iridescence (labradorite, right), and play of color (opal, below).*

FLUORESCENCE *Opal (L... under ultraviolet and norm...*

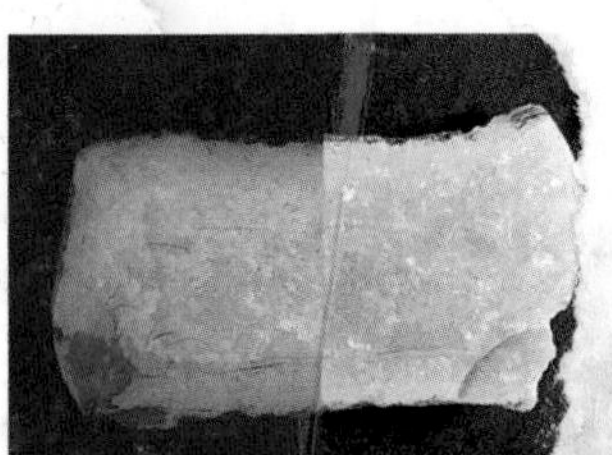

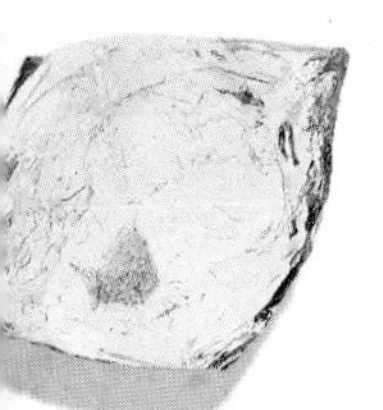

TRANSPARENCY *The three categories of transparency are (from left to right) transparent (quartz), translucent (chalcedony), and opaque (malachite).*

Identifying Minerals *by* Physical Properties

The visible form of a mineral is a demonstration of its invisible atomic structure. As such, physical properties provide excellent clues to a mineral's identity.

Many minerals can be identified by their external shape. First, look carefully at the crystal faces. Can you tell what shape they are and at what angles the faces meet? If so, you can determine the crystal system to which they belong. Crystals also have a distinctive appearance, or habit. Like people, they can be short and fat (tabular), tall and thin (acicular), or somewhere in between.

The crystal faces may have distinguishing features, such as lines (striations) or geometric pits and bumps, which are often specific to a particular mineral. Examination of broken surfaces also reveals clues. Have they fractured irregularly or broken along a series of parallel flat surfaces (cleavage planes)? With practice, you will learn to recognize these properties.

Crystal System

Minerals will always crystallize into one of six crystal systems reflecting their internal molecular arrangement. Perfect crystals are rare in nature, but even a broken fragment or distorted crystal is enough to give some clues to its identity.

The six basic crystal systems are cubic, tetragonal, hexagonal, orthorhombic, monoclinic, and triclinic. These are shown below, with characteristic examples.

AGGREGATES, *or interlocking crystals, form in confined spaces, as seen in this gypsum specimen.*

Habit

The size and shape of a mineral's crystals determine its habit. These range from isometric or equant (all faces equally developed), through pyramidal (crystal edges converge to a point), tabular, columnar, acicular (needle-like) to fibrous. Other habits include stellate (star-like), dendritic (tree-like), reniform (kidney-shaped), botryoidal (grape-like), and massive (with no distinctive shape).

Minerals in the same crystal system, or even the same mineral with different trace elements, can display vastly different habits. Ruby crystals, for example, are

THE SIX CRYSTAL SYSTEMS
h mineral examples.

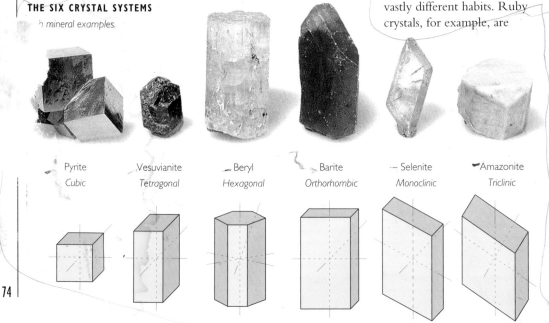

Pyrite *Cubic* — Vesuvianite *Tetragonal* — Beryl *Hexagonal* — Barite *Orthorhombic* — Selenite *Monoclinic* — Amazonite *Triclinic*

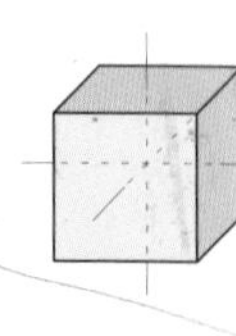
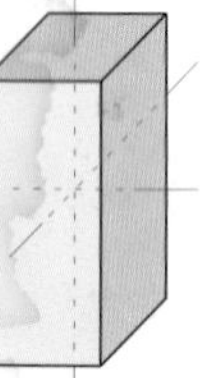
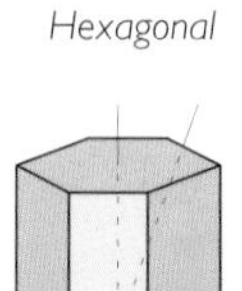
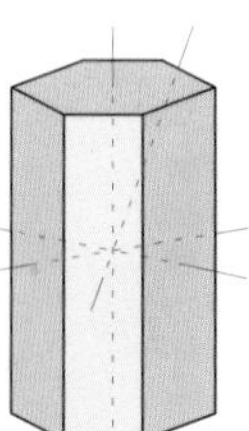
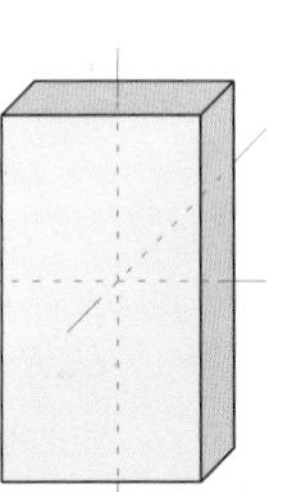
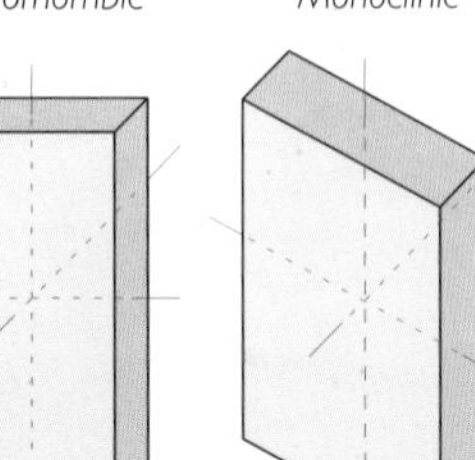
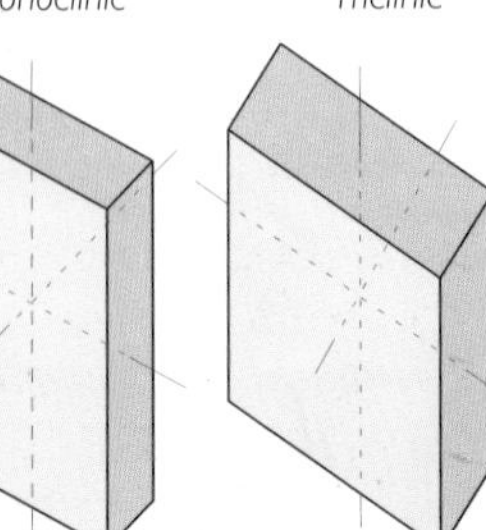

usually tabular, while sapphire crystals are generally pyramidal or prismatic.

Crystals often grow in confined environments in between already formed crystals. As a result, they may become distorted or they may form an assemblage of interlocking crystals, known as an aggregate.

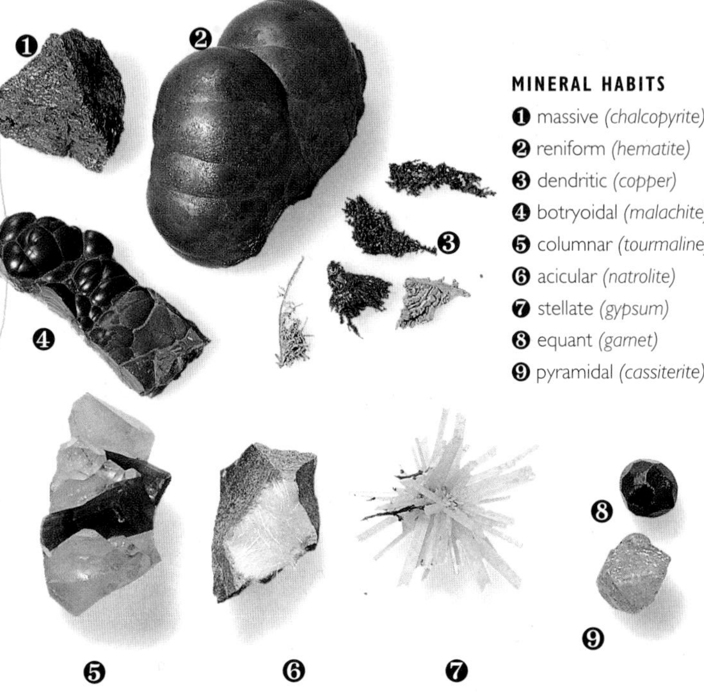

MINERAL HABITS
❶ massive *(chalcopyrite)*
❷ reniform *(hematite)*
❸ dendritic *(copper)*
❹ botryoidal *(malachite)*
❺ columnar *(tourmaline)*
❻ acicular *(natrolite)*
❼ stellate *(gypsum)*
❽ equant *(garnet)*
❾ pyramidal *(cassiterite)*

Crystal Faces

Crystal faces are rarely flat and featureless, often being striated, patterned or deformed in distinctive ways. Quartz is usually striated across the width of the crystal while striations in tourmaline run along the length of the crystal. The octahedral faces of diamond crystals often show triangular pits known as trigons. Similarly, sapphires and rubies may show raised triangular hillocks, in both cases due to corrosion damage from the magma that carried them to the surface.

Cleavage

Certain minerals, when struck with a hammer, will break, or cleave, along set planes of weakness related to their mineral structure.

MICA *(above) has perfect cleavage along one plane. A magnification of the face of a diamond (above right) showing triangular pits, or trigons.*

Cleavage may be poorly defined (as in bornite) to perfect (mica and diamond) and may occur in one plane (mica), two planes (gypsum, pyroxene, amphibole, feldspar), three planes (galena, calcite, halite), four planes (fluorite, diamond) or even five planes (sphalerite).

Minerals will therefore cleave into sheets (mica, talc), prisms (pyroxene, amphibole), cubes (halite, galena), octahedra (fluorite, diamond), and rhombs (calcite). The ease of cleavage is also important, and can range from easy (mica) to very difficult (diamond).

Again, don't take to your finest pieces with a hammer. By looking closely, it is often possible to find silvery, reflective planes within transparent minerals, known as incipient cleavages. These often parallel a flat outer surface, indicating cleavage directions and their relationship to the crystallographic axes.

Fracture

Minerals that do not cleave easily will fracture irregularly instead and the type of fracture can often be used to distinguish among mineral species. The most common type of fracture is conchoidal. This is the smooth, shell-like fracture seen in obsidian and quartz. Metals and tough minerals, such as jade, tend to have a hackly fracture, like broken cast iron. Arsenopyrite is an example of uneven fracture and kaolinite breaks with an earthy fracture, like clay or chalk.

COMMON FRACTURES *include (clockwise, from top left) uneven (arsenopyrite), earthy (kaolinite), and conchoidal (quartz and opalite).*

THE METALS

The discovery of metals led to great leaps forward for human civilization.

About 75 percent of the elements on the Periodic Table are metals. The most abundant are aluminum, iron, calcium, sodium, potassium, and magnesium, but these were not discovered by ancient civilizations because they are found only in compounds with other elements. The first to be discovered were the native metals, such as gold, silver, and copper, that occur naturally in a pure form. As each metal was discovered, societies were transformed.

Gold artifacts from Sumer date from 4000 BC, but gold was probably discovered at least 8,000 years ago. Strong and pliable, yet unbreakable, a native gold nugget could be honed to a fine edge or hammered into thin sheets.

Native copper, an attractive salmon-pink metal discovered about the same time, was stronger and harder than gold. It could be fashioned into tools and weapons as well as ornamental objects.

At some time before 4000 BC, a crucial discovery was made—molten metals could be combined (or alloyed) with one another when heated. The first metal alloy was bronze, and its discovery marked the birth of the Bronze Age. Made from roughly 90 percent copper and 10 percent tin, bronze is harder and more useful than either of its constituents. It facilitated the production of weapons and tools and expanded the potential for trade. People also realized that metals could be extracted from certain rocks by placing them in a very hot fire. This primitive form of smelting brought about the discovery of lead and mercury, as well as new sources of copper.

The last metal to revolutionize the lives of ancient people was iron. Although more difficult to extract than the other metals, its abundance and strength changed the face of civilization during the Iron Age. By 1100 BC, iron was in widespread use, creating vastly superior tools and weapons.

A BRONZE *statue of Hercules, from the second century AD. The green staining indicates the high copper content of bronze.*

MODERN METALS

Modern technology has given rise to a range of new metals. Aluminum, a strong, lightweight metal that resists corrosion, is used widely for construction and packaging. Although abundant, its early discovery was impossible because of the vast amount of power required to refine it from its ore, bauxite.

Another important metal is titanium. Its light weight and high melting point make it ideal for use in rockets and planes, and it can be used as an alloy to improve the quality of other metals, particularly iron. Chromium, tungsten, and platinum are also important modern metals.

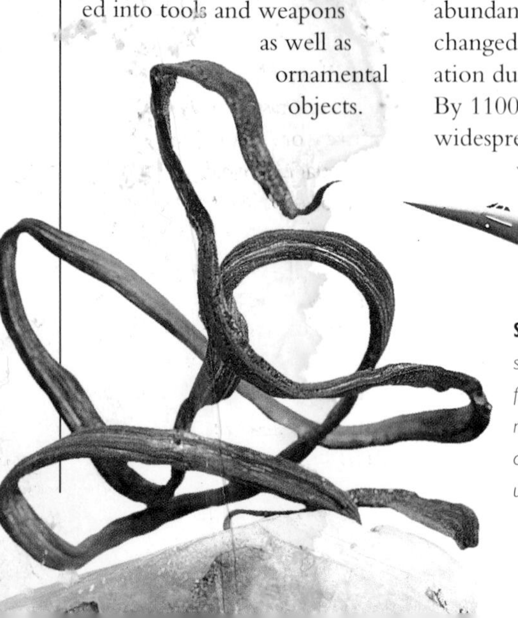

SOME METALS, *such as silver (left), are found in a pure form. Copper occurs naturally or is derived from the shiny compound "peacock ore" (top). Titanium is used in aircraft such as the Concorde (right).*

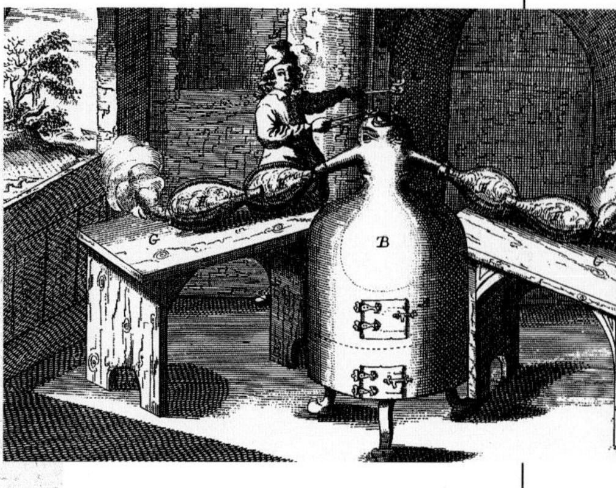

Metal-bearing Ores

The term "ore" refers to those minerals that form the raw product of a valued commodity. For example, the metal tin (Sn), is derived from the ore cassiterite (SnO_2). Ore deposits represent unusually high concentrations of raw minerals that are economically viable to extract and process. Most metals can be derived from a number of ore minerals, and are extracted by smelting in a carbon-burning furnace. The oxygen in the ores combines with the carbon to form carbon dioxide (CO_2), leaving behind the pure metal.

Colored staining on rocks can indicate ore deposits. Bright blue and green staining may indicate the presence of copper or nickel. Red, rusty-looking rocks indicate iron, while black staining reveals manganese. The metal ores tend to have a higher specific gravity than other minerals, and often show a metallic or sub-metallic luster.

Many of the minerals that make up the metal-bearing ores are spectacularly beautiful. Copper was first smelted from the attractive carbonates, azurite and malachite. Today, the principal ores of copper are chalcopyrite (a brassy yellow sulfide), bornite (called "peacock ore" because of its shiny purple tarnish), chalcocite and covellite.

Lead was first smelted from its carbonate ore cerrusite. It was used in the water pipes of Rome, and even for food and water vessels before its toxicity was understood. Now, the main ore of lead is the heavy, silver-grey mineral, galena.

Iron, our most important metal, is abundant and cheap. Hematite, the most common ore of iron, takes a mirror polish, making it a useful ornamental stone.

Silver occurs as ductile, twisted, branching masses of silvery native metal that tarnish to grey and black, or as the silver sulfide argentite.

Zinc was first smelted from its carbonate ore, smithsonite, and alloyed with copper to produce brass. Today, its primary ore is sphalerite, or "blackjack", usually found with ores of lead and copper.

Mercury, the only metal that exists as a liquid at room temperature, is extracted from the mineral cinnabar, although it is occasionally found in a native form as metallic drops. Used in pesticides and paints, mercury is highly toxic and persistent in the biosphere.

GOLD FEVER

During the Middle Ages, alchemists were obsessed with the possibility of making gold from lead (right), its near neighbor on the Periodic Table. Despite elaborate efforts, all attempts were fruitless. Today, nuclear facilities can actually create the synthetic gold that so frustrated the alchemists. However, the cost of doing so is far greater than the current market value of gold.

In the face of fluctuating currencies, gold is accepted by all nations as a common and reliable medium of exchange and an important reserve asset.

People find gold in fields, veins, river beds and pockets. Whichever, it takes work to get it out.

Art Linkletter
b. 1912, North American radio personality

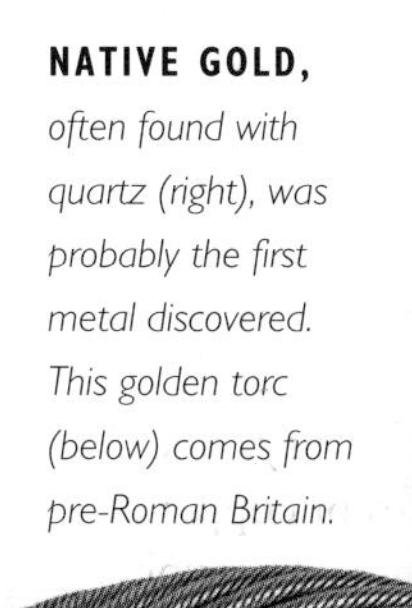

NATIVE GOLD, *often found with quartz (right), was probably the first metal discovered. This golden torc (below) comes from pre-Roman Britain.*

The Non-Metallic Minerals

The non-metallic minerals have provided many of the basic needs of our society since its beginnings.

The world's most common minerals belong in this group and without them we would not have bricks, tiles, cement, plaster, ceramics, insulation, filters, fertilizers, and other chemicals. These minerals may contain some metallic elements but they are not used as a source of metals.

A number of non-metallic minerals are evaporites, meaning that their crystals develop during the evaporation of water in desert lakes. Examples of these include calcite, gypsum, halite, and the borate minerals. Beautiful crystals of these minerals may be picked up from dry lake beds. Evaporite crystals form very quickly and you can even grow your own.

The clay minerals, dull, soft, and generally unattractive, were probably the first to be used, and remain the basic materials for building, pottery, and ceramics. A clay brick from beneath the city of Jericho has been dated from between 8000 and 7000 BC, and clay bricks were also used in ancient Egypt, China, and in the Americas. These early bricks were made by drying a mixture of clay and straw in the sun.

Calcite, the most common of the carbonate minerals, and gypsum, a soft evaporite mineral, were similarly important to the people of ancient Egypt and the Americas. It was discovered that if these minerals were crushed and mixed with water, they would harden into solid rock. This led to their widespread use in cement, plaster, and stucco. These carbonate minerals also make fine collection specimens. Gypsum often occurs as beautiful swallow-tail crystals or as massive alabaster which is equally interesting. Calcite, too, occurs as attractive crystals or massive non-crystalline forms, such as chalk.

In the days before refrigeration, the mineral halite, or common salt, was as valuable as gold itself.

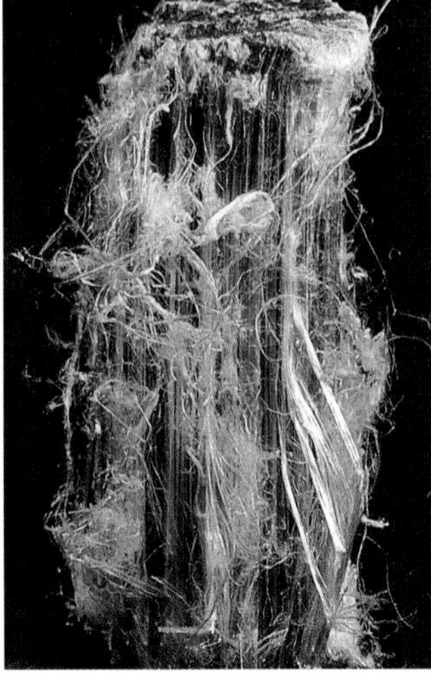

The preservative qualities of salt were vital for the transport and storage of meats. Salt is an evaporite, forming in layers after the evaporation of salty water. Because of its low density, salt may rise, punching through overlying layers and forming huge pillars or domes. These can be mined in the traditional way, or dissolved by pumping water in and out through boreholes, to recover the salt. Apart from being a popular food additive, salt is

SPECTACULAR *non-metallic minerals include selenite (left) and chrysotile (above), an asbestos. This soapstone pipe (top left) was made about 800 years ago by Native Americans.*

SALT CRUST *On this Australian salt lake (left), a dead shrub has been blown upside down and encrusted with salt.*

used in the manufacture of chlorine-based products and soda ash for glassmaking.

Another of the oldest known elements is sulfur, a soft, yellow, elemental mineral. When burnt, sulfur reacts readily with oxygen to produce the pungent gas sulfur dioxide, which reacts with water to produce sulfuric acid. Because of its association with active volcanoes, sulfur was known to the ancients as "brimstone" or "the fuel of Hell's fires". Most sulfur is converted to sulfuric acid and used in fertilizers. Its low melting point means that it can be mined from the surface by pumping superheated water underground through large pipes and drawing out the molten sulfur.

Graphite is an extremely soft (hardness 1.5) form of carbon. It is an invaluable industrial mineral because of its qualities as a lubricant, its resistance to corrosive chemicals, and its ability to conduct electricity. Graphite leaves a distinctive black streak and its name derives from the Greek word meaning "to write". The "lead" in pencils actually consists of a mixture of graphite and clay, molded and baked at high temperatures.

Among the most infamous non-metallic minerals are those collectively known as asbestos. Fibrous in appearance, they include crocidolite, or blue asbestos, and chrysotile, a form of serpentine. Because their fibers don't burn or conduct electricity, they were used for fireproofing and insulation for many years, until the fibers were found to cause permanent lung damage when inhaled.

Many other important non-metallic minerals make fine collection specimens. Barite, also known as "heavy spar" because of its high specific gravity, often occurs as large, transparent or whitish crystals. Another particularly attractive mineral, found in almost every color, is fluorite.

Talc, or "soapstone", is the softest of all minerals, with a hardness of 1. It has been used for carving since ancient times, but is now best known in the form of talcum powder.

The borate minerals are good examples of evaporites, and are found in desert salt pans. They include borax, the principal source of boron, and ulexite, also known as "television stone".

TELEVISION STONE *When ulexite (right) is polished, its crystals behave like optical fibers, transmitting images from one side to the other.*

GROW YOUR OWN SALT CRYSTAL

To grow your own salt crystals, stir salt into water in a small container, or glass, until no more will dissolve (using warm water will help). Simply allow this supersaturated solution to evaporate slowly and crystals will begin to form on the sides of the container. You may want to suspend a thread into the container, as the crystals will tend to grow on this.

To grow large crystals, continually remove the smallest crystals as they develop and re-dissolve them. This will ensure that your best crystals continue to grow.

SOURCES *of* ENERGY

Fossil fuels represent the warmth of the Sun locked within the Earth's crust. Uranium's energy stretches back to a time before the very formation of our Earth.

A lump of coal is compressed plant material from a prehistoric time, which may be burned today to release its stored energy. To appreciate the energy driving the technology around you, imagine a bright, sunny day millions of years ago. The scene is the edge of a swamp filled with strange-looking plants and insects. At that moment, a minuscule portion of the Sun's energy was captured and locked away to be released millions of years later, for the benefit of a species that had not yet evolved.

FOSSIL FUELS

Coal is classified as an organic sedimentary rock and is the fuel that fired the industrial revolution. As plants and trees are buried in an oxygen-starved environment and compressed, water and organic gases are driven off and the carbon content gradually increases. This process first produces peat, not considered true coal, containing about 80 percent moisture and gases. Further heating and compression produces a true coal, lignite (also known as brown coal). Continuing the process gives us bituminous, or black, coal. Yielding far more heat, bituminous coal represents 90 percent of all coal mined. Its higher grades store well and burn with an almost smokeless flame. Anthracite is the highest-ranking form of coal, containing only about 5 percent water and gases.

Petroleum is made up of the microscopic remains of plants and animals buried in the fine muds of the sea floor. The fluidity of the petroleum depends on the length of the chains of hydrocarbon molecules. The shorter chains form the lighter fluids, kerosene, gasoline, and diesel, while the longer chains form the more viscous tars and asphalts. Petroleum migrates through porous formations, such as sandstone, and becomes trapped in reservoirs when it reaches impermeable layers such as shale or salt.

Natural gas is a mixture of the lightest hydrocarbons

COAL FORMATION *(below, from left to right). In prehistoric times, dead vegetation was buried in a swamp, forming peat. As this was compressed and water driven off, higher grades of coal were formed. In the past, a "gusher" (top) was often the first sign of an oil strike.*

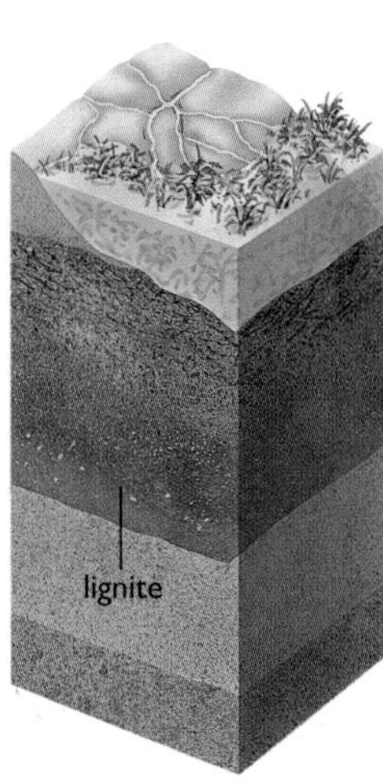

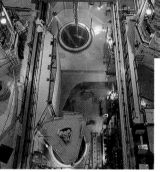

ENERGY NEEDS *Modern sources of energy include oil from North Sea oil rigs (above) and power generated by nuclear reactors (right). Large cities, such as New York (below right), consume enormous amounts of energy.*

found in the upper part of petroleum reservoirs. The pressure of natural gas in a reservoir formation is often enough to drive the oil and gas up the drill pipe, causing the famous "gushers" of old. These were dangerous and wasteful, and today's drillers pump high-density fluids (barite muds) into the hole to keep the gas pressure contained in the rock formation until the well is established.

Oil shales are organically rich, oil-bearing rocks that can be mined, then heated and processed, to yield petroleum and natural gas. At present, oil shales are not used, but they will soon become economically viable as liquid petroleum resources start to run out.

Uranium is an extremely heavy, unstable, naturally occurring metal, formed during the explosive evolution of stars. It breaks down to lighter, more simple products over time, releasing enormous amounts of atomic energy. Controlled in a nuclear reactor, the heat given off by uranium can be used to drive conventional steam turbine generators. When the mass of uranium in one place exceeds a critical threshold, known as its "critical mass", all the available energy is released at once. This is the basis of the highly destructive atomic bomb. The principal source of uranium is the mineral uraninite or pitchblende, a hard, black, heavy, radioactive ore.

Energy Dependence

These sources provide us with energy that we take for granted in our daily lives and it is virtually impossible to imagine living in the modern world without using any of the above-mentioned fuels. At present, however, we are burning up the Earth's fossil fuel resources at an unsustainable rate. Unlike metals, which may be partially recycled, fossil fuels are nonrenewable. While coal reserves will probably last for another 300 years, some estimates suggest that known oil reserves will be exhausted within about 50 years.

The burning of fossil fuels also releases a great deal of carbon dioxide into the atmosphere, which has been linked to global warming, as well as sulfur dioxide and nitrogen oxides, which produce acid rain. Nuclear power, too, has dangers that we have yet to deal with satisfactorily.

To meet these challenges, we must develop means to reduce our consumption of energy. We must also continue to encourage and financially support research into alternative, renewable sources of energy, harnessing natural forces such as the Sun, wind, and tides.

GEMSTONES, ORNAMENTALS, *and* BUILDING STONES

The Earth is a treasure chest of rocks and minerals that are sought for their durability and intrinsic beauty. Some are prized for personal adornment while others are used for building.

The word "gemstone" refers to any rock, mineral, or organic substance suitable for personal adornment. This may be any uncommon, transparent, colorful, or unusually brilliant material. The first use of gemstones dates from prehistoric times when the wearing of personal adornments preceded even the wearing of clothes.

A gem's weight was once measured by comparing it to that of a carob seed. While these seeds have a relatively constant weight, a standard unit of measurement was eventually agreed upon. Known as a carat, it is equivalent to 0.006 ounces (0.2 g).

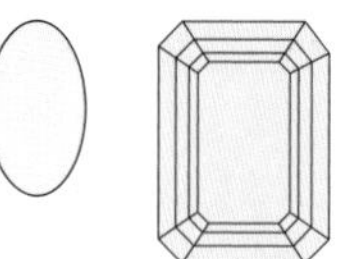

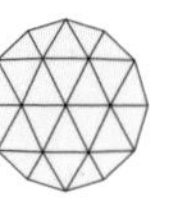

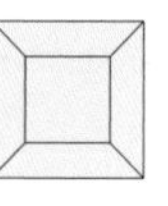

RAW GEMSTONES, *such as this tourmaline crystal in quartz (right), are often cut and polished to accentuate their natural qualities. Some popular cuts are (above, from left) cabochon, emerald cut, rose cut, and table cut.*

WHY ARE THEY PRECIOUS?

The three distinctive qualities that generally characterize gemstones are their beauty, durability, and rarity. Beauty is a gemstone's attractiveness to the beholder and is dependent on its optical properties. These are its color, transparency, luster, sheen, refractive index, and dispersion (the ability to split white light up into its spectral colors).

Durability is a gemstone's ability to resist damage over time, an important property if it is being worn or handled. This is dependent on the hardness and tenacity of the material. The durability of certain gemstones has invested them with mystical significance and allowed them to pass from generation to generation.

Rarity is an essential quality of a precious stone and, may be more important in determining value than physical properties. Amber, opal, and pearl, for example, are very precious, but not particularly durable.

Of the 3,600 mineral species known to science, only about 100 possess all the attributes required in gems. Traditionally, diamond, ruby, sapphire, emerald, and opal are considered precious gemstones. Other well-known gems include topaz, aquamarine, spinel, garnet, and tourmaline.

Some stones are notable only for their striking colors or patterns, not being transparent, durable, or rare. Stones such as these are known as ornamentals, and examples include jasper, onyx, jade, malachite, rhodonite, and lapis lazuli. Organic gemstones are those formed as a result of living processes. These include amber (fossilized tree sap), coral, pearl, jet (coal), and shell.

TIMELESS BEAUTY *A lion pendant (top) made from amber more than 2,500 years ago. A selection of cut topaz stones (left). Stonehenge (right) was constructed about 4,000 years ago from sandstone blocks. Texture details of granite and gneiss (insets, far right).*

BUILDING BLOCKS *The ancient Mayan Temple of the Inscriptions (above), in Mexico, was made from limestone blocks. A marble quarry (right) near Tuscany, Italy.*

Rocks for Building

Many of the Earth's common rocks are cut for building stones and look spectacular when polished or carved. However, the fundamental building blocks of our society are calcite, the basis of cement, and the humble clay minerals, pressed into bricks and fired. Other building stones have been largely replaced by steel, reinforced concrete, and glass, and are now mainly used for finishing and decoration. They must be attractive and long wearing, with no iron minerals that will oxidize and rust with time.

Blocks of unfractured material are quarried with cable saws, cracked out with wedges, or broken free using small amounts of carefully placed explosives. The enormous stones used in ancient structures, such as the pyramids of Egypt, or Stonehenge in England, had to be dug manually from quarries nearby because of the enormous difficulties involved in moving them.

Igneous rocks are often used as building stones because of their strength, ability to take a high polish, and resistance to weathering. These are generally known as granite (light-colored rocks, including true granites, porphyrys, and rhyolites), or black granite (all other darker-colored igneous rocks).

Metamorphic rocks, such as gneisses, are also favored for their attractive swirly, folded, or brecciated patterns. Slate, which splits easily into large, thin sheets, is used for roofing and floors. White, pink, black, or patterned marbles are particularly popular for carving. Softer than granites and gneisses, they are easier to cut and work but less resistant to weathering and the corrosive effects of pollution. Some well-known structures, such as the Taj Mahal in India, are made of marble.

Sedimentary rocks, such as sandstones, are often attractive, with delicate patterning. Sandstone is favored because its bedding makes it easy to quarry and stonemasons have few problems working it. Its only drawback is that porous varieties are less durable. Limestones, particularly those containing fossil corals, shells, and other marine organisms, are most attractive and interesting when cut and polished. The pyramids and statues of the great civilizations of Egypt and Mexico were constructed from limestone blocks.

Professional Geology

The equipment and technology that geologists use to aid their search for natural resources are becoming ever more sophisticated.

While out of the reach of most amateurs, advanced techniques allow professional geologists to analyze rocks in far greater detail than was ever previously possible. As a result, there has been enormous improvement in the accuracy and efficiency of mineral and petroleum exploration.

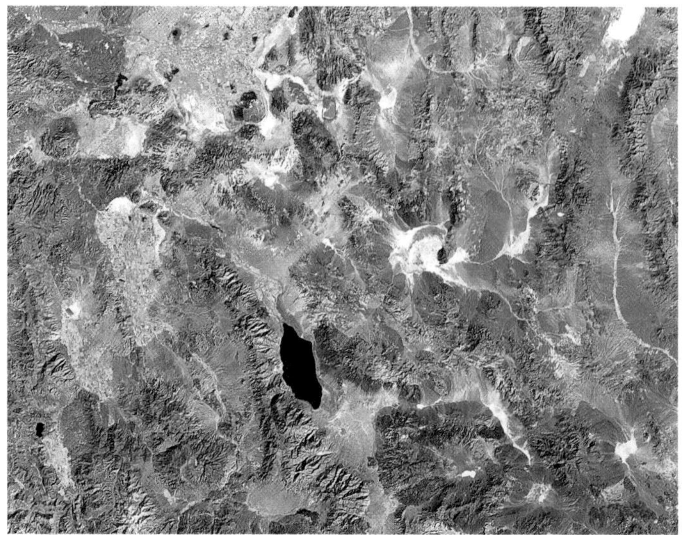

NEAR AND FAR *An enhanced satellite image of part of Nevada, USA (above). A thin section of quartz porphyry viewed through a petrological microscope (top left).*

A Close Look

The petrological microscope, used with thin, transparent slices of rock in normal and polarized light, is capable of revealing a great deal more about the rock than the naked eye. It can show exactly what minerals the rock is made up of, how they came to be there, and if they have been affected by heat or pressure.

Chemical analysis tools, such as the atomic absorption spectrograph, are also used to treat rock and soil samples in laboratories. They can detect the presence of 0.01 percent of any elements present in a given sample. Such tests are routinely carried out by geologists looking for hidden mineral deposits.

More sophisticated still, electron and proton microprobes can identify minerals, their compositions, and, in special cases, the time at which they crystallized. Age determination is possible where the minerals include unstable radioactive elements, as these decay to stable elements at a known rate. For example, uranium breaks down to lead, so, by comparing the ratio of uranium to lead in a mineral, its age can be determined. Such tools are far too expensive for amateur use, but they are crucial for the advancement of scientific knowledge.

A PROTON *microprobe (left). A photograph of gold formed bacterially (above), taken with a scanning electron microscope.*

SEISMIC SURVEYING *for oil (left). Shock waves from charges detonated in holes bounce off the rock strata below, providing a seismic profile of the area.*

The Big Picture

One of the greatest advances over past decades has been the availability of images of the Earth as viewed from satellites. These have revealed a wealth of information about our planet and its geological history. Satellite images can be recorded using any number of wavelengths and these can be computer enhanced. For example, soils and rocks may appear to the naked eye in various shades of brown, but by expanding this band of information to occupy the entire visible spectrum, soils with traces of different elements can be made to stand out as distinct colors.

Geophysical instrumentation enables professional geologists to look deep within the Earth. Earthquakes, being like large explosions, are recorded by numerous seismic stations around the world, and these records can be used to study the structure of the Earth's interior. Seismic energy from explosion sources bounces off different layers of rock, providing cross-sections that are used for picking up the rock traps within which hydrocarbons may be trapped.

Deposits of ore minerals are usually very dense, often magnetic, and conduct electric currents. As a result, a number of instruments have been developed to identify these properties in the rocks beneath our feet.

Portable gravity meters, consisting of a weight suspended on a sensitive spring, measure the strength of the Earth's gravitational field. Unusually heavy bodies below the instrument, such as gold, copper, iron, and uranium ores, will register as minute increases in gravitational pull. Magnetometers measure tiny variations in the Earth's magnetic field that can reveal the presence of iron-bearing ore bodies. Electromagnetic coils are laid out on the Earth's surface to register the presence of conductive bodies, as these are susceptible to carrying induced electric currents. Many of these instruments are designed to be towed by aircraft or ships, making it possible to explore large areas quickly and efficiently.

Drilling

Ultimately, all of this technology is put to the test when drilling is carried out to penetrate structures and confirm the suspected presence of underground resources.

Drilling machinery varies enormously from small, truck-mounted equipment to huge floating oil rigs. However, in all cases the principle is the same. Hollow diamond-impregnated drill bits are used to penetrate the rocks so that a solid core of rock passes up into the drill pipe for collection and analysis when the drill is retracted.

DRILLING *for oil in the North Sea. A floating rig (right) requires a large crew and is extremely costly to operate.*

MINING *in the* FUTURE

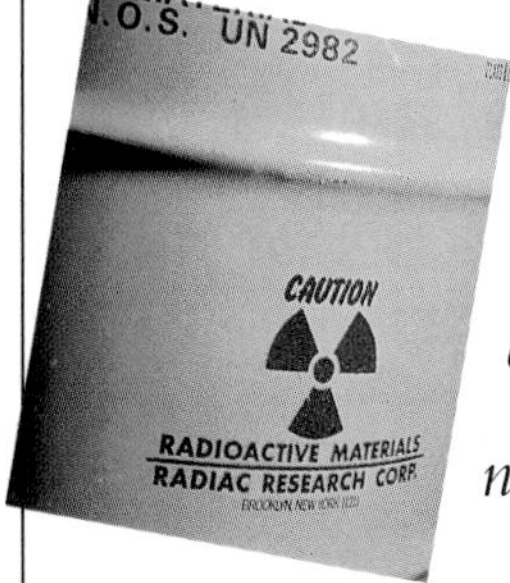

It is only recently that we have come to understand the consequences of our quest for the Earth's resources. We must now consider environmental, as well as economic, concerns.

Modern industrialized societies depend on the continued harnessing of the Earth's resources. We require increasing amounts of energy and raw materials to maintain our standard of living, and entire economies depend on the income and revenue generated by mining. However, every step of the mining process—exploration, extraction, processing, transportation, and utilization—can exact a heavy toll in human and environmental terms, and these problems will have to be addressed.

ENVIRONMENTAL RISKS
The disposal of radioactive waste (top), an oil spill off the Shetland Islands in 1993 (right), and the polluting of the Ok Tedi River in Papua New Guinea by mine tailings (below) all serve to highlight the concerns of environmentalists.

THE TROUBLE WITH MINING

No matter how carefully mining is monitored, it puts significant stress on the environment. In remote areas, the discovery of mineral resources often has tragic results for the local inhabitants. Geologists and miners move in, land is cleared, roads are built, and towns spring up. This often results in the displacement of the original inhabitants and the breakdown of established lifestyles and customs.

While the amount of land tied up by mining is generally small when compared with forestry and agriculture, the potential for environmental damage is great. In order to extract any valuable resource, a vast amount of material has to be processed. For example, 99.9 percent of the ore mined to produce silver is discarded as waste. As the richest deposits run out, mining companies are forced to mine less-concentrated ores. As a result, more material has to be mined to produce the same amount of the mineral, creating more waste.

These waste products, or "tailings", can disfigure the landscape and silt up rivers. Because of the chemicals and minerals often used in the mining process, these tailings may also be toxic. In the Amazon Basin, highly toxic mercury is used as an amalgam to extract gold and is then burnt off, finding its way into rivers and soil and the food chain. Native villagers have started to report unusual diseases and an anomalous number of birth defects, which may be linked to this buildup of mercury.

Once material has been mined, its transportation can also create problems. Oil tankers carry vast amounts of oil around the globe and, while accidents are rare, the results can be catastrophic. The grounding of the *Exxon Valdez* in March 1989, resulted in the release of

11 million gallons (50 million liters) of crude oil along the pristine Alaskan coastline.

Even when a mined resource reaches its destination safely, there are risks associated with its use. Fossil fuels provide most of the energy for our society and consumption continues to rise. This has led to increasing atmospheric concentrations of carbon dioxide and sulfur dioxide, which have, in turn, been linked to global warming and acid rain.

Nuclear power is often seen as a cleaner alternative source of energy, but there are dangers here, also. The meltdown of the Chernobyl nuclear reactor in 1986 released seven tons (7.1 tonnes) of radioactive material into the atmosphere, contaminating food and water throughout Europe. Unforeseen accidents, deliberate misuse of radioactive materials for arms production, and the need for safe disposal of the spent, but highly radioactive, reactor fuel are all very real concerns. Who can say with any certainty that a location chosen for a nuclear power plant or spent fuel dump will remain geologically stable for several thousands of years?

Sustainable Development

What, then, can we do, and whose responsibility is it to act? Increasingly, large mining companies are being forced to accept responsibility for their activities. Environmental impact statements (EIS) are now required in most countries before mining can proceed, and stringent restoration requirements generally apply.

There is a growing awareness of the need to recycle material and to develop safe, alternative sources of energy. The global implications of the continuing reliance on fossil fuels are beginning to foster greater cooperation among countries. As resources become more scarce, governments will have to act for their own economic, as well as environmental, survival. The responsibility for these changes rests with all of us, but the onus is on the developed countries to act first. After all, 25 percent of the world's population uses 80 percent of its resources.

We can no longer view the Earth's resources as unlimited. The environmental and social implications of mining must be considered along with economic concerns, with the aim of achieving sustainable development. How we deal with these issues will be the ultimate test of our maturity.

MINING *in Tasmania, Australia, has destroyed areas of rainforest along the King River (left). A solar-powered furnace in France (below left), and waste recycling in the USA (below) indicate growing environmental awareness.*

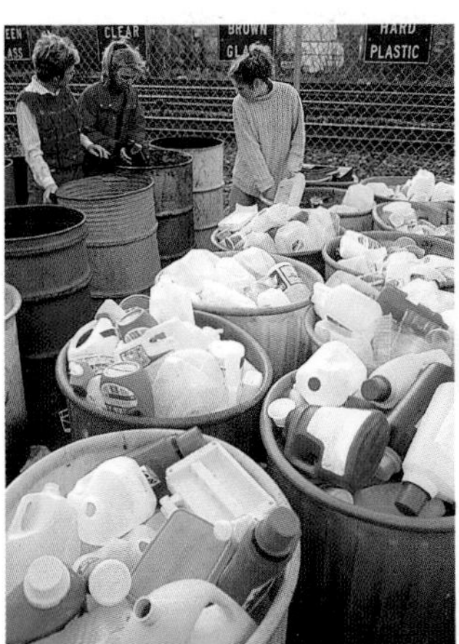

Touch the earth, love the earth, honor the earth ...

The Outermost House,
HENRY BESTON
(1888–1968), American writer

The book of Nature is the book of Fate. She turns the gigantic pages—leaf after leaf—never re-turning one.

The Conduct of Life,
RALPH WALDO EMERSON (1803–82), American essayist and philosopher

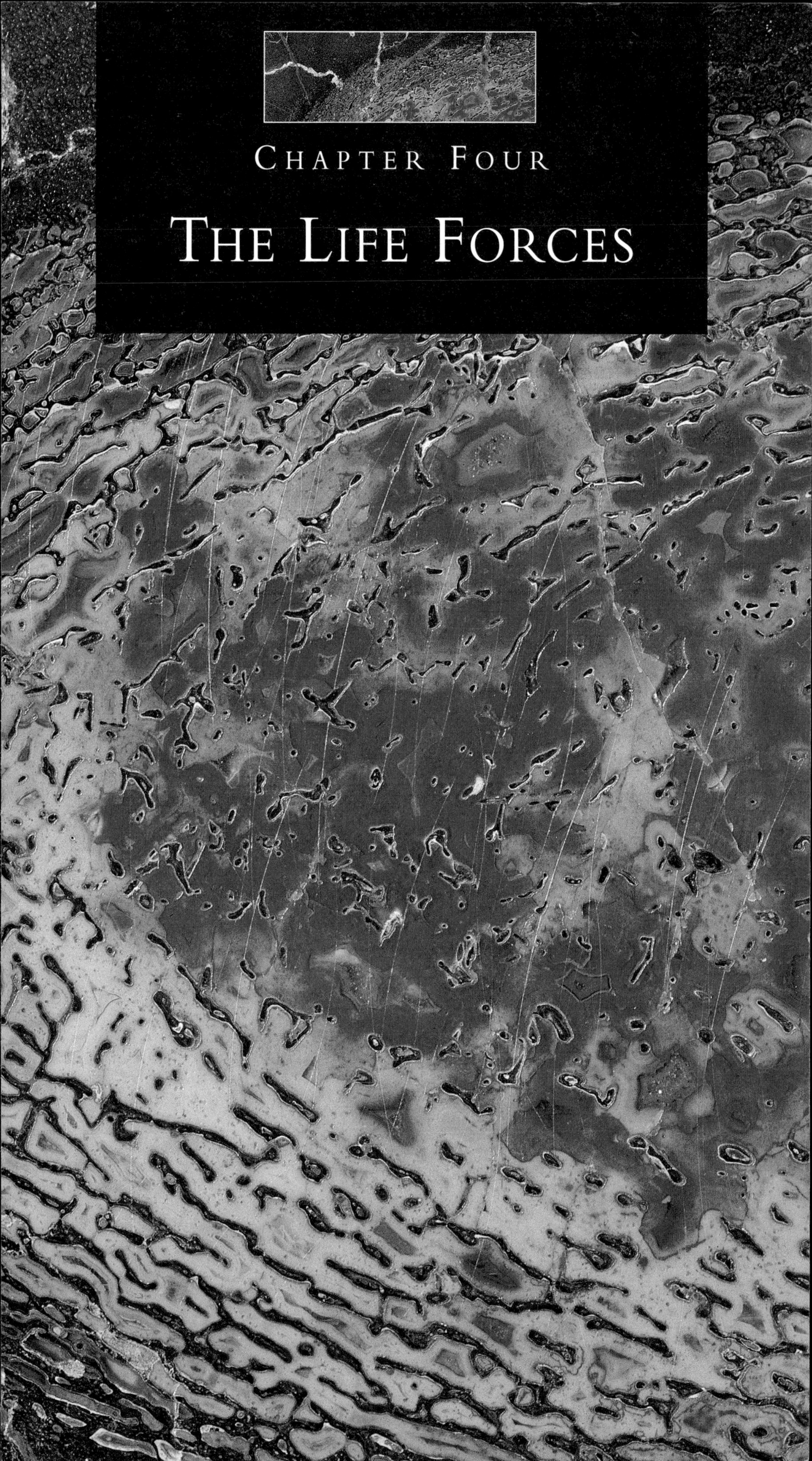

Chapter Four

The Life Forces

THE TREE *of* LIFE

To visualize the evolution of life forms through time, picture the spreading branches of a tree.

There is a complete unity to life. All the animals, plants, fungi, bacteria, and viruses that have lived on Earth are related by descent. Throughout the past 3,500 million years, life has been growing, dividing, and developing to produce the myriad forms we see around us today. We can call this concept the Tree of Life. This four-dimensional tree grows through the three dimensions of space, and also through the fourth dimension of time.

The diagram on the facing page shows part of the tree extending only through the past 600 million years—life began a long time before this. The oldest life forms are at the bottom of the tree and time progresses toward the present at the top. Only some of the major branches are shown.

By carefully studying fossils, we can trace back along the branches of the tree, find out when different branches divided, determine how the tree is structured, and also establish how organisms are related.

TWO BEETLES, *one living (top) and the other a 50-million-year-old fossil. Both show striking iridescent blue coloring, illustrating the continuity of life through time.*

THE HISTORY OF LIFE

As far as we can determine, life first appeared on Earth about 3,500 million years ago and it started only once. While the exact mechanism by which the first life form came into existence is still unclear, once it had appeared, the processes of evolution began acting on it, resulting in a changing and complex diversity of plants and animals.

For most of its history, life has been very simple. For the

STROMATOLITE FOSSIL SECTION *(above) and living stromatolites (right) at Shark Bay in Western Australia. Their structure has remained basically the same since the early Precambrian.*

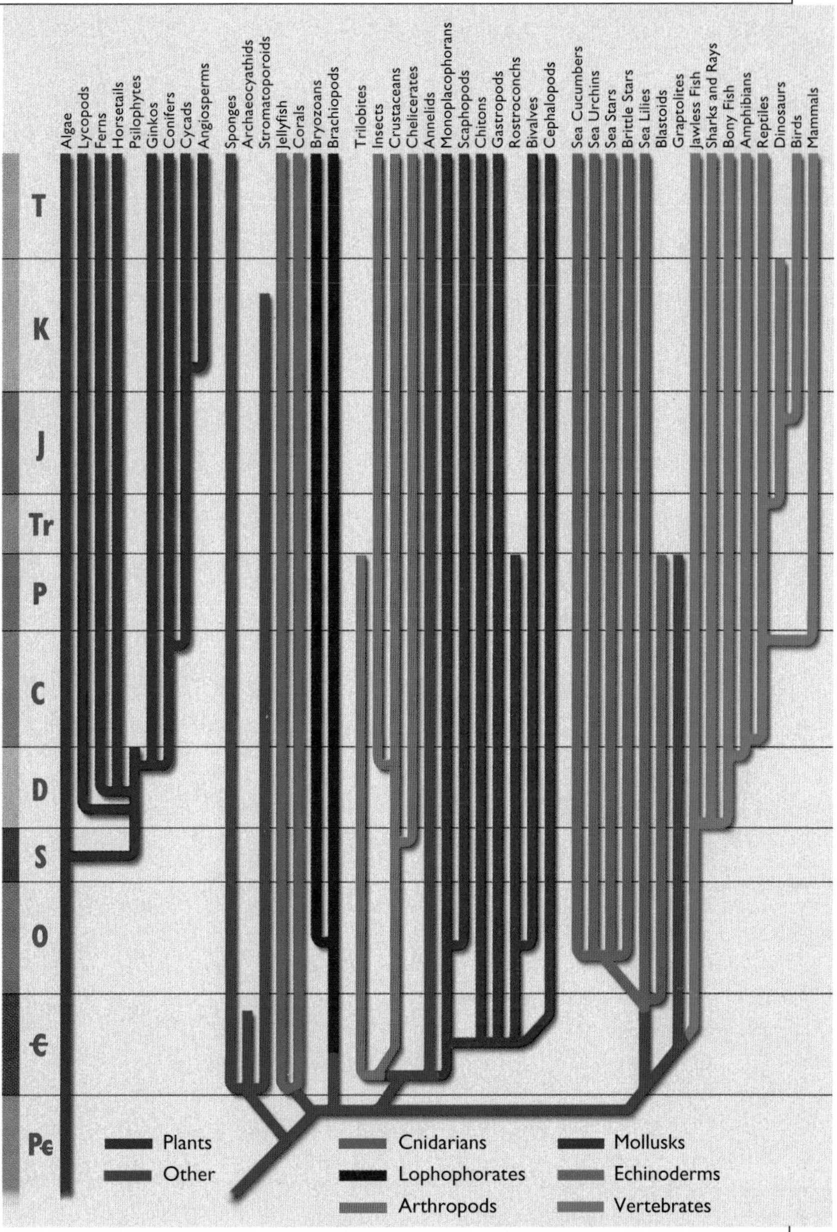

THE TREE OF LIFE

Each line on the diagram (right) represents a group of organisms and shows its distribution through time and its relationship to other groups of organisms. Plants, animals, and even fungi and bacteria (not shown here), are truly our siblings. The geological periods are shown on the left and proceed from the oldest at the bottom to the present at the top. Their abbreviations are as follows:

- *T Tertiary*
- *K Cretaceous*
- *J Jurassic*
- *Tr Triassic*
- *P Permian*
- *C Carboniferous*
- *D Devonian*
- *S Silurian*
- *O Ordovician*
- *€ Cambrian*
- *P€ Precambrian*

Algae
Lycopods
Ferns
Horsetails
Psilophytes
Ginkos
Conifers
Cycads
Angiosperms
Sponges
Archaeocyathids
Stromatoporoids
Jellyfish
Corals
Bryozoans
Brachiopods
Trilobites
Insects
Crustaceans
Chelicerates
Annelids
Monoplacophorans
Scaphopods
Chitons
Gastropods
Rostroconchs
Bivalves
Cephalopods
Sea Cucumbers
Sea Urchins
Sea Stars
Brittle Stars
Sea Lilies
Blastoids
Graptolites
Jawless Fish
Sharks and Rays
Bony Fish
Amphibians
Reptiles
Dinosaurs
Birds
Mammals

T
K
J
Tr
P
C
D
S
O
€
P€

Plants
Other
Cnidarians
Lophophorates
Arthropods
Mollusks
Echinoderms
Vertebrates

first 3,000 million years, it was no more complex than single-cell organisms that sometimes built colonies called stromatolites. About 600 million years ago, complex life forms evolved with multi-cellular organisms, similar to jellyfish, flourishing. The next important step, about 570 million years ago, was the development of animals with shells or hard skeletons.

Life spent most of its time (the first 3,000 million years or so) in the sea, and only comparatively recently, in geological terms, have life forms ventured onto land.

The first land plants occurred 410 million years ago; the first insects crawled on land 360 million years ago; and the earliest terrestrial vertebrates appeared some 20 million years later.

OLD SURVIVORS, BRIEF APPEARANCES

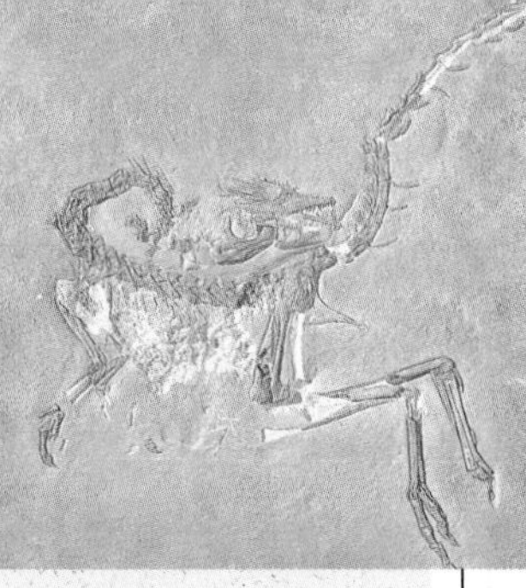

Even though life is always changing, some life forms have remained relatively unaltered for long periods. Such groups are commonly known as "living" fossils. Stromatolites in Shark Bay, Western Australia, are almost identical to those that lived thousands of millions of years ago. The small brachiopod *Lingula* has remained unchanged since its appearance 550 million years ago.

Most forms of life, however, appear and disappear relatively quickly. For example, the smallest dinosaur, *Compsognathus*, just 16 inches (40 cm) long, is known from a single specimen (above) found at Solnhofen, Germany. This is the only representative of a species that probably existed for a million years and whose total number would have been counted in the hundreds of thousands.

HISTORY *of* EVOLUTION

Fossils show that life has a complicated history and that organisms have become increasingly more complex.

Throughout the vast expanse of time life has existed, it has altered in response to environmental pressures. The changing of life through time is known as evolution.

There are many unscientific explanations for the origins of life, generally based on the existence of a divine creator, or many creators. The first suggestions that life forms were related by descent occur in early Greek texts. The ancient Greeks noted certain physical similarities among organisms and assumed that these resulted from their relationship to one another, but they were unable to explain the relationship.

CREATION THEORIES, *such as the story of Adam and Eve (below), were essentially unchallenged until the nineteenth century. French naturalist Georges Cuvier (right).*

Little advance was made in understanding the relationships of organisms until the mid eighteenth century, when Swedish naturalist Carl Linnaeus (1707–78) published his *Systema Naturae*, in which he set out a method of classifying plants and animals that is still in use. The Linnaean system places every species within a group, known as a genus, of similar plants or animals. Genera (the plural of genus) are placed, in turn, in larger groupings based on the sharing of similar structures by their members. Linnaeus considered the order that he found in nature to be God's "ladder of creation", rather than being the result of an evolutionary history.

Heaven knows what seeming nonsense may not tomorrow be demonstrated truth.

Science and the Modern World, ALFRED NORTH WHITEHEAD (1861–1947), English philosopher and mathematician

Advances in geology in the late eighteenth and early nineteenth centuries revealed that the Earth was a great deal older than had previously been believed. The study of fossils also revealed a number of perplexing facts. Modern organisms had not been around since the beginning of the fossil record and most organisms that had lived in the past were not alive in the present. Most types of organism appeared at some point in the fossil record, persisted for a period of time, then disappeared. New organisms appeared throughout Earth's history rather than being concentrated in a single appearance at the beginning, as would have been expected from a divine act of creation. Organisms also disappeared throughout time, although there seem to have been at least seven instances in the fossil record when huge numbers of organisms disappeared at the same time.

NATURAL SELECTION, *expounded by Charles Darwin (far left) in 1859 in* The Origin of Species *(left), is still regarded as the basic, most widespread mechanism by which evolution occurs. Variations evolve (below left) among closely related species, such as finches.*

By the mid nineteenth century, the accumulating evidence of biology (the similarity of organisms) and paleontology (the complex history of life) was directing the minds of intellectuals toward the concept of evolution. Erasmus Darwin (1731–1802), grandfather of Charles, wrote about the possibility of organisms being related by ancestry. French naturalists Etienne Geoffroy Saint-Hilaire (1772–1844) and Georges Cuvier (1769–1832) studied and documented similar structures in different organisms and Geoffroy believed these similarities were the result of an evolutionary history. Despite such support, evolutionary theories were not generally accepted because no one had yet proposed a viable mechanism by which they could operate.

Natural Selection

Jean-Baptiste de Monet Lamarck (1744–1829) was the first biologist to propose such a mechanism: he suggested that organisms changed during their lifetimes to meet particular pressures or demands and that these changes could be passed on to their offspring. He suggested that the accumulation of such minor changes through many generations would lead to new species being formed, but his theory was widely discounted.

It was not until the work of the English biologists Charles Darwin (1809–82) and Alfred Russel Wallace that a viable mechanism for evolution was proposed. Working independently, both Darwin and Wallace came up with the theory of natural selection. The theory drew on extensive scientific evidence relating to competition for resources among organisms, with their success or survival being dependent on their varied forms and the transmission of favorable traits from one generation to the next.

With the acceptance of natural selection as a valid scientific theory, evolution became the core concept of modern biology. The theory has since been supplemented by the science of genetics.

ALFRED RUSSEL WALLACE

While Darwin is most often cited as the originator of the theory of natural selection, the role of Alfred Russel Wallace (1823–1913) in the development of this theory should not be overlooked. Like Darwin, Wallace had spent long periods in Asia and South America collecting specimens, and his familiarity with nature led him to similar conclusions. Ironically, Wallace sent his manuscript describing natural selection to Darwin for comment. Darwin, who had been working in secret, was startled at the similarity of their ideas. The theory was eventually proposed under their joint authorship at a meeting of the Linnaean Society with neither present.

PROCESSES *of* EVOLUTION

Millions of species of plants and animals, all with a common ancestry, have evolved on Earth through many processes.

Although it is widely accepted that evolution has taken place, the mechanism by which it operates is much less certain. Evidence of evolution comes in many forms. Related species of animals or plants can be shown to have the same or similar structure.

EVOLUTION: THE FACT

Deoxyribonucleic acid (DNA) is the fundamental chemical in life. It is a complex molecule found in every living being and is the "blueprint" or code that tells an organism how to build itself. DNA is inherited from the parents so there is great similarity between the structure of the DNA of parents and their offspring. Over a greater distance of relationship, those organisms that are closely related will have a greater similarity in their DNA than those whose relationship is more distant. By comparing the degree of similarity, we can determine how closely two organisms are related. In this way, it is possible to map the history of relationships among organisms.

FEATHERED DINOSAUR? *As much a dinosaur as a bird,* Archaeopteryx *(right), with its modern-looking feathers, provided a convincing link between the two groups and confirmed evolutionary theory. The molecule DNA (above) controls the form of all organisms.*

Fossils provide evidence in support of evolution by tracking the progress of life-forms as they change through time. The classic example comes from Jurassic rocks of southern Germany.

When Darwin published *The Origin of Species*, he noted that one problem with the theory of evolution was the apparent lack of transitional forms in the fossil record. In 1861, miners near Solnhofen in southern Germany unearthed the first specimen of *Archaeopteryx*.

This specimen would have been described as a small dinosaur because of its dinosaur-like skeleton, except that the fossil clearly had impressions of feathers covering its body. Here was a link between birds and dinosaurs, two groups that had previously been thought to be separate.

The similarity of body parts (homology), the similarity of DNA sequences, and the changes in the fossil record make sense only in an evolutionary perspective.

MECHANISMS OF EVOLUTION

The most convincing mechanism for evolutionary change is natural selection, as proposed by Darwin and Wallace.

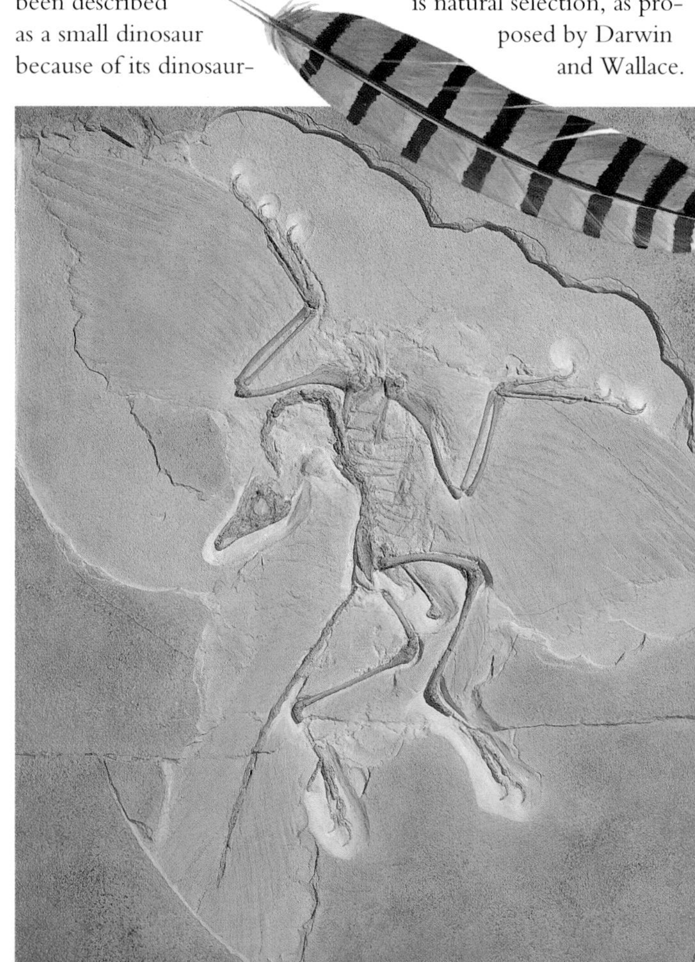

THE SCORPION *(left), like other organisms, produces more offspring than can possibly survive to adulthood. Offspring, such as the penguins below, show individual variation, even within the same family.*

ARMS FOR EVOLUTION

The front leg of a crocodile, the wing of a bat, the flipper of a whale, and the wing of a bird superficially look quite different from each other. On closer examination, however, these differences begin to disappear. For example, the arrangement of bones in the front limbs of these four animals is identical, even if the shape of those bones varies greatly. This common arrangement of bones, which extends also to humans, indicates that these limbs have all evolved from the same basic plan.

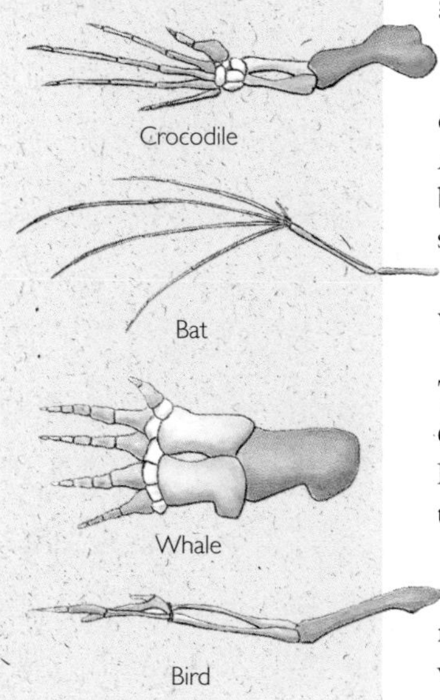

The theory results from three basic observations. First, each generation produces far more offspring than can possibly survive to adulthood and this leads to competition for resources. Second, all individuals vary in form, even siblings within a family showing slight differences. By chance, some variations within a population will be better suited to the prevailing environment and have a better chance of surviving to adulthood. Last, favorable traits of parents are passed on to their young.

The struggle among individuals for survival, the selection of the most fit individuals, and the passing on of favorable traits from one generation to the next allows a species to change through time to adapt to changing environmental pressures. Additional mechanisms have been proposed, but natural selection appears to be the most common means by which evolution proceeds.

The Tempo of Evolution

Darwinian evolution was thought to have been a gradual process that affected an entire species. By this model, called gradualism, one whole species was thought to evolve gradually into a new one. This view went largely unchallenged until the 1970s when Harvard professor Stephen Jay Gould and fellow paleontologist Niles Eldredge proposed that evolution had progressed in short bursts followed by long periods of little or no change, a model called punctuated equilibrium.

Debate between the two camps raged throughout most of the late 1970s and early 1980s. Gradualism found particularly strong support from genetics and biochemistry. Punctuated equilibrium was best supported by the fossil record, where new species appear "suddenly", and persist for a time without change before going extinct. This pattern, which fits the punctuated equilibrium tempo of evolution, is the result of a new species evolving in small populations in isolation from its ancestors. It is now generally accepted that evolution has proceeded in both of these ways.

FORMATION *of* FOSSILS

Preserved in a variety of ways, the remains of plants and animals that lived millions of years ago reveal the secrets of life on Earth.

There is no precise definition of a fossil. Generally, fossils are the remains of animals or plants, usually turned to stone, but this is not always the case. Mammoths frozen solid or insects trapped in amber are also considered to be fossils, even though their original composition has not been altered. While fossils are usually the remains of extinct prehistoric animals or plants, there are also fossil specimens of many animal and plant species that still survive. Pollen, excrement, tracks, eggs, and even the cast-off shells of living animals can also be fossilized.

TRAPPED IN MUD, *the dragonfly (below) may eventually become fossilized like the beautiful specimen shown at left with a modern relative. Shells, such as this gastropod (far left), fossilize readily.*

The vast majority of fossils are impressions of a once-living organism preserved in stone. While there are a number of processes that will convert a living organism to a fossil, they generally conform to a set series of events. Initially, after the animal or plant dies, the soft parts (guts, muscles, leaves, and so on) rot away, leaving the hard parts (bones, teeth, or wood).

There are exceptional situations where the soft parts are also fossilized, but these are rare. Soon after the organism has been reduced to its hard parts, it is buried by sediment—usually in association with water—in soft mud on the bottom of a lake, or in silt dropped by a flooding river. Once buried, the remains are impregnated with water carrying dissolved minerals that slowly replace the original organic material, turning the remains to stone.

After a fossil has been formed, one more essential process must occur before it can be found: the rocks that entomb the fossil must be uplifted and eroded back to reveal the fossil's location.

"INSTANT" FOSSILS

A relatively common fossil found on many parts of the coast of northern Australia, *Thalassina* is a crustacean, a kind of lobster, that lives in burrows on tidal mudflats. Like all other crustaceans, it has to cast off its shell in order to grow. Throughout its life, a single *Thalassina* may produce a dozen or more shells. Each time it molts, the shell is buried at the bottom of the burrow and a new living chamber is excavated. Buried in mud, the cast-off shell can be fossilized in a very short time, perhaps less than a year. Some of these fossils are so young that the animals that shed them may still be alive.

The chess-board is the world; the pieces are the phenomena of the universe; the rules of the game are what we call the laws of nature.

A Liberal Education,
THOMAS HENRY HUXLEY
(1825–95), English biologist and writer

WHERE FOSSILS OCCUR

The distribution of living organisms is governed by many factors: whales are found in the open sea, dogs live on land, and some apes inhabit forests. After the death of an organism, its carcass is rarely transported very far, so if the remains are fossilized, these will usually be found where the organism died, most often where it lived.

The location of fossil sites is determined by three factors: the environment where the organism lived, the rock type in which it is preserved, and the current exposure of that rock. Only rocks of certain types are suitable for preserving fossils. Volcanic rocks or rocks formed deep within the crust of the Earth will not contain fossils. Similarly, rocks with a very coarse structure, such as conglomerates, will usually destroy the remains of any organisms before they can be fossilized. The best types of rock for preserving fossils are limestones or such fine-grained sedimentary rocks as shales and mudstones.

Regardless of the fossiliferous nature of a rock unit, the fossils can be impossible to collect if the unit is not exposed. Mountainous areas where rocks have recently been uplifted and eroded back are good places to look. Some human-made rock exposures, such as quarries or road cuttings, can be prime sites for collecting. But the forces of weathering and erosion that expose fossils will also, eventually, destroy them.

FOSSILIZATION *sometimes happens on a grand scale, such as occurred at Dinosaur National Monument, Utah, USA.*

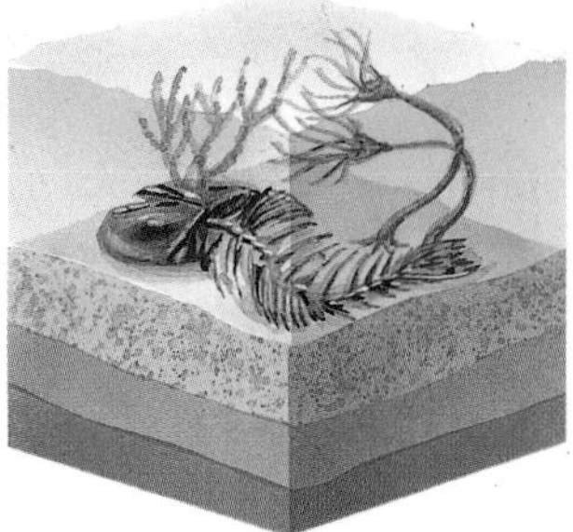

HOW A FOSSIL FORMS *In stage one, the soft tissues of the animal are broken down, leaving only hard parts such as teeth and bones.*

OVER A LONG PERIOD, *the bones become buried under layers of sediment, and the original bones are replaced by minerals from groundwater.*

WITH THE SLOW MOVEMENTS *of the surrounding terrain, the rock holding the fossil is uplifted and gradually makes its way to the surface.*

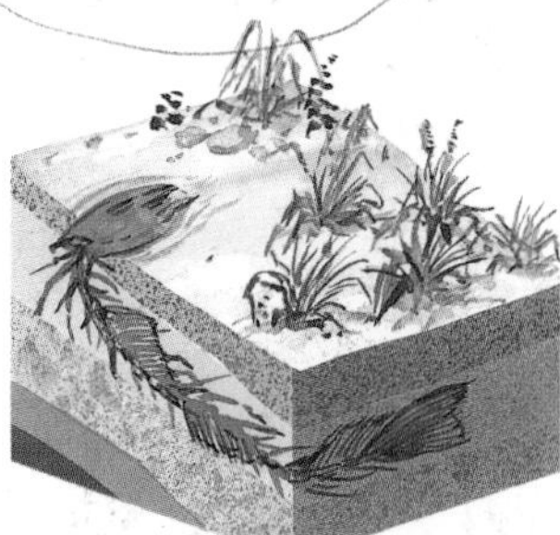

ONCE BACK NEAR THE SURFACE, *the fossil may be eroded out of its rocky tomb, or released by a fossil hunter, to disclose its information from the past.*

VARIETIES *of* FOSSIL

The range of fossil forms is astonishing, not only in the diversity of organisms represented, but also in the numerous types of preservation.

Most fossils are impressions of the original organism preserved in rock. The same effect can be created by taking a leaf, lightly pressing it into wet cement, and removing it. When the cement has dried, a "fossil" of the leaf remains.

In nature, the cavity left behind after the organism has rotted inside its rock tomb is often all that is recovered as a fossil. On rare occasions, the organic compounds from the organism decay to a carbon film that is left inside the natural mold. Sometimes the cavity, or mold, is filled with minerals deposited by water seeping through the rock.

In other cases, minerals are deposited from the water while the original organic matter is decaying, which results in petrification, or cell-by-cell replacement of the organic matter by minerals. In some cases, the resulting cast may be harder than the surrounding rock and the fossil will erode out of the rock and be found loose on the Earth's surface. Usually, replacement minerals are common elements from the surrounding environment, but sometimes more exotic minerals form fossils. For example, in Australia, fossils are sometimes formed from opal.

OTHER FORMS OF PRESERVATION

There are numerous other types of preservation. Since the movie *Jurassic Park*, amber fossils have become very well known. Amber is a semiprecious mineraloid formed from the sap of ancient trees. In rare cases, as the sap oozed out of the tree, a spider, insect, or other small animal would become entombed in the sticky goo. When the sap hardened, the creature was perfectly preserved in exquisite detail.

Some fossils retain their original organic composition. The most common of these are calcareous shells, where the calcium carbonate shell of the living animal forms the fossil shell, but often

FORMS OF PRESERVATION *Fern fronds (above) have been turned into a carbon film on a rock surface, while a tree trunk (above left) has been replaced by opal.*

PEARLY LUSTER *This Jurassic ammonite has its original mother-of-pearl coating, like that of modern oyster shells.*

PAST AND PRESENT *Although preserved only as a two-dimensional impression, this fossil seedpod is clearly similar to its modern relative.*

with some reorganization of the individual molecules. In some cases, the mother-of-pearl luster of the living animal is preserved on the fossil. Bone, too, can retain parts of its original composition, but because the more volatile fatty compounds quickly degenerate after the death of the animal, any bony remains preserved this way are usually chalky, brittle, and crumbly.

Another exceptional form of preservation occurs when bodies are buried in peat bogs. The high acidity of peat prevents bacteria and other decay-causing organisms from attacking the carcass and the organism is preserved in a process similar to that used to tan leather. In such unusual circumstances, the soft tissue, bones, teeth, and even the stomach contents may be preserved.

The ultimate preservation of ancient life is when bodies have been frozen solid, although it is debatable if these can be classed as fossils—they are usually termed sub fossils. Frozen "fossils" of mammoths, woolly rhinoceroses, and even people have been found high on mountains, or in the permanently frozen parts of northern Europe and Asia. Some specimens preserved in this way may be up to a million years old but there is no replacement of the original organic material and the specimen has not left an impression in rock.

Usually, only the hard parts of an organism, such as shell, bones, or teeth, will be fossilized. In some rare sites, however, because of a variety of factors, soft tissue may be preserved as well. These sites are so important to our understanding of the evolutionary history of various groups that they have a special name, *Lagerstätten,* from the German word for "mother lode".

Although fossils show how life has changed through time, the record is imperfect —99 percent of all the different life forms that have ever existed have left no fossils of themselves. Of the life forms that have produced fossils, many are known from a single example and most of the rest are represented by only a handful of specimens. Clearly, this is an inadequate record, but it is the only one we have. While many details of the story of life will never be revealed, we can at least deduce the major themes.

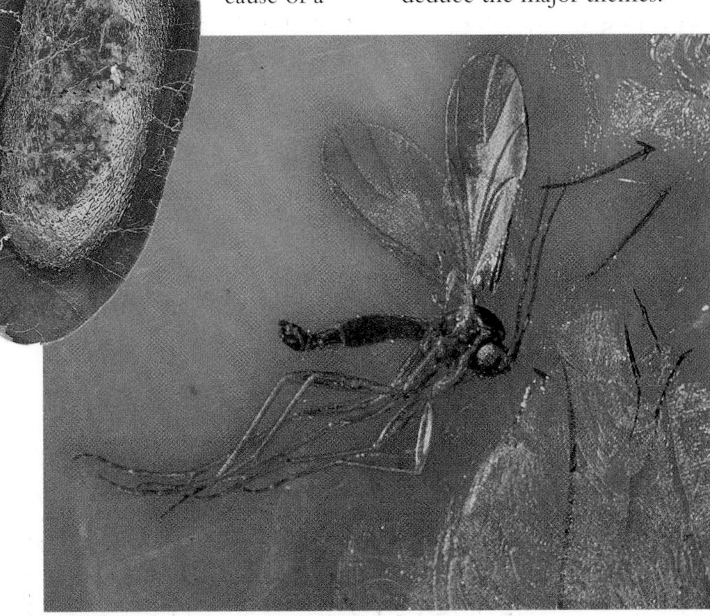

BONES AND INSECTS *A section of dinosaur bone (above) has been petrified and replaced with agate, while a tiny mosquito (right) has been preserved in perfect detail in amber.*

FOSSILS *in* LIFE

The only record we have of most of the animals and plants that have ever lived on Earth is from their fossils.

Paleontologists have developed sophisticated and complicated techniques to decode the fossil record and learn something of how animals and plants lived in the ancient world. Usually, it is only the hard parts of an animal or plant that become fossilized. Particularly in the case of vertebrate animals, only parts of a skeleton may be found, but a small part of a skeleton is often enough to make a reconstruction of the whole animal.

From shells or a few fossilized teeth or bone fragments, the paleontologist first identifies the type of animal. This is usually straightforward because different types of animal each have certain types of shells, teeth or bones. Remains of a species will resemble those of close relatives so even when a new type of animal is discovered, it can often be recognized by its similarity to other species.

Once the fossil fragments have been identified, the complete skeleton can be reconstructed by filling in the missing portions, using the structure of its closest relatives as a guide. Imagine that you were trying to reconstruct a previously unknown '57 Ford from only the bumpers and a small piece of bodywork. Your approximation would be more accurate if you based your reconstruction on '56 and '58 Fords rather than on a '57 Chev or, even worse, on a '91 Toyota.

❶

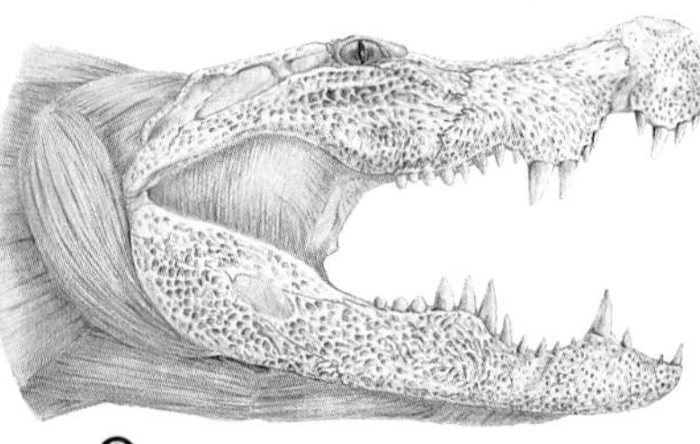

❷

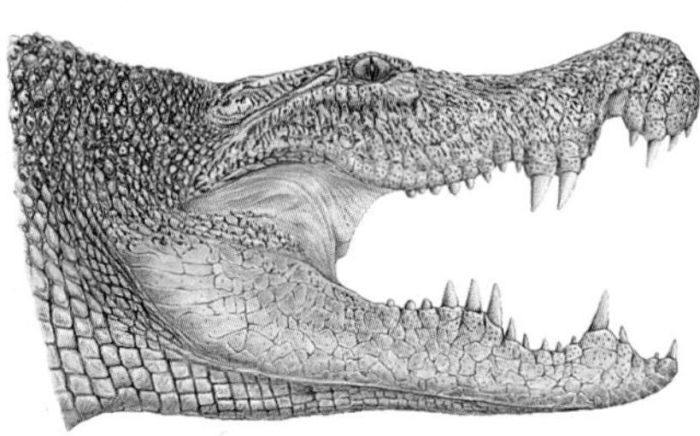

❸

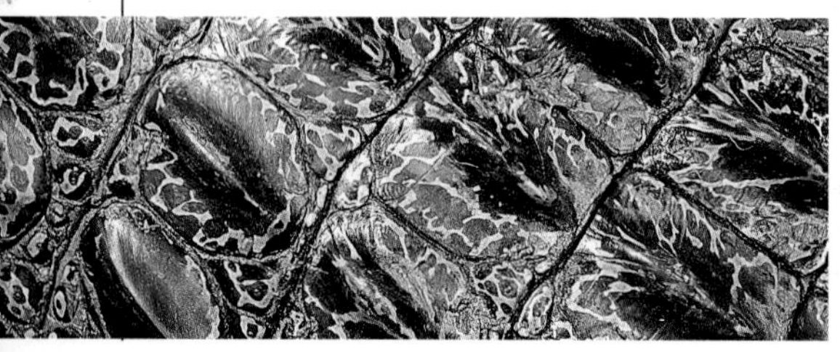

ASSEMBLING THE EVIDENCE
Reconstructing an animal from an original fossil ❶ involves completing the bones and musculature ❷, then covering with appropriate skin ❸. Each of these steps is aided by comparison with details from living relatives, such as the crocodile skin (below left).

From a complete skeleton, a paleontologist can reconstruct the size, shape, and posture of the animal. Muscles are located in similar positions in most vertebrates, and by looking for sites of muscle attachment on the bones, the paleontologist can reconstruct the musculature that gave the animal its form.

The last step is the addition of skin and color. Skin is rarely fossilized, so the paleontologist makes an educated guess as to what kind of skin an extinct animal might have had by looking at that of close relatives. Similarly, the color of extinct animals is seldom preserved, so this also has to be an educated guess. While the skin, fur, feathers, scales, or color may be the most noticeable aspects of a reconstructed animal, they are of minor significance compared with its size and shape. These aspects can be deduced fairly accurately from fossils.

EDUCATED GUESSWORK

A reconstruction of an extinct animal is only a model of what it might have been like. We can never be really sure, but a good reconstruction will present what we do know about the animal. Knowledge of extinct animals and plants changes with each new find

FROM A FRAGMENT
such as this shark tooth, a whole animal can be reconstructed. Many fossil sharks are known only from their teeth.

and subsequent reconstructions reflect this. For example, the first finds of the dinosaur *Iguanodon* were a few bones, teeth, and a mysterious spike. The teeth indicated that this animal was similar to the modern iguana but much bigger, so *Iguanodon* was reconstructed as a gigantic iguana with the spike on its nose.

When complete skeletons were found, it was clear that this animal could move on all four legs but probably spent a large amount of time standing on its hind legs and tail like a kangaroo. They also showed that the spike was not on the nose, but on the thumb. So *Iguanodon* was reconstructed in a pose similar to that of a giant kangaroo with spiked thumbs. We now know from more detailed studies that *Iguanodon* did not rest on its tail, but carried it straight out behind. It is also thought that the animal probably spent most of its time on all four legs, but could run on its back legs when necessary.

Fossils tell us more than what an extinct organism looked like. Analysis of teeth, claws, and even the contents of the gut (which are sometimes fossilized) can tell us what the animal ate. The association of different fossils in a single deposit allows us to reconstruct their environment and their positions in food chains. Fossil trackways (a series of footprints) can tell us a great deal about how animals moved. A series of fossils representing different growth stages of an animal reveals much about the development of individual members of that type of organism. In many cases, a relatively complete picture of an ancient ecosystem can be reconstructed from the fossils left behind.

> *The universe is not only queerer than we imagine, it's queerer than we can imagine.*
>
> *Possible Worlds,*
> J.B.S. HALDANE (1892–1964),
> British geneticist

RESTORATION OF THE IGUANODON.

CHANGING UNDERSTANDING
Reconstructions of extinct animals are modified with each new find. Early models of Iguanodon *(above) pictured it as a giant lizard with a nose spike. Later illustrations (above left) improved* Iguanodon's *posture and moved the spike from its nose to its thumb. Even the most recent interpretation (left) may be modified after future finds.*

IDENTIFYING *and* CLASSIFYING FOSSILS

Being the remains of living organisms, fossils are placed into groups in the same way that living organisms are classified.

All cultures have systems for naming familiar plants and animals, but these common names vary among cultures and geographic areas. The biological sciences require a naming system that is universal and consistent, so all living organisms that have been described scientifically have been given a two-part name, or binomial. The first part is the genus to which the organism belongs and the second is the species. This is like the Western system of people having a given and a family name, but in biology the order is switched so that we first find out to which group the animal belongs and then the species name.

In addition, all plants and animals are classified according to kingdom, phylum, class, order, and family. This system is the legacy of Carl Linnaeus. The concept was later expanded to include all known living organisms.

Like their living counterparts, fossils are identified as species, the smallest group in classification. Species are grouped in genera (the plural of genus), which are grouped in families; families are grouped in orders, classes, phyla (the plural of phylum), and kingdoms. For example, *Tyrannosaurus rex* is classified:

Kingdom	Animalia (animals)
Phylum	Chordata (animals with a dorsal nerve cord)
Class	Archosauria (dinosaurs)
Order	Saurischia (lizard-hipped dinosaurs)
Family	Tyrannosauridae (large, carnivorous dinosaurs with small arms)
Genus	*Tyrannosaurus*
Species	*T. rex*

TRACE FOSSILS *such as these 400-million-year-old eurypterid tracks (above) are given different names to the animals that made them. Beetles (above left) exhibit a huge range of diversity, making identification a challenging task.*

SCIENTISTS *had no satisfactory system for naming living things until 1735, when Swedish naturalist Carl Linnaeus (right) published* Systema Naturae, *his system for classifying plants and animals. His basic heirarchy is still used.*

Because binomials are derived from Latin, Greek, or other languages, the convention is for them to be written in italics or to be underlined. The genus always starts with a capital letter but the species never does. Another convention is that a generic name can appear by itself (for example, *Tyrannosaurus)* but a species name must always be accompanied by the genus name (*Tyrannosaurus rex*).

LIVING ORGANISMS, *such as the* Nautilus *(left), and fossil organisms, such as the ammonite (below), are named using the same classification system.*

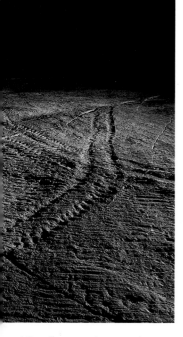

After the generic name has been used in full once, it can then be abbreviated to the first letter in all subsequent references to species of the same genus (*T. rex*).

A species is defined as a group that can reproduce with other members of that group but not reproduce with members of other groups. This is the principle of reproductive isolation. Experiments on the reproductive isolation of fossils are impossible, so another method of defining fossil species is required.

In paleontology, species are identified by their shape (or morphology) so a species of fossil is a group of fossils morphologically similar to each other but different from all other fossil groups.

Confusing the Issue

The species concept presents paleontology with another set of problems. Juveniles can look quite different from adults, and eggs and trackways bear no resemblance to either. It can thus be very difficult to relate juveniles, adults, eggs, and trackways to the one species. In cases where juveniles and adults of the same species have been accidentally named as different species, there is a comprehensive set of rules that allows us to determine which name is correct. A well-known case of juvenile–adult confusion is where the dinosaur *Brontosaurus* (an adult) and *Apatosaurus* (a juvenile) were described as different species. When the mistake was realized, it was determined that *Apatosaurus,* the name given first, had priority.

Identifying trace fossils, such as eggs and trackways, is more difficult, so a different naming system has been devised for them. They are given a binomial, genus and species, but because we can rarely associate a trace fossil, or ichnite, with its maker, it is classified an ichnotaxa with ichnogenus and species. For example, large, three-toed footprints with sharp claws were almost certainly made by carnivorous dinosaurs, but they are given the name *Anchisauripus* rather than *Tyrannosaurus* or *Allosaurus*. Consequently, it is quite possible for an extinct animal to have been given four names: one for the juvenile, one for the adult, one for its footprints, and one for the type of eggs it lays. Without evidence to link them, we may never discover that these four names belong to fossils of the same animal.

TYRANNOSAURUS REX *is probably the most widely known binomial. It means "tyrant-lizard-king".*

Tracks, Traces, *and* Microfossils

While the fossil of an organism tells us much about the creature it once was, trace fossils help establish its patterns of behavior.

In broad terms, fossils fall into two groups: body fossils (the remains of the organism itself), and trace fossils (signs and remains of an animal's activities, such as footprints, trackways, bite marks, feeding and dwelling burrows, nests, eggs, droppings, and stomach stones). Because conditions for preservation differ, the two types are rarely found in the same localities. Since we can seldom identify which animal made which trace fossil, the names of trace fossils differ from those of their makers.

Tracks and Traces

Fossil footprints can tell us the animal's weight, and a trackway can suggest how fast it moved. Usually, the normal walking speed is indicated, because comparatively little of an animal's time is spent running. Trackways can also reveal if the animal lived alone, in small groups, or in large herds. Sometimes, we may even glimpse how those groups behaved. In Lark Quarry, near Winton, central Queensland, Australia, there is a trackway of more than 190 small dinosaurs stampeding at the approach of a single, large carnivorous dinosaur. This is a moment preserved from life 98 million years ago. Such sites are fossilized in what were sandflat or floodplain deposits beside rivers and lakes. Other important places for dinosaur tracks include Glen Rose, Texas, and the Connecticut Valley, USA.

Trace fossil include the feeding burrows of invertebrate animals, each species of organism making a distinct type of feeding track. The burrows of underground dwellers may also be preserved. Some sediments become very mixed up by the actions of animals feeding and burrowing through them and are termed "bioturbated" (literally, "mixed-by-life").

Sometimes, the eggs of an animal are fossilized in the nest. Although rare, such sites can tell us much about the reproductive and parental behavior of the animal. From a dinosaur hatchery in Montana, USA, we can say that at least one kind of dinosaur nested in enormous rookeries, that the young remained in the nests for a considerable time, and that the parents brought them food and warded off predators.

Fossilized animal droppings, or coprolites, tell us about the diet of an animal and something of its digestive system.

TRACE FOSSILS *Tubular structures (left) in a matrix of fibrous dolomite, are thought to be worm burrows. "Worm tracks" and wave marks (below) on a shallow bottom, in Lower Cambrian sandstone. The radiolarian (top left), highly magnified to show its intricate form, is one of many types of microfossil.*

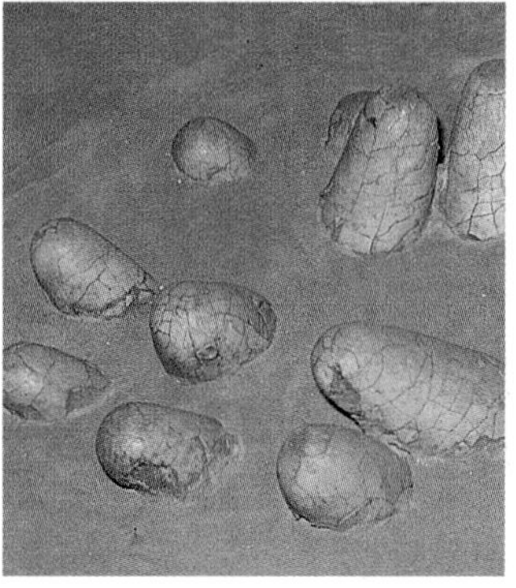

A RARE FIND, *dinosaur eggs (above), fossilized in the nest, have been discovered on several continents. These small stones (below), preserved alongside the bones of a giant bird, were once held in the bird's crop.*

FOOTPRINTS OF GIANTS

There are many sites worldwide where the fossilized footprints of dinosaurs and other animals are preserved. Among the first areas in North America found to have footprints were the brownstone quarries of New England. Much of the early work on footprints was conducted by Edward Hitchcock (1793–1864), of Amherst College, Massachusetts, who thought the footprints were those of giant birds.

A more prolific site is the area around Glen Rose, Texas. Here, footprints, including many extensive trackways, were known only to local residents until a visit to the area in 1938 by Roland T. Bird (1899–1978).

Bird found a trackway in which a meat-eating dinosaur appears to be stalking a long-necked dinosaur. This significant find is now on display in the American Museum of Natural History, New York.

A number of animals ingest stones to help with digestion or for ballast in water. These stones, which become smooth and rounded inside the animal's gut, are called gastroliths or stomach stones. Gastroliths are sometimes found in the body cavity of the fossil or alone in sediments where they were regurgitated after use.

Teeth marks of a predator or scavenger are sometimes seen on the fossilized bones of an animal and indicate which predators were eating them. In some cases, it's the teeth marks not of another species but of a member of the same species, possibly the result of a clash over a mate or territory.

MICROFOSSILS

Most people think of fossils as being visible to the naked eye, but the most abundant and useful types of fossil are too small to be seen without a powerful lens or microscope. The variety of microfossils representing the plant and animal kingdoms is vast.

A number of different types of microscopic, single-cell organisms have hard shells that can become fossilized. Typical among these are dinoflagellates, radiolarians, and diatoms. Dinoflagellates have both "animal" and "plant" characteristics and leave fossils of a "shell" or cyst, typically between 20 and 150 thousandths of a millimeter long.

Radiolarians are larger, at between 100 and 2000 thousandths of a millimeter. Diatoms are single-cell algae that produce a "shell" of silica. These "shells" can be so abundant that they form the whole rock, called diatomite. Diatomite is composed of compacted diatom shells, regular in size and shape. This makes it commercially useful for filtration in brewing and other industries.

Because a single tree can produce billions of grains of pollen and these can be scattered over a wide area, pollen grains are the most likely tree parts to survive as fossils. Plant spores and pollen can both be fossilized—they are very useful for reconstructing ancient climates and floras.

Parts of animals, such as the bony scales of some fish, are also classed as microfossils. Other types of microfossil, including conodonts, foraminifera, and coccolithophores, appear in Chapter 7, the fossil field guide.

PLANT FOSSILS

When plants emerged from the water and spread onto land about 410 million years ago, they had many problems to overcome.

Plants have a long history—algae are known from Precambrian rocks. However, until about 410 million years ago, plants were represented only by algae found in aquatic environments. Because of their fragility, algae are largely inconspicuous in the fossil record, except where they formed stacked algal mats called stromatolites, or where they produced calcareous skeletons similar to corals. Stromatolites are the most abundant macrofossils of the Precambrian, and various calcareous algae are reasonably common in Ordovician and Silurian deposits.

To survive on land, plants had to stand without the support of water surrounding them. They evolved in two ways: by developing a vascular system that transported water and nutrients around their tissues and also acted as a pressurized skeleton; and by developing woody stiffening compounds. Because these woody fibers were generally harder than other plant material, plants that developed them were much more likely to be fossilized.

The earliest land plants, Psilophyta, were little more than creeping root systems with vertical shoots and no

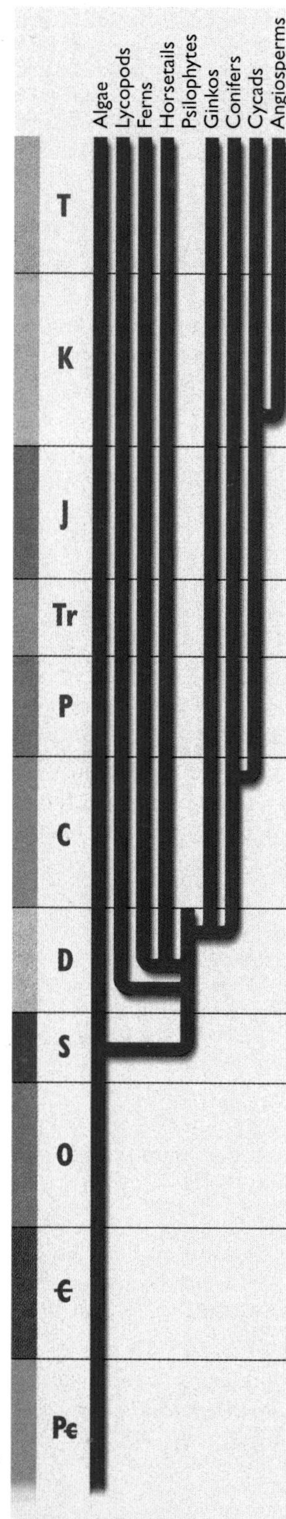

PLANTS *(dark green on diagram) diversified rapidly after they spread onto land in the Silurian. The dominant plants of today are the angiosperms, flowering plants, which first appeared in the Cretaceous.*

FOSSILIZED CONES *(top left) from the monkey puzzle tree—the one on the right has been sectioned and polished. The delicate form of a leaf (above) that grew in the Eocene.*

evidence of leaves. Ferns, Pterophyta, appeared in the Devonian. Their close relatives, the seed ferns (Pteridospermophyta), and lycopods, a group of tree-like plants, grew in such abundance by the Carboniferous that the earliest coal deposits were formed from their remains. Seed ferns are now extinct but modern ferns are familiar to us all. Despite their early success, lycopods are represented today by only three genera.

Cycads, a group of palm-like plants, became relatively common in the Mesozoic era, but today are represented by only a few species growing in tropical to warm temperate regions. Conifers, the group that includes fir and pine trees, dominated most of the Mesozoic and, together with the cycads, probably formed the bulk of the diet of plant-eating dinosaurs.

The plant world today is dominated by angiosperms, the

flowering plants, which include all grasses, and most trees, palms, and shrubs. Angiosperms appeared early in the Cretaceous, but did not become common until toward the end of that period.

Plant parts (leaves, seeds, roots, trunks, and so on) may be found as separate fossils and identified as different organisms. Until the various parts are found in association, this confusion cannot be addressed. For example, a single lycopod had one name for its trunk fossils (*Lepidodendron*), another for fossils of its roots (*Stigmaria*), a third for those of its leaves (*Lepidophyllum*), and a fourth for its cones (*Lepidostrobus*). *Lepidodendron*, the first name given, is now assigned to all parts.

Plant fossils are usually found in rocks where the environment was slightly acidic. Mudstones and shales, particularly where they are associated with coal deposits, are good locations for finding fossils of plant parts. In these fine-grained rocks, flat objects such as leaves have been laid down and preserved in detail. The accumulation of rock on top of the entombed fossil will usually have reduced it to an imprint, but often the original carbon content of the plant material is preserved as a film on the rock surface. In special cases, the waxy leaf covering, or cuticle, can be preserved and removed for examination.

FERNS HAVE THRIVED *for over 300 million years, retaining the same essential form. Compare the fossil ferns (above) and their living relative (right). The petrified logs (below) are from the late Triassic. Petrified wood is prized by collectors for its beauty and color.*

The more robust parts of a plant, such as the trunk, branches, cones, or seeds, can withstand more rigorous environments, and fossils of these plant parts can be found in sandstones. In such cases, the original plant material has been completely replaced by minerals. This process, proceeding slowly, cell by cell, is called petrification. The result is a faithful reproduction of the plant material in minerals, the structure inside the individual cells being retained. Whole forests preserved in this way have been found in the USA in Yellowstone National Park, Wyoming, and Arizona. In Australia, the minerals in some petrified wood have been replaced by opal.

AN ETRUSCAN SACRED OBJECT

When one Etruscan chief was buried some 2,500 years ago, a section of the fossilized cycad *Cycadeoidea* was interred with him. This fossil is particularly well preserved, still showing scars where the fronds once grew from the main trunk. Although it is unlikely that the chief recognized the paleontological significance of the specimen, it was obviously important to him. We can surmise that the owner believed that some magical process had produced such an unusual form, giving the object mystical power. This specimen is now held in the Bologna Museum, Italy.

SIMPLE INVERTEBRATES

Fossils from groups of invertebrate animals with the simplest level of organization demonstrate the developing complexity of animal life.

The abundance of invertebrate fossils is the result of a number of factors: shells are hard enough to withstand fossilization; the animals usually live in high concentrations; and they often live in environments where fossilization is likely to occur.

Multicellular animals, or metazoans, first appear in the fossil record in the late Precambrian, although fossil tracks and burrows of these or similar animals are known from earlier rocks. These early animals are enigmatic—they are possibly early representatives of later groups such as jellyfish, annelid worms, and crustaceans, or they could belong to an early group that evolved, flourished, and became extinct before the beginning of the Cambrian.

Certainly, there are fossils found in the late Precambrian that appear to belong to no known metazoan group. These include the circular *Tribrachidium,* with three-fold symmetry that is seen in no other known group.

BRACHIOPODS *(above), a group of bottom-dwelling marine animals with two hinged shells, are abundant as fossils. Cambrian limestone (right) containing fossils of archaeocyathids.*

CAMBRIAN EXPLOSION

There are very few fossils from the beginning of the Cambrian, and these are from only a few types of animal. By the mid Cambrian, animals were abundant and diverse. This apparent explosion of life, known as the Cambrian Explosion, could be due to a number of factors. It could be that, because there were many environmental niches, organisms evolved rapidly to fill them. It could also be that the ability to form a hard shell (thus increasing the chance of producing a fossil) appeared simultaneously in a number of different groups in response to changed conditions.

Whatever the cause, and it is likely to be a combination of factors, by the mid Cambrian, there were representatives of most major groups and even of a few animals that have since become extinct.

Sponges (phylum Porifera) and their allies, the archaeocyathids and stromatoporoids, are multicellular animals, but they are not regarded as metazoans. All three groups first occur in the Cambrian and sponges survive today. The sponges are mostly represented by microscopic spicules—slender, pointed, crystalline structures—but all three groups leave macroscopic calcareous structures as fossils. These are usually either cup-shaped or cylindrical.

... [time is] the process of decay and transformation, just as it always has been.

Timebends: A Life,
ARTHUR MILLER (b. 1915),
American playwright and author

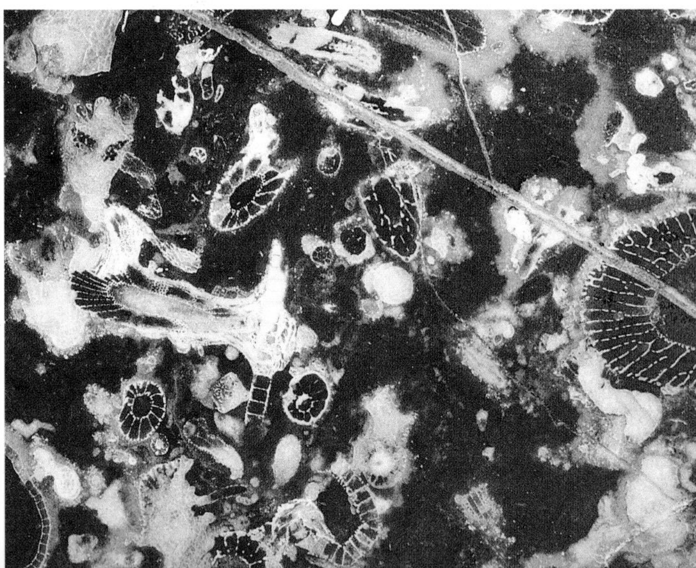

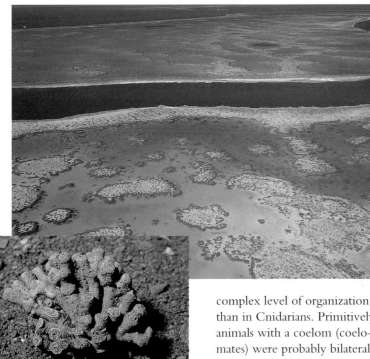

SPONGES AND CORALS *are among the oldest multicelled animals. A fossil sponge (below left) is from the Cretaceous. Great Barrier Reef, Australia (left).*

Stromatoporoids, particularly, were important reef-builders from the Ordovician to the Devonian.

Cnidarians

Probably the most primitive group of metazoans is the Cnidarians, including jellyfish, corals, and hydroids. Cnidarians have two layers to their body walls and are arranged in a radial pattern. They have two forms in their life cycle, a free-swimming medusoid and an attached hydroid.

In different groups, different forms are emphasized or lost. In jellyfish, it is the free-swimming medusoid that is most familiar to us, but in corals, the attached hydroid, or polyp, is the only form, the medusoid stage having been lost. Cnidarian fossils are among the oldest fossil metazoans and, while jellyfish are rare as fossils, corals are abundant and widespread throughout the fossil record.

All remaining animals, vertebrate and invertebrate, have three layers in the body wall. The middle layer, the mesoderm, can form a body cavity or coelom. This allows for a far more complex level of organization than in Cnidarians. Primitively, animals with a coelom (coelomates) were probably bilaterally symmetrical and divided into segments along their length. Many coelomates retain this form of organization and, even in groups that have lost their segmentation, we suspect that they are derived from segmented ancestors.

Lophophorates

A relatively small group of coelomates that is well represented in the fossil record is the lophophorates. This is a group of filter feeders that includes bryozoans and brachiopods. These animals strain microorganisms from the water with a pair of coiled furry arms called lophophores.

Bryozoans, commonly known as cold water corals or lace corals, are a group of mostly marine colonial animals. Today there are more than 3,500 known species.

Brachiopods are known as lampshells because of their resemblance to Roman lamps. There are only 70 surviving species, most of which are living in marginal environments.

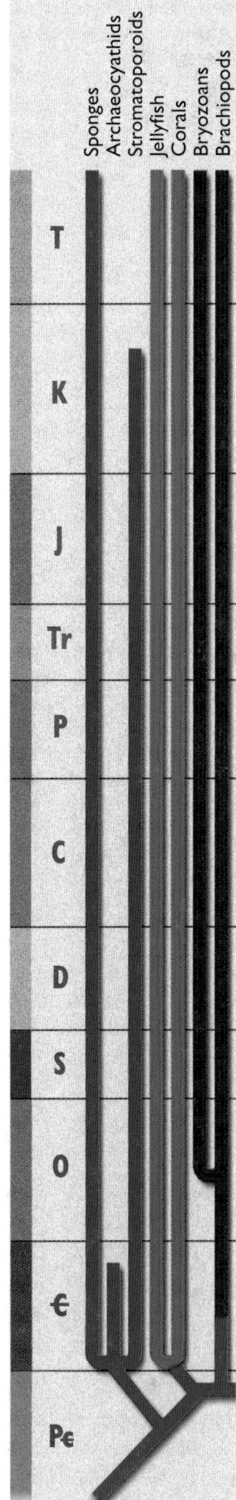

THE GROUPS *of invertebrate animals (sponges dark blue, Cnidarians light green, lophophorates burgundy) shown in the diagram (right) have relatively simple levels of organization. The ancestral link to all these groups also appears in dark blue.*

More Invertebrates

Among animal species, those without backbones are by far the most numerous and they are widely represented in the fossil record.

A major group of invertebrates includes the annelid worms (segmented worms and leeches), the arthropods (crabs, shrimps, insects, spiders, and trilobites), and mollusks (snails, clams, and squid). All are united by having similar larvae and development patterns, and most are bilaterally symmetrical. Annelids and arthropods are clearly segmented but mollusks have lost most traces of segmentation, which suggests that they split from the group early in their history.

Annelid worms, which include the familiar garden and marine worms, are rarely fossilized because they have no hard parts.

Arthropods

Arthropods have a good fossil record because they have hardened shells that fossilize easily. Further, because arthropods regularly cast off their old shells and grow new ones, a single individual may produce many shells in its lifetime, multiplying the opportunities for fossilization.

NATIVE AMERICANS *found trilobites (above) so appealing that they gave them a name meaning "little water bug in the rocks".*

Arthropods are easily identified by their jointed legs, which are moved by a series of internal muscles. It is this thin suit of armor that has allowed the arthropods to become so successful. It affords protection and supports the animal's weight while still allowing a wide range of movement.

Within the arthropods there are four classes of varying importance to the fossil record: *Trilobita*, *Uniramia*, *Crustacea*, and *Chelicerata*.

Trilobites are one of the most important fossil groups of the early to mid Paleozoic and their attractive fossils are much sought after by collectors. They are characterized by a multisegmented body of three basic parts: a broad head

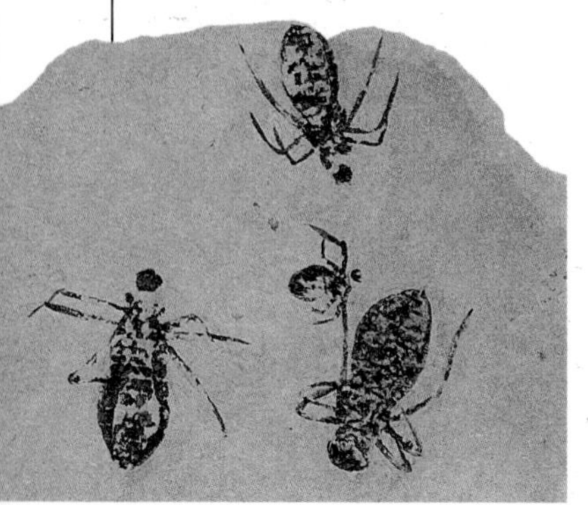

RELATED GROUPS *The seemingly disparate groups (right) of arthropods (orange), mollusks (red), and annelid worms (dark blue) can be shown to have a common ancestor (also dark blue). Insects (left) are the most diverse group of living metazoans, but they are rarely fossilized.*

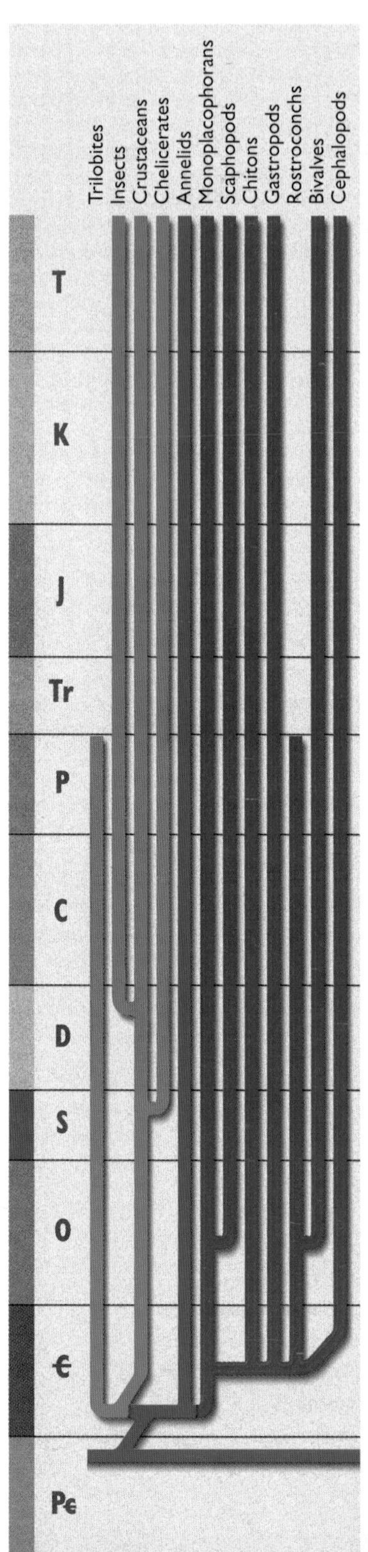

CRUSTACEANS, *such as this lobster (left), are arthropods with legs that divide into two branches.*

or cephalon, a body or thorax, and a tail or pygidium. Each body part can be divided into three sections (lobes). Trilobites became extinct at the end of the Permian.

Uniramia is a group of arthropods with unbranched limbs. It includes hexapods, or insects, onychophorans (small arthropods that superficially resemble caterpillars), and myriapods (centipedes and millipedes).

Insects, the most diverse group of metazoans in the world today, do not have a fossil record to match their current massive diversity. However, rare preservations, such as those in amber, indicate the major evolutionary paths of this group.

Crustaceans (shrimps, crabs, and lobsters) are mostly marine and have limbs that branch into two. The earliest representatives are about 550 million years old. Most crustaceans have relatively thin shells that readily disarticulate after death so they have been preserved only in exceptional situations.

Mollusks

Mollusks are mostly marine invertebrates with some freshwater groups and a single terrestrial group, the land snails. Mollusks have a very good fossil record because most have hard shells. The success of mollusks is due largely to the protection given by their heavy shield-like shells. There are six main groups: monoplacophorans, chitons, gastropods, scaphopods (tooth-shells), bivalves (clams), and cephalopods (squid, octopus, nautiloids, and ammonoids).

Gastropods are familiar as snails—marine, freshwater, and terrestrial. The helically coiled shells of gastropods are common fossils from the Cambrian to the present.

Bivalves are readily recognized as clams, oysters, and mussels. These two-shelled mollusks first appeared in the Ordovician, but their possible progenitors, the rostroconchs, are known from the late Cambrian through to the Permian. Bivalves are also known as pelecypods and lamellibranchs, both of which are old names now considered to be incorrect.

The most advanced mollusks are the cephalopods, including octopus, nautiloids, squid, and ammonoids. They have a single shell that can be straight, spiral, or some variation of these. In some groups, the shell is reduced or completely lost. The largest shelled cephalopods were some types of ammonite that could grow to more than 10 feet (3 m) in diameter but the largest cephalopod is the living giant squid, which reaches 50 feet (15 m) in length.

MOLLUSKS *come in a variety of forms, including gastropods (above) and cephalopods (below).*

ECHINODERMS *and* VERTEBRATES

More complex animals, such as the echinoderms and vertebrates, represent some of the most fascinating organisms in the fossil record.

CRINOIDS *(left) are upside-down echinoderms attached to the sea floor by a stem with "roots". Sea stars (above) are one of the most important invertebrate predators of marine life.*

The last major group of fossil animals on our Tree of Life consists of the echinoderms (including sea stars, sea urchins, and sea lilies), the protochordates (including graptolites), and the chordates, which include vertebrates (fish, amphibians, reptiles, birds, and mammals).

These animals are placed in one group because of numerous similarities. They have similar larval forms, show segmentation, and have either bilateral or five-part symmetry around a central axis.

ECHINODERMS

Echinoderms are distinct in the animal world in having five-fold radial symmetry (five similar parts arranged around a central axis) instead of bilateral symmetry, although they are probably descended from a bilaterally symmetrical ancestor. The echinoderm skeleton is composed of many small, interconnected plates made of calcite. These animals also have an internal water-vascular system—a series of fluid-filled canals and reservoirs through which water and food circulate. Externally, they are covered with spines. There are four main groups, all of them marine.

Echinozoa, including sea urchins (echinoids), and sea cucumbers (holothuroids), are mostly rounded or discoid. Echinoids gather their food in a variety of ways—there are active hunters moving over the sea floor, and burrowing forms that process huge amounts of sediment to extract microscopic organisms. Some echinoids have jaws powerful enough to allow them to burrow into solid rock. The earliest echinoids and holothuroids are from the Ordovician.

Asterozoa includes sea stars (asteroids) and brittle stars (ophiuroids). Both have a familiar form with five or more radiating arms, and both have a reasonable fossil record dating from the Ordovician.

Crinoids, or sea lilies, mostly live in deep water and have not been studied in great detail. Blastozoa is an extinct group of attached echinoderms that superficially resemble the crinoids. These two groups first appeared in the Cambrian.

CLEAR DETAILS *of the skeleton and fin rays are visible in this small fossil fish. Bony fish first appeared in the Silurian.*

PROTOCHORDATES

Graptolites are colonial organisms that floated through the Paleozoic seas. These protochordates are most widely known from thin impressions of branches of colonies, like hacksaw blades drawn on rocks. Other members of this group are rarely fossilized.

CHORDATES

Chordates are bilaterally symmetrical animals with segmented body muscles in V-shaped bands (myotomes). They also have a stiffened rod (notochord) running along the back and a nerve cord above the notochord. In this basic form, primitive chordates possess no hardened parts that are likely to be fossilized and the fossil record of early chordates is sketchy. The earliest recognized chordate is the tiny *Pikaia,* from the middle Cambrian.

After chordates developed hard parts, they became better represented in the fossil record as vertebrates. The earliest fossils of vertebrates are microscopic "scale-bones" from late Cambrian rocks, probably belonging to early jawless fish (agnathans), which proliferated in the Silurian and Devonian.

The cartilaginous skeletons of sharks and rays (chondrichthyans) do not fossilize well, but early chondrichthyans are known from the Devonian.

Other groups of fish have appeared and disappeared throughout evolutionary history. The group that contains the bulk of modern fishes—the ray-finned fish or actinopterygians—had its origin in the late Silurian.

The earliest terrestrial vertebrates were amphibians that resembled salamanders. The next big step in vertebrate evolution was the development of an egg that could be laid on land, thus freeing the parents from remaining close to water to reproduce. This appears to have occurred during the Carboniferous, and resulted in the evolution of reptiles, birds, and mammals.

FOSSIL FROG *(left). These amphibians date from the Jurassic. Graptolites (below) first appeared in the Ordovician and went extinct by the end of the Permian.*

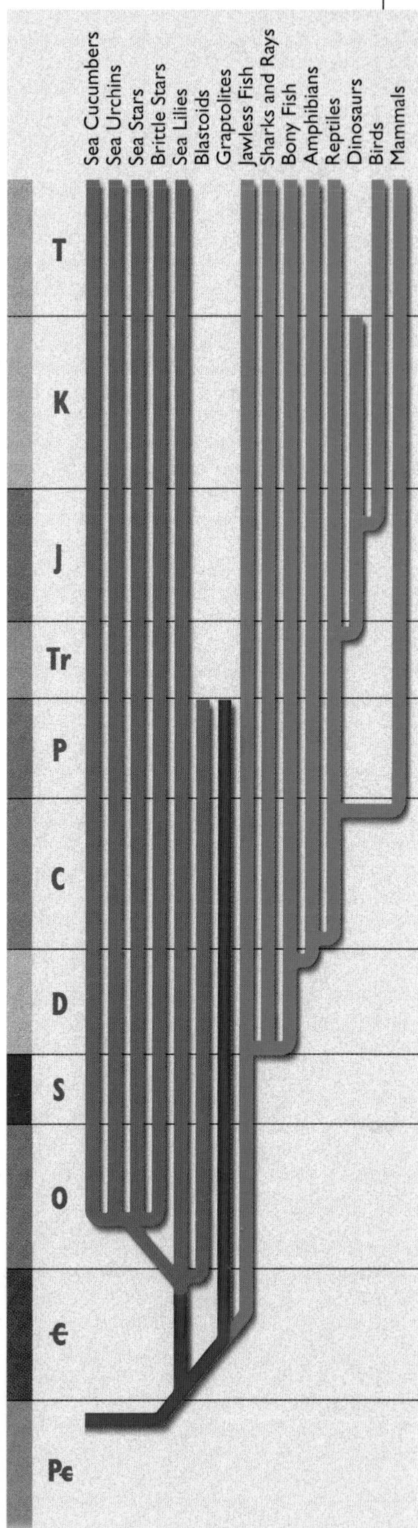

THE SEEMINGLY DIVERSE GROUPS *of echinoderms (purple) and chordates (dark blue), including the vertebrates (light blue), can be shown, from studies of their development and morphology, to have a common origin in the early Cambrian.*

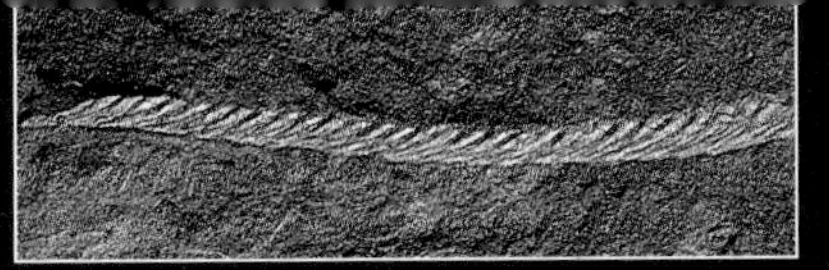

Chapter Five
Unlocking *the* Secrets *of* Fossils

Unless you expect the unexpected you will never find [truth], for it is hard to discover and hard to attain.

Fragments,
HERACLITUS (c. 500 BC), Greek philosopher

Studying Fossils

This fascinating science not only reveals the secrets of the past, but has a variety of practical applications.

FIERCE RIVALRY *between Othniel Charles Marsh (above) and Edward Drinker Cope (left) led to the retrieval of huge collections of dinosaur specimens.*

Paleontology is the study of fossils, a scientific discipline that has emerged from the long history of association between humans and relics of the past. Today, as the result of long intellectual struggle, fossils are readily recognized as the remains of ancient life forms.

Fossils have always been difficult to interpret. Native North Americans thought that the petrified tree trunks scattered around the landscape in Arizona were the bones of monsters, or weapons from battles among the gods. The ancient Egyptians used some fossils for road-making and carving, but they do not seem to have appreciated their origins.

As early as the seventh century BC, Ancient Greeks, including Xenophanes, Pythagoras, Xanthus, and Herodotus, observed fossils of marine shells and even a small fossil fish found inland in the mountains and realized that these areas of land must once have been under the sea. Although Pliny the Elder (AD 23–79) recognized that amber was ancient pine resin, he also thought that fossilized shark teeth had magical powers helpful for those who court fair women.

During the Dark Ages and Middle Ages, fossils received little attention and were generally regarded as curiosities, oddities, or freaks of nature. While some, such as the Moslem philosopher Avicenna (980–1037), held more accurate ideas about fossils and the Earth's history, such views were far from universal. Leonardo da Vinci (1452–1512) recognized that fossils might once have been living organisms and went some way to explaining how they could have been formed by burial in sediment. Anatomist Gabriello Fallopio (1523–62) thought that fossils were the result of movements and activities inside the Earth. This view, and the idea that fossils grew within the Earth by various processes, were widely held throughout the Middle Ages. Another concept that retarded a scientific understanding of fossils was that they were the remains of organisms killed in the Biblical Flood.

PLINY THE ELDER *(left) was familiar with some types of fossil, including shark teeth. Native Americans thought petrified logs (above) were bones of monsters.*

By the Renaissance, a more rational view of fossils was becoming prevalent. Scholars began to make fossil

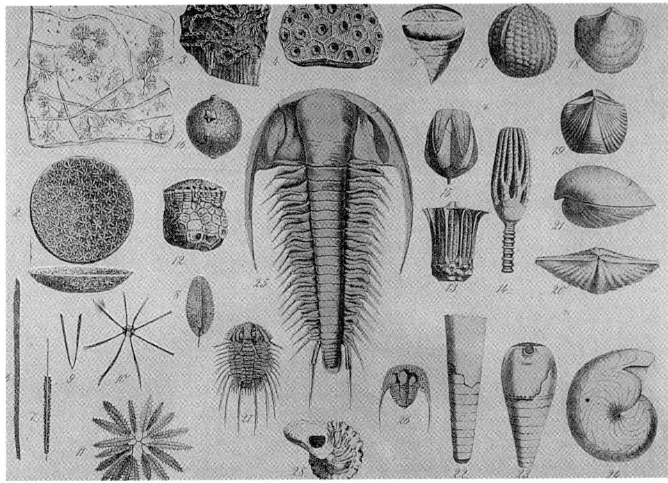

collections and to illustrate and describe them in printed works. The Scottish physicist and mathematician Robert Hooke (1635–1703) was the first person to make detailed studies of the fine structure of petrified wood with a microscope and to comment on the similarities, even at the most detailed levels, between fossilized wood and living trees. He was convinced that fossils must be the remains of once-living organisms.

The eighteenth century marked the foundation of the sciences of geology and paleontology, and by the nineteenth century the two were firmly established. Fossils showed that the world had changed and that numerous organisms had appeared and disappeared. They were instrumental in establishing evolution as a scientific fact—

BY THE LATE *eighteenth century, fossils were being collected for study by both scientists and amateurs.*

no longer were they poorly understood oddities but a record of the history of life.

Every great and commanding moment in the annals of the world is the triumph of some enthusiasm.

Nature Addresses and Lectures, RALPH WALDO EMERSON (1803–82), American essayist

During the nineteenth century, the great museums of the world competed for the best collections of natural objects and prized among these were fossils. In the latter half of the nineteenth century, two great American adversaries, Othniel Charles Marsh and Edward Drinker Cope, battled fiercely for the latest and largest dinosaur specimens for display in their respective museums. Both were wealthy and employed professional fossil hunters to scour the badlands of the western United States for specimens.

Today, paleontology has gone far beyond the esoteric collecting of fossils and interpreting the history of life. It is an active and vigorous science practiced worldwide by people working for universities, museums, geological surveys, and mining companies. It is of particular use in the mining industry, especially for the petroleum exploration industry, and is widely used in the assessment of modern environmental problems by providing evidence from environments of the past.

SHE SELLS SEASHELLS ...

Mary Anning (1799–1847) lived in Lyme Regis, Dorset, England. Her father owned a souvenir shop and, finding that the abundant fossil shells from the area fetched high prices, Mary began collecting them to sell. She inspired the tongue-twister "She sells seashells by the seashore".

In 1811, she found her first skeleton of an ichthyosaur (a dolphin-like reptile), and later dug up many other skeletons of Jurassic marine animals from the Age of Dinosaurs. Many fine specimens on display in the British Museum of Natural History and other museums around the world were found by Mary Anning.

From Evolution *to* Extinction

From fossils, we learn that we are intimately related to all other life on Earth and, like all species, are vulnerable to extinction.

HINDU ICONOGRAPHY *sees the cosmos as a turtle, with four elephants supporting the world.*

Historically, it was thought that the Earth and all life upon it was created by a god or gods in the recent past, certainly within the past 100,000 years and more likely only 10,000 years ago. One notable exception is the Hindu religion, which conceives the universe as being the dreams of the gods, recurring in cycles of tens of billions of years.

The concept of a universe beginning billions of years ago, with or without divine intervention, found its way into Western philosophy only recently. Along with this view came the realization that the universe, the Earth, and all life, have changed over time.

Before the scientific revolutions of the eighteenth and nineteenth centuries, people in the West considered humans the species chosen by God to preside over all life on a planet at the center of a universe that He had created. However, our future depends on a better understanding of Earth's place in Nature. Astronomy has taught us that the universe is ancient and that the Earth is a minor planet orbiting one of billions of stars on the edge of one of billions of galaxies in an ever-expanding universe. Geology has taught us that the Earth is also very old and that humans evolved only recently.

Wild Cards

A surprising lesson from the history of life is the possibility of periodic exterminations or mass extinctions. Early in the systematic study of fossils it was realized that there have been at least seven moments in the history of the Earth when the enormous variety of life was suddenly reduced to relatively few forms. In the worst case, at the end of the Permian, up to 95 percent of all life on Earth was wiped out. The second-largest mass extinction, 65 million years ago, destroyed the dinosaurs and about 80 percent of all species that lived with them.

Although we have known about mass extinctions for nearly 200 years, they were largely considered to be mere hiccups in the progress of life from ameba to human. Despite this, and the dearth of direct evidence, their causes had long been the subject of speculation. More than 200

METEORITE IMPACT *This crater at Wolf Creek, Western Australia, is just one of numerous spectacular impact sites around the world.*

A THIN WHITE LAYER *of iridium-rich clay separating Cretaceous and Tertiary rocks (right) indicates a meteorite impact about the time dinosaurs went extinct. The layer contains extra-terrestrial magnetite (above).*

extinction scenarios had been proposed but all lacked credible evidence to support them.

Since the early 1980s, however, paleontology has had to reconsider both the nature of mass extinctions and their effects. At that time, workers studying rocks in northern Italy found high concentrations of iridium, a rare element, in a layer of clay that marked the end of the Cretaceous. It has been argued that this iridium was concentrated in a thin layer by the impact of a meteorite.

In this scenario, a large meteorite, about 6 miles (10 km) in diameter, hit the Earth, causing a massive explosion far greater than the detonation of all the world's nuclear weapons. The result, including the blocking of sunlight to the Earth for months or possibly years, would have been enough to cause a mass extinction.

Mass extinctions tell us two things. First, the effects of large explosions, such as meteorite impacts or multiple nuclear detonations, are sufficient to erase most life. Second, the fact that humans are here today was not predetermined by evolution. Mass extinctions are the wild cards of evolutionary history. If a meteorite impact had not wiped out the dinosaurs, they would probably still be the dominant terrestrial vertebrates in the world today.

EXTINCTION OF THE DINOSAURS

Fossils often tell surprising stories about the history of life. Until recent times, it was assumed that life was a continuous process of refinement, winnowing out the unfit to produce new and better life forms. Current studies of early organisms, however, indicate that there was an amazing diversity of life very early in the story. Apparently, the selection of those forms that survived was not entirely dependent on Darwinian selection. This area of paleontology, called the Contingency Theory, is best illustrated by the extinction of the dinosaurs. After 160 million years of evolution, the dinosaurs were perfectly adapted to their environments, but no amount of Darwinian selection could have prepared them for the effect of something like a large meteorite's impact on the Earth. This extremely rare and unpredictable event is probably what eliminated the dinosaurs from their long-held dominance of terrestrial niches and made way for the rise of the mammals.

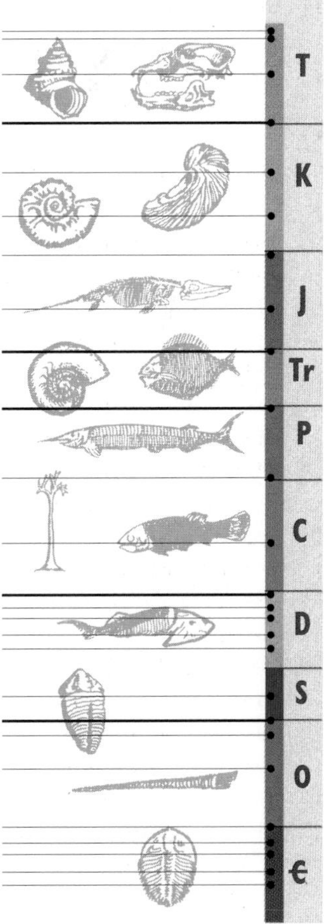

NUMEROUS EXTINCTIONS *have occurred throughout the Earth's history.*

60% or more of species lost

30%–60% of species wiped out

THE PAST REVEALED

If we are to preserve the Earth's wonders and resources for future generations, we need to examine all the evidence and learn as much as possible from past calamities.

Fossils can be thought of as snapshots from the past, giving indications of what earlier environments were like. Such relics provide a chronicle of the evolution of the biosphere. And, like many historical documents, the fossil record holds important lessons for the present.

GLOBAL WARMING

One of the major crises confronting the world today is global warming, the raising of the temperature of Earth's atmosphere. Known as the greenhouse effect, this phenomenon is being caused by various human activities, chiefly the burning of fossil fuels. One of the important questions that confronts us is how to predict what will happen if the temperature of the atmosphere rises. We can invoke complicated computer models, which are only as good as the assumptions made in them, or we can look at the Earth's past, to a time when the atmosphere was hotter than it is now.

Measuring the past temperatures of Earth's atmosphere presents considerable difficulties, but the temperature of the oceans is a little easier to gauge. There is a correlation between the temperatures of oceans and the atmosphere. Chemical analysis of certain minerals that were present in the shells of once-living creatures and that are now trapped in their fossils can be used for this purpose. Living animals trap oxygen molecules in their shells and the ratio of the two most common isotopes of oxygen are correlated to temperature. So, analysis of the oxygen isotopes of a shell indicates the temperature of the ocean at the time the animal was alive.

This type of analysis has been conducted on the fossil shells of animals dating back

CHEMICAL ANALYSIS *of fossil shells (top) reveals the temperature of the ancient world at different times.*

THE MAKING OF A DESERT

There are some frightening lessons from fossils, both with respect to natural changes and those wrought by human activity. Fossils reveal that much of northern Africa was covered by savanna woodands until comparatively recent times. Animals, such as elephants and hippopotamuses, roamed much of the area and their presence was recorded by early human inhabitants.

The nature of this region has been gradually changing over the past few million years, becoming drier and less hospitable to life. These changes have been accelerated by human activities such as grazing, fire management, and land clearing. As a consequence, there are now large areas of desert.

IN THIS CHANGING WORLD, *ancient rain forests (top) that once covered vast areas of Australia and other southern continents have been replaced by desert (left). Extensive glaciers (above) locked up enough water to lower the sea level, creating "bridges" over which species crossed to new lands.*

to the early Cambrian, giving us a fairly accurate measure of the temperature of the Earth's atmosphere since then. There are other ways to take the temperature of ancient atmospheres and oceans, and these are used for corroboration.

As well as indicating the temperature of ancient worlds, fossils also tell us about the life occurring at a particular period. In past times, when the world was warmer than it is now, forests extended over larger areas than they cover today. Warm-habitat plants and animals spread to higher latitudes because the prevailing conditions were favorable to them. In general, a warmer world is a wetter world and the sea level rises as water is released from the melting of polar ice caps. Such conditions prevailed in the mid Cretaceous and again in the Eocene. During these times, rain forests covered areas such as Australia and Antarctica, now the two driest continents on Earth.

On the other hand, fossils tell us about life when the Earth's temperature was cooler than now. During past ice ages, for example, glaciers covered large parts of the Northern Hemisphere, extending as far south as the latitude of New York. Because more of the Earth's surface water was bound up in ice sheets, the sea level was lower and precipitation decreased. When the sea level fell, land bridges appeared where once there were shallow seas. These bridges allowed animals and plants to disperse into areas that were previously inaccessible to them.

It was the extensive land bridges formed during the last ice age (100,000 to 15,000 years ago) that allowed our species to spread to most parts of the globe, including the Americas and Australasia. In this cooler, drier world, deserts were much more extensive than they are now.

It is worth remembering that the radically different world of the last ice age was an average of 9° Fahrenheit (5° C) cooler than the world today. Most prognostications about the effects of global warming are based on an increase in global temperatures of a similar magnitude.

STUDIES OF PAST *climatic changes may help us to deal with current environmental problems, such as air pollution.*

PRACTICAL APPLICATIONS *of* PALEONTOLOGY

Fossils are present in many of the things we take for granted in the modern world, from building materials to the fuels we depend on.

It is easy to dismiss fossils as curiosities, or as merely the basis of an arcane science, but the study of fossils produces a vast body of information that has both direct and indirect economic value.

Fossil fuels, such as gas, oil, and coal deposits, are all formed from once-living organisms. The accumulated debris of a forest that was growing for many thousands of years before burial eventually produces coal. It is because of its organic content that coal and some other deposits, particularly some sea-floor sediments, are high in complex hydrocarbons.

When coal deposits are heated, volatile components are driven off and migrate through the sediments to be deposited elsewhere as oil or gas. The hydrocarbon chains in the gasoline that powers our cars, in the natural gas that heats our stoves, and in the oils that lubricate our engines were all put together by plants and animals millions of years ago. Most of the world's electricity is produced by coal-fired generators that burn the compacted residues of ancient forests. In a very real sense, our energy needs are filled by the fossilized remains of ancient worlds.

READING ROCKS

The rocks of the world were laid down, one atop the other, in grand sequences called stratigraphic columns. No rock unit, however, is distributed worldwide and most are very restricted geographically. So, while stratigraphic columns can be constructed for a particular area or region, it is difficult to correlate the layers of one stratigraphic column to those from another region except through their fossil content.

Unlike rock units, some plant and animal species are distributed over large areas, many even globally, so their fossils are potentially useful in correlating distant rock units. If the fossils are of organisms that evolved rapidly and of species that occurred in short, well-defined periods, the correlations should be quite accurate. If the species existed for a more extended period, precision may be compromised. It helps greatly if the fossil organisms are abundant and easy to find.

The rapid evolution, wide distribution, and abundance of many microfossils, such as

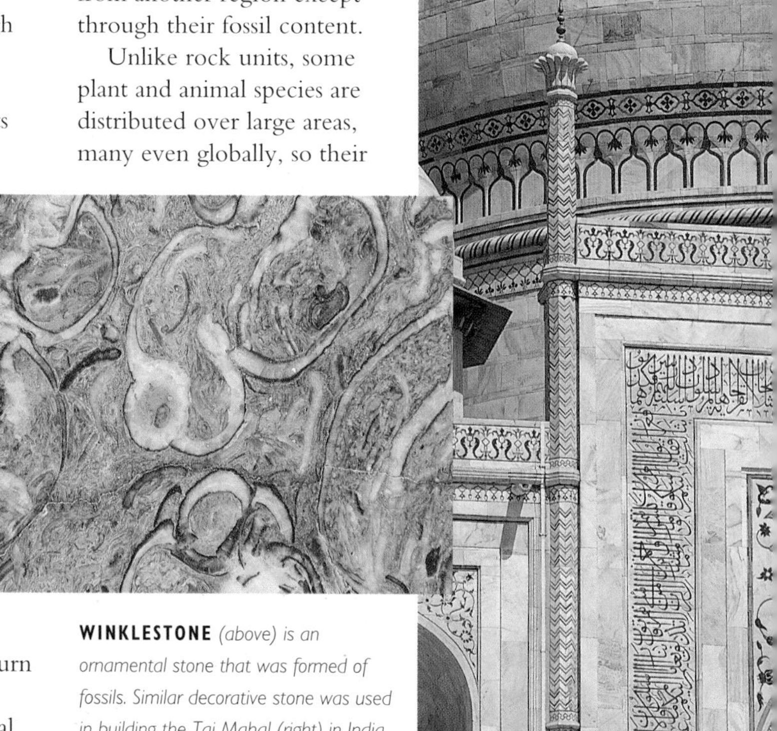

WINKLESTONE *(above) is an ornamental stone that was formed of fossils. Similar decorative stone was used in building the Taj Mahal (right) in India.*

THE LUXURIANT SWAMPS *of the Carboniferous (above) eventually became extensive coal deposits. Geological processing of coal produces oil deposits. In Kuwait (right), these were wasted by war.*

radiolarians, foraminifera, and conodonts, make them useful indicators for correlating rock units over large areas. Fossilized plant pollens and spores reveal a great deal about stratigraphic correlation in rocks younger than about 290 million years.

Correlation of rock units is economically important in the search for deposits of minerals, oil, gas, and coal. If a particular type of mineral deposit is found in rocks of a particular age, it is useful to investigate rocks of a similar age when seeking similar deposits.

Building with Fossils

There must be few buildings in the world that do not, in some way, rely on the fossils of ancient organisms to keep them standing. Cement (powdered limestone) is perhaps the most commonly used material, either alone or mixed with aggregate to form concrete.

Most of the world's limestone deposits suitable for cement production are fossilized reef systems. Ancient organisms extracted calcium carbonate from the surrounding seawater to form their shelters, and we recycle their ancient dwellings to construct our own homes.

Decorative stone, principally marble, is also used extensively, particularly in bathrooms and kitchens. Marble is limestone that has been heated and subjected to enough pressure to reorganize the crystal structure of the original rock. Certain types of marble retain the fossilized remains of ancient reef builders. It is always worth inspecting marble for signs of fossilized animals.

MICROSOPIC DIATOMS *(top far left) form diatomite, a rock used to filter beer.*

Other Uses

While collectors pay well for spectacular specimens, fossils have many more practical applications. The most widely used material for the filtration of beer and similar liquids is diatomite, a porous rock composed of the fossilized shells of microscopic diatoms. The main source of superphosphate for fertilizers is sea-bird excrement known as guano. In some places, the guano is so deep and old that it is best thought of as fossilized bird droppings.

Black ivory, the tusks of mammoth fossils collected in Siberia, had numerous uses, including black keys for pianos. This resource was largely exhausted by the middle of the twentieth century, as were commercial deposits of particular types of amber that had been used in the production of varnishes and lacquers.

Limitations of Fossil Resources

All fossil resources are finite. They have taken thousands, often millions of years to form and, once depleted, they will take a similar period to be replenished. We are using fossil resources, particularly our limited fossil fuel deposits, at a much higher rate than they can be produced, so we need to develop other solutions to the problems these resources currently solve.

FINDING FOSSILS

With minimal equipment, proper care, and a little planning, searching for fossils can be a satisfying hobby.

The search for and recovery of fossils is a rewarding pastime that requires physical effort, some basic research and knowledge, some simply acquired skills, and a minimum of special equipment.

MAKING PREPARATIONS

Fossils are not found in every rock, so some basic research is needed before embarking on a fossil-collecting expedition. It is important to find out not only where to search, but also what kind of fossil can be expected in a particular area, who owns the site, whether access is permitted, and if there are any particular dangers, such as traffic at roadside sites, or tides at the seaside. Prime collecting localities are often listed by State mining departments and local geological surveys.

Amateur collectors who know the area are usually helpful and can often be contacted through rock-collecting and lapidary clubs. Tourist information centers, souvenir and rock shops, and regional museums also sometimes have local information.

Once you have identified a likely site, it is worth surveying it for recent exposures—road cuts or landslips may have exposed some fresh material. Quarries have great potential because the ongoing work of the quarry continually turns up new rocks. Remember that quarries can be particularly dangerous places and that safety aspects should be thoroughly checked with the property owner before you venture in.

A HAMMER AND CHISELS *(top left) and eye protection are all that is needed in the field, but sledge hammers (left) and other specialized equipment are also options.*

FINDING FOSSILS INSIDE ROCKS

It is relatively uncommon to find fossils exposed on the surface. Better specimens are usually obtained by breaking rocks open with hammers, cold chisels, or crowbars. When at a new site, experiment with some of the rocks to see how they break.

Usually, fossils are found on bedding planes, like bookmarks between the pages of a book. The trick is to work out the best way to open a particular type of rock along its bedding planes. In some cases, it is easier to cut across the bedding planes; in others, it is

DETAILED FIELD NOTES *should be kept for all field activities. Keep information about site locations and access for future reference.*

BRUSHES *can be used (left) to remove the loose debris that clings to your precious specimen. Often the best technique for finding fossils is to persevere, systematically splitting the rock at one particular site (center).*

better to cut along. Only trial and error will show the best approaches to take for particular types of rock.

Don't be too ambitious in the field. It might be better to take a fossil home still embedded in lots of rock so that it can be removed carefully later. Many good specimens have been destroyed by impatient collectors removing too much rock in the field.

Taking Your Specimens Home

Although fossils are made of rock, they should be treated like glass. Once you have cut away as much rock as is wise in the field, write a number on the fossil or surrounding rock with a felt-tipped pen for easy identification later, then wrap your prize in plenty of newspaper to protect it on the journey home.

Most important: don't be greedy. Never take more than you need for your collection. The more you collect, the less there will be for other enthusiasts. As a courtesy to others, leave the site in the condition you originally found it.

WHAT TO TAKE ON A FIELD TRIP

The most important equipment is protective gear such as eye goggles and, for powdery or dusty conditions, a face mask. Fossil-collecting sites are usually in rough terrain, so sturdy footwear is essential—steel-capped boots have saved many a toe from being crushed by falling rocks. Check weather reports and ensure that you have adequate protection from the Sun and the elements (sunscreen, broad-brimmed hat, long-sleeved shirt, and rainwear). A small first-aid kit is also a good idea.

A notebook and pen (take spares) allows you to record what you have found and where you found it. A felt-tipped pen will write on most rocks or fossils so you can immediately number specimens for later identification. Don't rely on memory and don't write important collecting information on separate scraps of paper that can become detached from the specimen and get lost. You might take maps of the area you are looking at to plot your finds and progress.

You will need packing material, such as newspaper, bubble wrap, and plastic bags, to transport your finds, and a sturdy knapsack will be useful.

Of course, you will also need tools to break the rocks to get at the fossils. A geologist's hammer and a range of cold chisels in different sizes is really all that is required. Crowbars, sledge hammers, picks, and shovels can all be useful, but remember that they are heavy and tiring to carry.

Sometimes the facts are of value in themselves, ... now and then someone finds a key that opens a whole new field.

The Molds and Man,
CLYDE CHRISTENSEN
(b. 1905), Professor of plant pathology

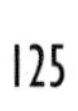

PREPARATION *and* CARE *of* FOSSILS

Fossil collecting begins in the field with the retrieval of specimens, but this is only the start of your adventure with paleontology.

Fossil specimens are usually brought home embedded in the rock matrix to protect them. The amateur paleontologist must clean and preserve the specimens collected, then identify, catalog, and store them.

The first step is to remove the surrounding rock with fine chisels, small hammers, and probing tools, such as those used by dentists, or even an electric engraving tool. This takes skill and patience.

The tools and techniques used will vary with the nature of the fossil and the matrix. Proceed slowly, using a variety of tools and techniques to determine the most appropriate for the type of fossil. The experience and guidance of other fossil collectors is invaluable at this stage. The golden rule is not to do too much—it is better to leave a specimen partly embedded in matrix than to damage it by taking risks in the preparation.

Once the specimen has been prepared, it may need to be hardened. One technique is to dissolve some water-soluble wood glue in water with a drop or two of detergent—10 percent solutions of glue are usually fine for most fossil work. This solution is dribbled onto the surface of the specimen where it forms a clear, hard skin. For porous, crumbly specimens, acetone-soluble glue dissolved in acetone is better—again, a 10 percent solution will be satisfactory. This solution will soak into the specimen and harden it throughout. Don't mix different types of glue as this can result in the formation of an insoluble cement that cannot be removed if further cleaning is required.

A SELECTION OF TOOLS, *including brushes and fine probes, is essential. Sometimes it may be necessary to glue specimens (right).*

Often a specimen is broken when found or breaks during excavation or preparation. Once the individual pieces have been cleaned and hardened, they can be glued together again. Be sure that surfaces to be joined are clean and fit together properly prior to gluing. Use an acetone-soluble glue so that mistakes can be undone if necessary.

Epoxy glues are useful for large specimens that need strong joints, but if a mistake is made, these glues are almost impossible to dissolve. A box of sand makes an excellent "pair of hands" to hold a fossil while the glue is drying.

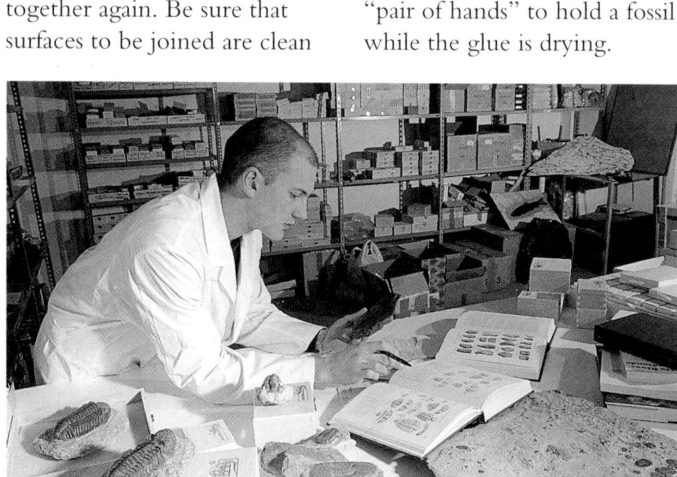

TO IDENTIFY *your finds, compare them with illustrations in reliable books.*

CLEANING SPECIMENS *with a fine, pointed tool (below). Larger fossils may need to be protected in a plaster cast (left).*

Identifying Fossils

The levels to which hobby collectors wish to identify their finds will vary greatly. With this book, an amateur will be able to identify the major group to which a fossil belongs and, in some cases, the genus. For many amateur collectors, this level of identification is satisfactory.

To identify finds more accurately, buy or borrow books with illustrations and photographs of excellent specimens with detailed descriptions of each type of fossil. These are available through rock and fossil shops, local geological surveys, or mining departments. Again, the knowledge and resources of more experienced collectors will prove invaluable.

Storing Fossils

There are two basic types of storage: display storage for your more spectacular and interesting fossils, and collection storage for more mundane specimens. For both groups, humidity and temperature must be kept as low and as constant as possible. This means being careful when choosing lighting for a display cabinet—it should not heat the specimens. Wood is the perfect material for these cabinets because it helps to buffer the humidity around the fossils as well as providing some insulation against temperature extremes. Display cases can be fitted with locks to protect your collection against theft or damage from inept handling.

Collection cabinets are usually stacks of shallow drawers. Consider lining the bottoms of drawers with soft material, or placing your specimens inside small cardboard boxes within the drawers. This stops them from rolling around when drawers are opened and closed.

Cataloging

Since collectors rapidly acquire many different types of fossil from a number of locations, it is important to establish a system for cataloging and identifying them—don't trust your memory. All fossils in a collection should carry a unique number, which can be written on the specimen. This number, plus the fossil's identification and collecting information, should be recorded on cards to be kept with each specimen. It is also a good idea to keep a central register of specimens detailing the number of each, its identification, collecting information, and details of its preparation history and any preservation treatments that have been carried out.

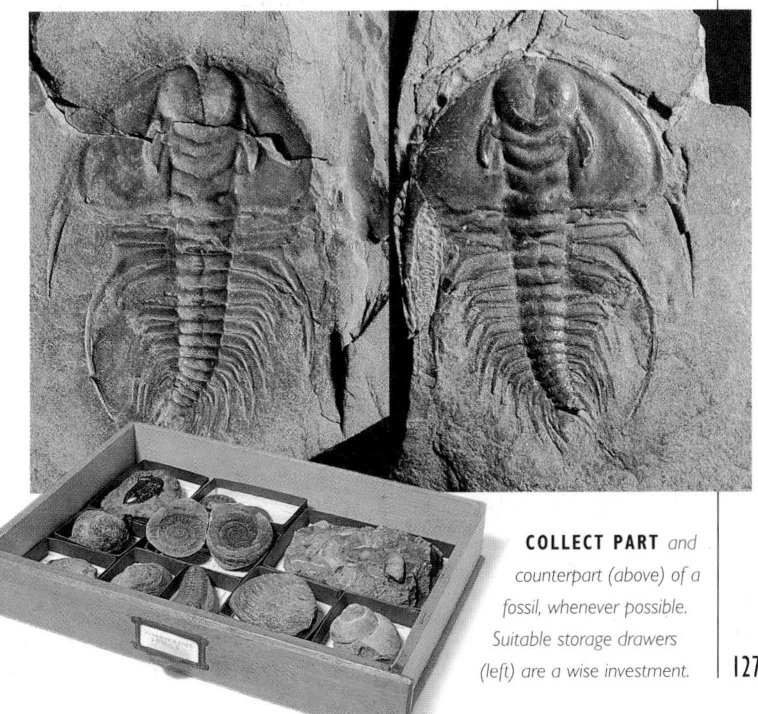

COLLECT PART *and counterpart (above) of a fossil, whenever possible. Suitable storage drawers (left) are a wise investment.*

Special Techniques

Although the amateur can prepare many types of specimen, some require special equipment, patience, skill, and more demanding techniques.

Many types of fossil can be prepared at home by amateur collectors with simple, readily available tools, but other preparation techniques are more complicated and call for equipment to which the average collector doesn't usually have access.

Thin Sections

When fossils are sliced thinly with a diamond saw, paleontologists can study the internal structure of the organism. Further information can be retrieved from thin sections carefully ground to a thickness through which light can pass and mounted on microscope slides. Studying these by means of microscopes that polarize light can reveal details of the organism's structure and composition.

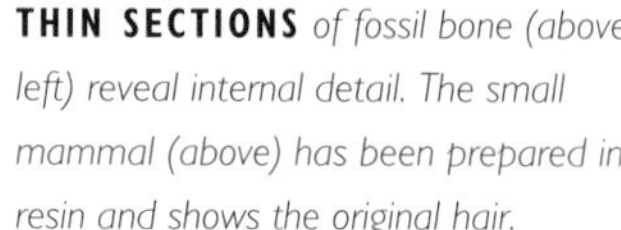

THIN SECTIONS *of fossil bone (above left) reveal internal detail. The small mammal (above) has been prepared in resin and shows the original hair.*

Acid Preparation

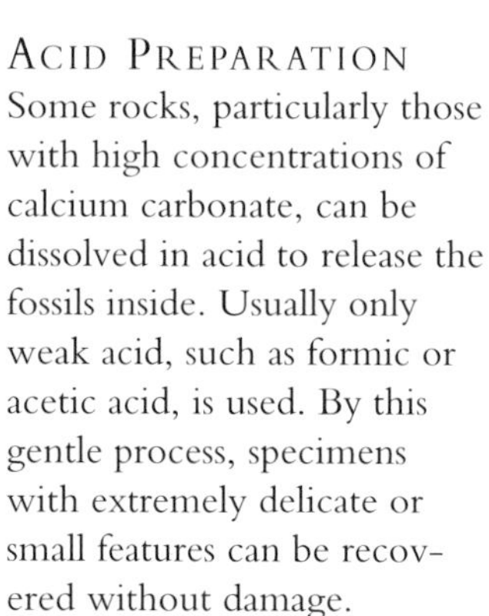

Some rocks, particularly those with high concentrations of calcium carbonate, can be dissolved in acid to release the fossils inside. Usually only weak acid, such as formic or acetic acid, is used. By this gentle process, specimens with extremely delicate or small features can be recovered without damage.

Acid preparation is also the easiest way to release microfossils from rock. To retrieve microfossils such as radiolarians, which have shells that resist acid, stronger acids are used.

Conversely, acids can be used to dissolve the fossil. This is particularly useful where the fossil is fragile and the surrounding rock is impervious to acid attack and otherwise too hard to remove. By soaking the rock in an acid that dissolves only the fossil, the cavity where the fossil lies is cleaned out, then filled with latex, or some other molding compound, and the rock smashed away. The result is a perfect latex copy of the original fossil.

ACID PREPARATIONS *involve carefully dissolving the rock surrounding the specimen (left) in weak acid and hardening the emerging fossil.*

Copying Fossils

There are numerous techniques for copying fossils. Basically, a mold is made of the fossil, usually with latex or silicon rubber. The fossil is removed from the mold, which is then filled with plaster of Paris or a hard-setting resin. A more basic technique used by many amateurs is to press the fossil into modeling clay to form a mold. Copies are then made with plaster of Paris. In recent years, the prices of silicon rubber molding compounds and casting resins have fallen to a level that is affordable by the average amateur. Experiment with making copies of your favorite fossils, but remember to use a suitable releasing agent between the fossil and the molding compound (soapy water or petroleum jelly are good), and try your technique on disposable objects first.

Embedding in Resin

Particularly delicate fossils that rely on the surrounding rock matrix for support can be liberated by embedding them in resin. Although a simple procedure, it takes practice and patience. Once the specimen is prepared from one side, a clear casting resin is poured over the surface and allowed to set. The block is then turned upside-down and the matrix carefully removed from the other side. Eventually, all that is left is the fossil, embedded in clear resin.

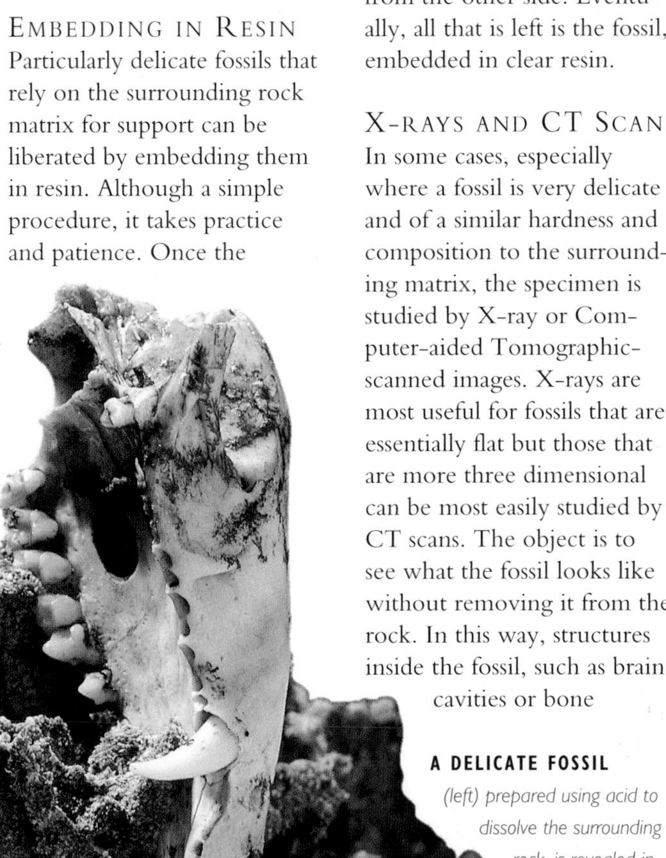

A DELICATE FOSSIL *(left) prepared using acid to dissolve the surrounding rock, is revealed in exquisite detail.*

X-rays and CT Scans

In some cases, especially where a fossil is very delicate and of a similar hardness and composition to the surrounding matrix, the specimen is studied by X-ray or Computer-aided Tomographic-scanned images. X-rays are most useful for fossils that are essentially flat but those that are more three dimensional can be most easily studied by CT scans. The object is to see what the fossil looks like without removing it from the rock. In this way, structures inside the fossil, such as brain cavities or bone thicknesses, can be looked at and measured without damage. It is now possible to take CT-scan data and make a three-dimensional resin reconstruction of the fossil. In effect, this technique produces a perfect copy of a specimen that is still locked inside rock.

COPIES OF SPECIMENS *can be made by first coating them in latex (left). Some fossils, such as the Devonian arthropod below, are best studied using X-rays.*

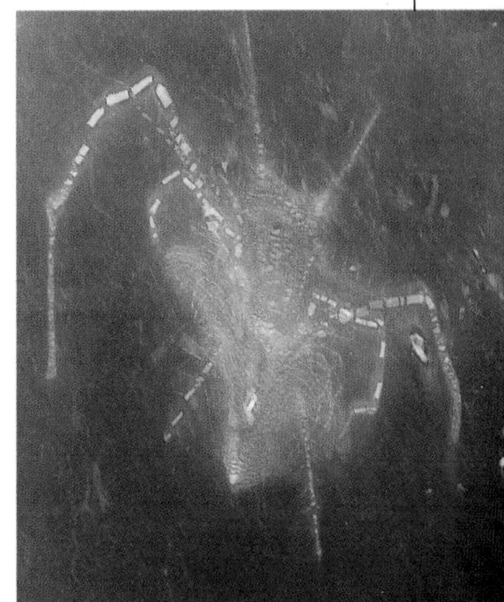

Ultraviolet and Fluorescent Fossils

Some fossils are preserved in minerals that glow under UV light. The fossil may be invisible to the naked eye until viewed under UV light. One example of this phenomenon is the remains of the dinosaur *Seismosaurus*. Uranium-rich minerals have been deposited in the fossilized bones, causing them to glow under UV light. These minerals were not present in the surrounding rock. The fossilized bones look similar to the surrounding rock, but under UV light, the difference between the two materials was obvious. This greatly assisted in preparing the specimen.

CLUBS, SOCIETIES, *and* INFORMATION

As with any hobby or pursuit, guidance and advice from more experienced people is invaluable to the amateur.

Meeting other hobby paleontologists takes only a little initiative. There are many rock-collecting clubs, lapidary clubs, as well as fossil-collecting clubs and societies, all of which cater, to some degree, to those interested in fossils. These are scattered all over the world, so it should be easy to locate a suitable one close to your home.

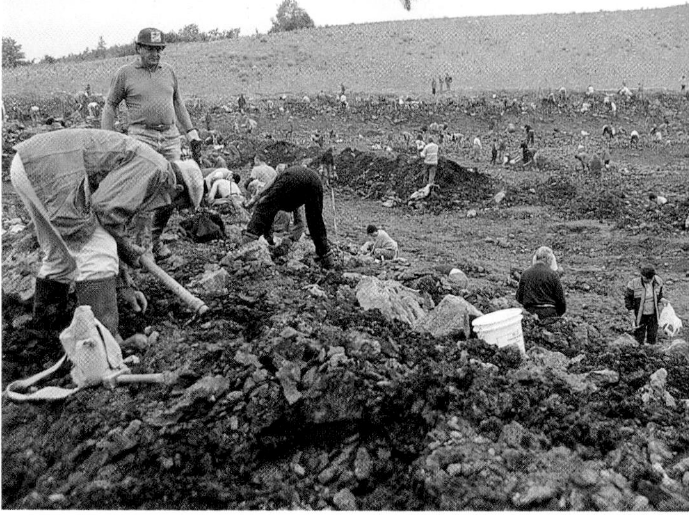

JOINING AN ORGANIZED GROUP *on a dig can be a fun way to learn about fossil-collecting as a hobby.*

CLUBS

Clubs are usually concerned with the more active side of rock and fossil collecting, and many clubs organize field trips to collect in particular localities. Through a club, you will meet other, more experienced hobbyists, who are usually only too happy to share information on where to find fossils, and how to collect, prepare, and take care of them. Clubs also have extensive fossil-trading networks, which are a wonderful way to increase and diversify your collection, as well as to make new friends with similar interests around the world.

VOLUNTEERS *for scientific excavation can use the task to gain knowledge and experience from the experts.*

SOCIETIES

Generally, societies tend to deal more with the theoretical side of paleontology. They are less concerned with expanding private fossil collections and more involved with gathering and disseminating technical information.

Societies often have news of scientific excavations, particularly those that need volunteers to help at the diggings. If you would like to help excavate a dinosaur skeleton, particular societies will be able to put you in touch with the right people.

OTHER SOURCES OF INFORMATION

In this age of computer-information services such as the Internet, a useful way to learn about paleontology as a hobby is to search cyberspace. Of particular interest would be the discussion group rocks-and-fossils@world.std.com,

But there is one single moment that is never quite duplicated … when the first specimen is secured.

Galapagos: World's End,
CHARLES WILLIAM BEEBE
(1877–1962), American scientist

AMMONITES *(above left) are sought-after specimens. Many natural history museums have information centers, like the one above, that amateurs can use.*

which will put you in contact with thousands of hobbyists and professionals around the world. Similarly, the news group sci.bio.paleontology has regular postings on all aspects of the subject, many of which are of interest to the hobbyist.

Another easy way to explore paleontology as a hobby is to visit rock and fossil shops. Look in telephone directories under rock shops, fossil shops, lapidary supplies, geological supplies or nature shops. Visit a few of these and talk with the owners. Usually, such shops are run by people who are absorbed in the hobby and they will know all the local clubs and societies. It is also worthwhile to look over their stock in case there are particularly appealing specimens.

There are a number of magazines readily available that publish news and articles on fossil collecting, as well as advertisements for clubs, societies, and businesses that may be of interest to you. Look for titles that cover the obvious areas, plus those on lapidary and earth sciences.

Books are a valuable source of information on all aspects of the hobby so check out new titles. Fossils tend to be a specialist subject and usually the better books are found in specialist science bookstores, university bookstores, or hobbyist shops. Don't forget to check your local library.

Natural history museums are a good source of local paleontological information and they usually provide a service for identifying your finds. However, use the time of busy professionals sparingly. While they are usually very accommodating, you will gain more satisfaction and learn more if you make the effort to identify specimens for yourself.

SHARING *information is of mutual benefit. Museum shops (below) and specialist shops are a good source of books and fossils.*

Many large natural history museums have resource directories that will put you in touch with clubs and societies in your area. Similarly, local geological surveys and departments of mines are worth approaching for information on good collecting localities and the activities of amateurs.

FINDING SUITABLE CLUBS OR SOCIETIES

The Resources Index at the end of this book provides a list of some of the larger clubs and societies that may be of interest to you.

The true scientist never loses the faculty of amazement. It is the essence of his being.

Newsweek, March 31, 1958
HANS SELYE (b. 1907), Austrian-born American physician

Chapter Six

Rocks Field Guide

USING *the* ROCKS GUIDE

This field guide introduces a variety of landscapes and treasures of the Earth, revealing many of the fascinating secrets behind its long history.

Rocks abound in every setting and this field guide can be used to seek out some interesting landscapes or to look for some elusive gems hidden beneath the Earth's surface.

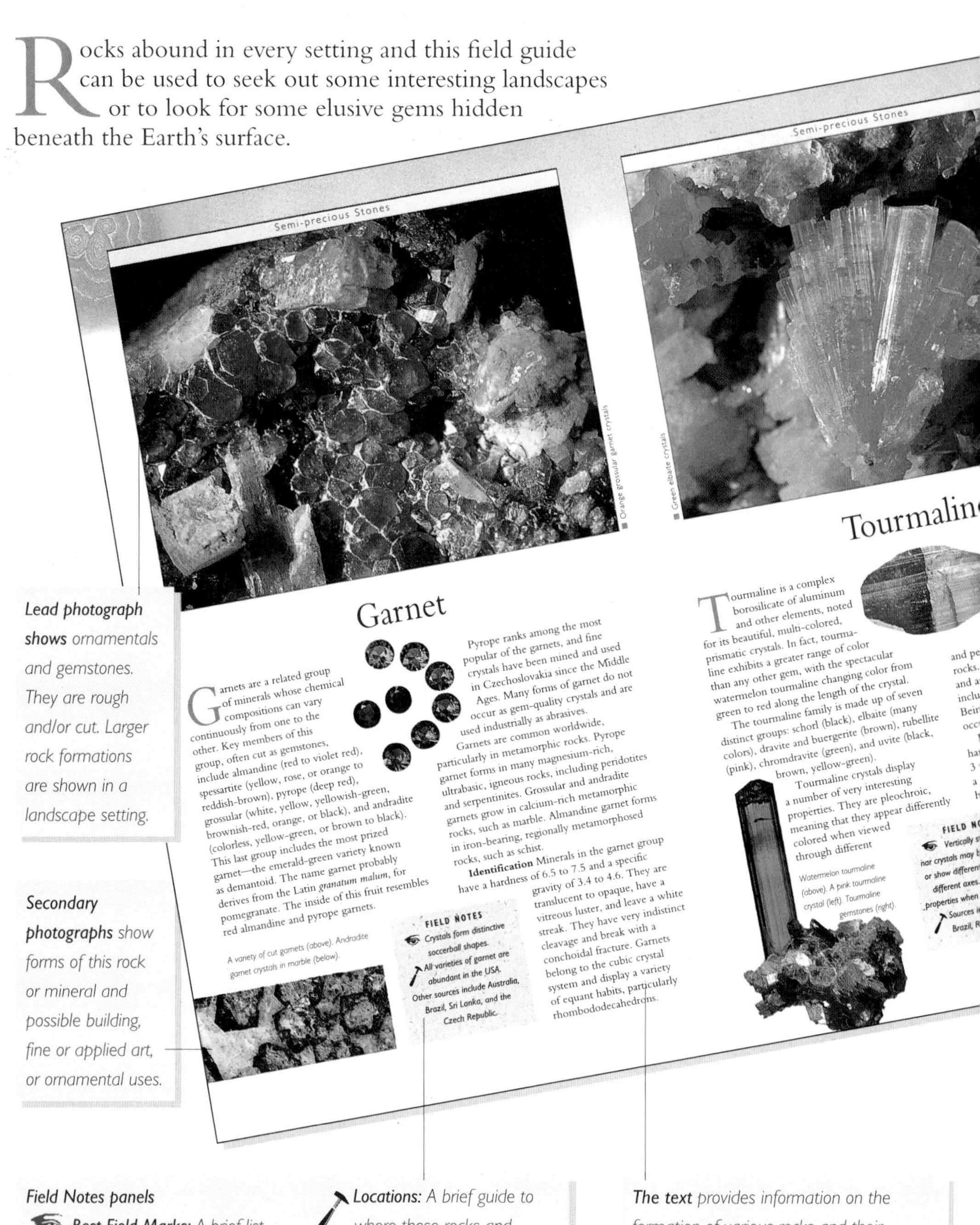
Semi-precious Stones

Orange grossular garnet crystals

Semi-precious Stones

Green elbaite crystals

Garnet

Garnets are a related group of minerals whose chemical compositions can vary continuously from one to the other. Key members of this group, often cut as gemstones, include almandine (red to violet red), spessartite (yellow, rose, or orange to reddish-brown), pyrope (deep red), grossular (white, yellow, yellowish-green, brownish-red, orange, or black), and andradite (colorless, yellow-green, or brown to black). This last group includes the most prized garnet—the emerald-green variety known as demantoid. The name garnet probably derives from the Latin *granatum malum*, for pomegranate. The inside of this fruit resembles red almandine and pyrope garnets.

Pyrope ranks among the most popular of the garnets, and fine crystals have been mined and used in Czechoslovakia since the Middle Ages. Many forms of garnet do not occur as gem-quality crystals and are used industrially as abrasives.

Garnets are common worldwide, particularly in metamorphic rocks. Pyrope garnet forms in many magnesium-rich, ultrabasic, igneous rocks, including peridotites and serpentinites. Grossular and andradite garnets grow in calcium-rich metamorphic rocks, such as marble. Almandine garnet forms in iron-bearing, regionally metamorphosed rocks, such as schist.

Identification Minerals in the garnet group have a hardness of 6.5 to 7.5 and a specific gravity of 3.4 to 4.6. They are translucent to opaque, have a vitreous luster, and leave a white streak. They have very indistinct cleavage and break with a conchoidal fracture. Garnets belong to the cubic crystal system and display a variety of equant habits, particularly rhombododecahedrons.

A variety of cut garnets (above). Andradite garnet crystals in marble (below).

FIELD NOTES

Crystals form distinctive soccerball shapes.

All varieties of garnet are abundant in the USA. Other sources include Australia, Brazil, Sri Lanka, and the Czech Republic.

Tourmaline

Tourmaline is a complex borosilicate of aluminum and other elements, noted for its beautiful, multi-colored, prismatic crystals. In fact, tourmaline exhibits a greater range of color than any other gem, with the spectacular watermelon tourmaline changing color from green to red along the length of the crystal.

The tourmaline family is made up of seven distinct groups: schorl (black), elbaite (many colors), dravite and buergerite (brown), rubellite (pink), chromdravite (green), and uvite (black, brown, yellow-green).

Tourmaline crystals display a number of very interesting properties. They are pleochroic, meaning that they appear differently colored when viewed through different

Watermelon tourmaline (above). A pink tourmaline crystal (left). Tourmaline gemstones (right).

Lead photograph shows ornamentals and gemstones. They are rough and/or cut. Larger rock formations are shown in a landscape setting.

Secondary photographs show forms of this rock or mineral and possible building, fine or applied art, or ornamental uses.

Field Notes panels

Best Field Marks: A brief list of the most characteristic features of the rock or mineral to aid identification in the field.

Locations: A brief guide to where these rocks and minerals occur. In the case of rocks, the geological environment is given. For minerals, a list of the best source countries is featured.

The text provides information on the formation of various rocks and their constituent minerals. In the case of minerals, each entry ends with the mineral's identifying characteristics (see Mineral Identification pp. 70-75).

The **illustrated banding** at the top of the page identifies either general categories of collecting rocks or rock sites—see the key below.

Volcanoes

Mt Augustine, Alaska, USA

Volcanoes

Mt Kilauea, Hawaii, USA

Explosive Volcanoes

After an explosion, a lake may form in the resulting crater. Subsequently, a small cone forms in the center of the lake, as in Lake Toya, Japan, and Crater Lake, Oregon, USA. Eventually, the small cone grows to the size of the original volcano, and the next explosion occurs. This pattern is recorded around Mt Fuji, Japan. The growth and explosion of andesitic volcanoes continues until subduction stops feeding them magma.

An 1831 sketch (right) of a volcano, Julia Island, Sicily. Sunset Crater (below) Arizona, USA.

FIELD NOTES

Volcanoes along subduction lines, as in Russia, Japan, ...ska, Indonesia, New Zealand, North and South America. Crater lakes can be seen Lake Toba, Indonesia; Crater Lake, USA; ...ake Toya, Japan.

Shield Volcanoes

Hot-spot volcanoes form from fast-running, low-silica basalt magma. On land, they have low-angle cones and are often referred to as shield volcanoes, after ancient Roman shields. When they form underwater, they start with a steeper shape because the lava freezes much faster and does not travel as far. The volcanic pedestal is formed from fractured pieces of pillow basalt (see p. 189). The shape flattens to the shield form as the cone builds above sea level, as in the Hawaiian Islands.

A hot-spot volcano has a limited life, forming as a plate passes over a hot spot deep in the mantle (see p. 43). The volcano builds in size as the plate moves steadily on. Once the plate has moved sufficiently, a new volcano will appear over the stationary hot spot and the old one will become inactive and eventually erode. Continental hot-spot volcanoes erode to subdued hills, as can be seen all along the eastern coast of Australia. Oceanic hot-spot volcanoes erode to sea level to produce flat-topped pedestals called guyots. These subside below sea level as their load depresses the ocean floor. This series of events is clearly recorded in the Hawaiian Island chain and surrounding reefs.

An eroded oceanic hot-spot volcano in Tahiti, Polynesia.

FIELD NOTES

Hawaiian Islands, USA
Yellowstone National Park, USA
Mt Erebus, Antarctica
Mt Kilimanjaro, Tanzania
Tristan da Cunha, Atlantic Ocean
Réunion Island, Mauritius
Lord Howe Island, Australia
Barrington Tops, NSW, Australia

187

161

Field Notes panels

Locations: Places where these significant geological formations can be found are listed. Plate tectonic addresses are given in some cases to give a broader range of sites, particularly where common.

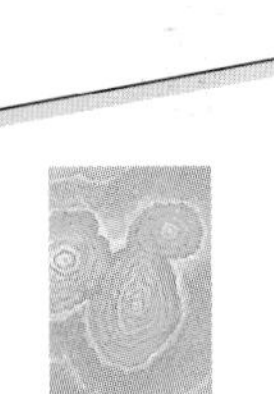

Collecting Rocks 136

Granite, Rhyolite, Obsidian, Andesite, Basalt, Breccia, Conglomerate, Sandstone, Mudstone, Limestone, Chert and Flint, Evaporites, Marble, Quartzites, Slate, Schist and Gneiss, Diamond, Ruby and Sapphire, Emerald and other Beryls, Opal, Chrysoberyl, Topaz, Garnet, Tourmaline, Peridot (Olivine), Quartz, Spinel, Feldspars, Zircon, Hematite, Agate, Other Chalcedony, Rhodochrosite, Rhodonite, Lapis Lazuli, Turquoise, Malachite and Azurite, Jade, Veined Gold, Alluvial Gold, Silver, Rocks from Outer Space, Silica Replacements, Jet, Pearls, Shells.

Rock Sites 184

Explosive Volcanoes, Shield Volcanoes, Columnar Jointing, Pillow Basalt and Basalt Flows, Sedimentary Remnants, River Bends, Caves, Lava Tubes, Folded Rocks, Gorges and Canyons, Glacial Valleys, Rift Valleys, Deserts, Crossbeds, Geysers, Meteorite Craters.

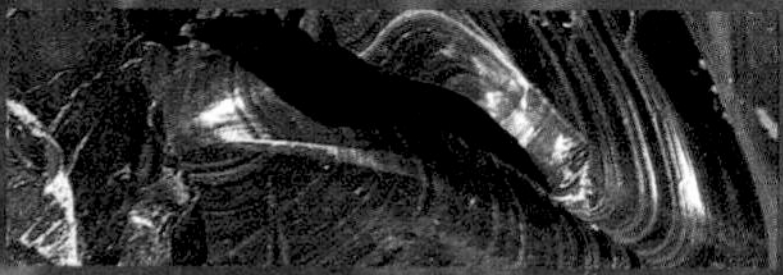

Collecting Rocks

The Granite Dells, Arizona, USA

Granite

Granite is a light-colored igneous rock, rich in quartz and orthoclase feldspar but poor in the ferro-magnesian (iron and magnesium) minerals. Because it cools for a long time underground, its large crystals are visible to the naked eye. Granite is structureless so weathers to spheres, called tors, in otherwise rolling countryside. These rocks form deep underground and as they rise, there is a great reduction in pressure and temperature. As a result, granite easily breaks down. Surface heating and cooling causes it to flake off, or exfoliate, in large sheets parallel to the ground surface.

Granite forms above subduction zones where two plates are colliding, and where heat melts the crustal rocks, (see p. 40). Convection holds the melt there. The granite cracks as it cool, and the cracks fill with the last of the melt. The resulting dikes may be either fine-grained aplite, or coarse-grained pegmatite. Pegmatite, which is produced from water-rich melts that allow easy movement of elements, forms crystals that can be larger than a person. Granite frequently occurs in the cores of old, folded mountain chains, often in association with metamorphic rocks. This is the signature of an old subduction zone.

A very tough rock, granite polishes to a fine mirror finish and is a popular building stone. Two of the more spectacular occurrences of eroded granite are Devil's Marbles, Australia, and Half Dome, Yosemite National Park, California, USA.

Granite detail (above). Carving on the tomb of Ramses II, in Egypt.

FIELD NOTES

Light-colored with large crystals. Often block-jointed. May form smooth spheres or erode to sharp spires and crags.

Forms at continent-continent collision sites, so is found in the cores of old, folded mountains.

Volcanic flow-banding, Sheep Creek Range, Nevada, USA

Rhyolite

Rhyolite is commonly pale gray, pink, or yellow. It usually forms from the melting of deep crustal rocks of granitic composition. In North America, such rocks are common in western Texas, New Mexico, Arizona, and Nevada. In western Texas, unusually fluid rhyolites flowed from fissures to cover thousands of square miles of land.

Rhyolite has basically the same composition as granite—its fine-grained structure is due to rapid cooling. It contains some large crystals, phenocrysts, formed early within the magma chamber. The many small crystals in rhyolite respond to flowing by aligning themselves in bands, an effect called flow-banding. Such patterned rocks are keenly sought after as a stone for building.

If charged with steam, rhyolite erupts noisily, ejecting clouds of ash and steam. If little steam is present, the magma will be far too stiff to flow easily and so rises directly out of the ground as mounds of hot rock.

When the magma cools too quickly for crystallization to occur, glass-like obsidian is produced. If the cooling lava froths, a rock called pumice is formed. This variety of rhyolite is so porous and light that it can float on water—the only rock that is able to do so.

Gas cavities in rhyolite sometimes fill with silica precipitates, such as agate. When the rock is cut and polished, the resulting mosaic of banded rhyolite and agate makes pieces of great beauty that are popular with collectors.

Rhyolite detail (above). Lava spire (right) in Arizona, USA.

FIELD NOTES

Fine-grained, light-colored rock, often pink or gray. Frequently exhibits a flow-banding pattern.

Forms when molten granite erupts at the surface or forms intrusive bodies.

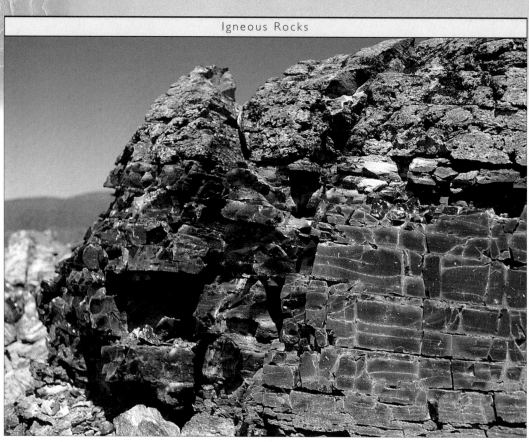

Newberry Crater, Oregon, USA

Obsidian

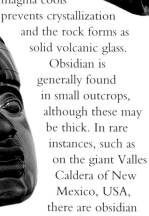

Like rhyolite, obsidian forms from the melting of deep, crustal, granitic rocks. In the case of obsidian, the speed at which the magma cools prevents crystallization and the rock forms as solid volcanic glass. Obsidian is generally found in small outcrops, although these may be thick. In rare instances, such as on the giant Valles Caldera of New Mexico, USA, there are obsidian flows hundreds of feet thick. The Glass Buttes, in Oregon, USA, are composed entirely of obsidian.

With its glassy luster and sharp conchoidal fractures, obsidian is a distinctive rock. There are several varieties of obsidian, distinguished by their color and sheen: pure black, light brown, brown mottled with black, and black with a beautiful golden or silvery sheen. Snowflake obsidian is dotted with white patches where parts of the rock have begun to crystallize. The most prized obsidian of all is the rainbow variety, with its purple, green, and gold bands of sheen.

Because of its absence of crystals and conchoidal fractures, obsidian can be easily worked to razor-sharp edges, even sharper and finer than those of flint. It is also able to take an extremely high polish. As a result of these properties, obsidian was revered by ancient cultures. It was one of the major materials of barter, and was prized for the manufacture of sharp and delicate heads for arrows and spears. Today, obsidian is used throughout the world as a semi-precious stone for jewelry or decorative items.

FIELD NOTES

Generally black and glass-like, containing no crystals. Tends to form small outcrops, although these may be thick.

Forms where magma erupts at the surface and cools very rapidly.

Rainbow obsidian Olmec head replica (left). Obsidian detail (center). Aztec obsidian arrowhead (right).

The three main craters of Mt Bromo, Java, Indonesia

Andesite

Andesite is intermediate in composition between granite and basalt, so it has some small component of iron and magnesium, giving it a speckled, salt-and-pepper appearance. Andesite magmas are very viscous and set quickly, forming steep-sided volcanoes, such as Fuji, in Japan, and Anak Krakatau, Indonesia. These magmas tend to block volcano vents, causing explosions as the pressure builds.

Ignimbrite is a rock of andesitic composition that has formed in an unusual way. It usually contains some quite large crystals, but the major component is extremely fine particles of volcanic glass. It forms when gas dissolved in the lava suddenly boils as the roof of a magma chamber collapses. This causes the andesitic lava, which has been slowly crystallizing, to explode. The pressure blows some molten rock into tiny fragments that cool rapidly to fine dust or ash. Occasionally, fragments reach the ground while they are still aglow, and the particles are cemented into a solid rock called ignimbrite (meaning, literally, fire rock).

Andesite and ignimbrite are common where oceanic crust subducts under continental crust. In these conditions, melting basalt rises and mixes with melting rocks of granitic composition, as in mountain chains such as the Cascade Mountains of Washington and Oregon, USA, and the Andes of South America, from which andesite took its name.

Andesitic volcanoes are very destructive. In the Yellowstone area in Wyoming, USA, for example, several eruptions since the Pleistocene threw huge volumes of volcanic ash into the atmosphere that are each thought to have changed world climate. Small ignimbrite explosions have covered historic sites, such as the Roman city of Pompeii.

FIELD NOTES

Speckled, usually brown or dark gray. Ignimbrite may contain large crystals.

Extrusive igneous rock that forms at ocean–continent subduction zones, such as the South American Andes.

Victim of the Vesuvius eruption in Pompeii (below). Hornblende andesite detail (center).

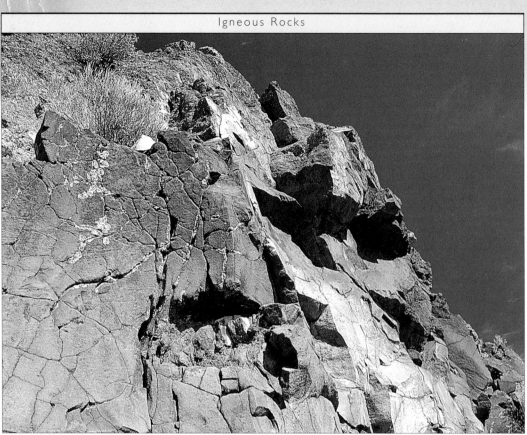

A typical basalt outcrop

Basalt

Basalt is the most common rock on the Earth's surface. It is a fine-grained, black, igneous rock, rich in silica and the ferro-magnesian minerals. Basalt occasionally contains large crystals of the green mineral olivine. In wet climates, ferro-magnesian minerals are susceptible to chemical weathering, and basalt weathers rapidly to clay.

Some black beach sands, such as those in Hawaii, are formed when hot basalt lava flows into the sea and is shattered into fragments as it suddenly chills. Basalt is typically the first lava to issue from any volcano. The majority of basalt is extruded at mid-ocean ridges from where it has formed 70 percent of the Earth's surface. Most basaltic ocean volcanoes, such as the Hawaiian Island chain, result from hot spots deep in the Earth. Islands closer to continents may be basaltic, although there is a transition to andesitic composition near continents and above subducting slabs.

As hot, flowing basalt cools, it forms a skin that distorts and stretches into rope-like shapes called *pahoehoe*. With further cooling, it slows and may crack into blocks called *a-a*. Basalt is also extruded through fissures on continents when they begin to rift. Basalt lava is very fluid and can flow for hundreds of miles, forming vast sheets. On continents, it is mostly found in areas that have undergone extension, or pulling apart, such as the Rio Grande Rift in New Mexico, or the Columbia River Plateau in the States of Washington and Oregon, USA.

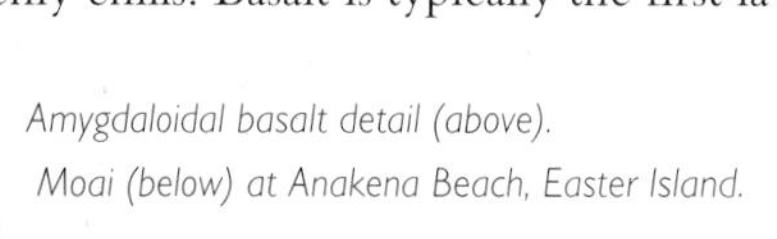

Amygdaloidal basalt detail (above).
Moai (below) at Anakena Beach, Easter Island.

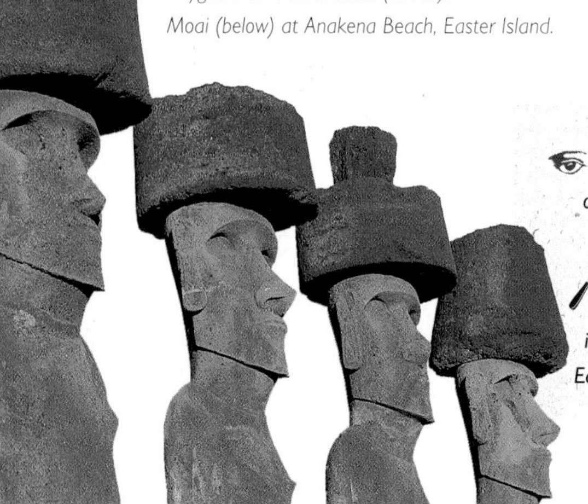

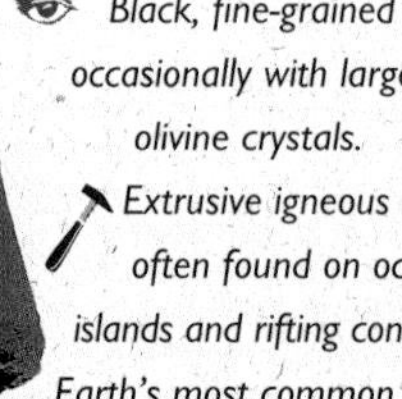

FIELD NOTES

Black, fine-grained rock, occasionally with large olivine crystals.

Extrusive igneous rock often found on oceanic islands and rifting continents. Earth's most common rock.

Breccia at the summit of Castle Mountain, Colorado, USA

Breccia

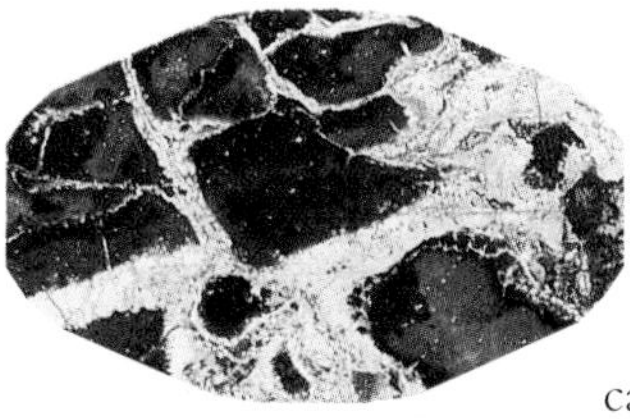

Breccia is a rock composed of generally large, sharp fragments cemented together. These fragments may be produced by volcanic explosion, faulting, or sedimentary deposition. The sharpness of the fragments indicates that they did not travel far from where they fractured. Conglomerate, on the other hand, has rounded fragments, indicating significant travel. Breccia forms among volcanoes at convergent zones and wherever faulting occurs. It may also form where desert or alpine conditions cause fast erosion and the collection of debris in alluvial fans.

Mudflows are breccias made up of small particles mixed with water. They often result from volcanic debris falling on the steep flanks of snow-covered volcanoes. They also occur at the toes of alluvial fans in desert regions. Mudflows travel downhill mostly as slurries, and may set to the consistency of cement. They may be so dense that they can pick up and carry cars and large boulders. These sometimes move at such high speeds that they strip the bedrock clean of trees. A significant mudflow occurred in 1980 when Mt St Helens, in Washington, USA, erupted. Hot ash landing on snow caused a massive avalanche of debris and mud down the Toutle Valley, smashing everything in its path. Mudflow deposits are characterized by the chaotic mix of particles of all sizes and can turn V-shaped valleys into flat-floored valleys. Where the surrounding topography is steep, a flat-floored valley may indicate the presence of a mudflow.

Breccia detail (above). House (right) engulfed by a mudflow.

FIELD NOTES

Breccia contains a mix of rock fragments of all shapes and sizes.

Breccias are found in fault zones, mudflows on the flanks of volcanoes and in mountainous deserts.

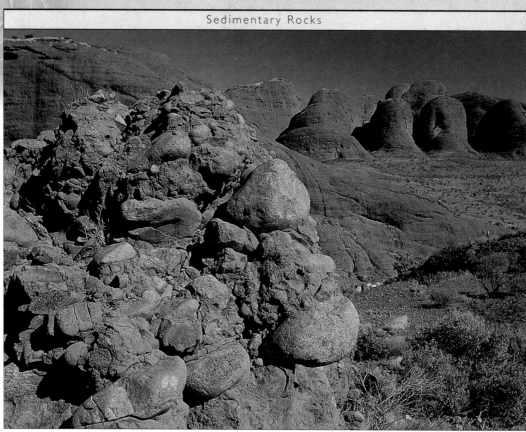

The Olgas, Central Australia

Conglomerate

A conglomerate is a sedimentary rock composed mostly of coarse fragments, but including a range of grain sizes. The fragments are well rounded, which indicates transportation by water. They are often deposited near mountains where gradients decrease, river velocity suddenly drops, and the river is unable to carry the coarse sediment any further. Conglomerates are common along present and ancient continental edges.

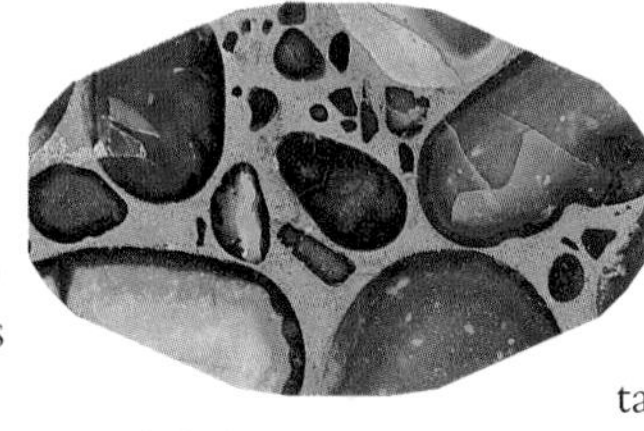

Conglomerates are common and accessible in desert alluvial fans, such as Death Valley, in California, USA. In ancient, giant alluvial fans, such as the Olgas of Australia and the Van Horn, West Texas, USA, conglomeratic deposits hundreds of feet thick can be found.

Hertfordshire pudding stone (center). Particles of various sizes in a conglomerate (above).

Large areas of conglomerate are produced by continental collision, where uplifting has resulted in mountain building. New rivers rise in these mountains and loose debris is swept into the river channels and carried downstream to the flat coastal areas. Later, this coarse material is covered by thick sand deposits. Many ancient sedimentary basins, such as the Sydney Basin, in Australia, record this history.

Other ancient environments represented by conglomerates include rocky beaches and the front of coral reefs, where coral that has broken off the living reef may form a slope of coral debris. As a result, fragments of shallow-water organisms may be found in conglomerates that have formed at great depths.

Look for conglomerates in desert areas where roads may skirt over alluvial fans and where recently deposited conglomerates have been cemented by minerals in the groundwater. These are quite common and easy to collect along the stream channels on the fan surface.

FIELD NOTES

Composed of coarse, rounded fragments in a range of sizes.

Sedimentary rock that forms in rift valleys and near mountains. Common in rocky desert areas.

Navajo Sandstone, Colorado Plateau, Utah, USA

Sandstone

Sandstone is any rock made of particles that are sand-size, up to 1⁄12 of an inch (2 mm) in diameter. The vast majority is made of rounded particles of quartz, but it can contain feldspar and even fragments of rock. Sandstone is a very common sedimentary rock and is easy to find. It forms landscapes that reflect the orientation of its layers. Flat layers erode to cliffs as in the Grand Canyon, Arizona, USA, while folded sandstone erodes to peaks such as seen in the Rocky Mountains in the west of North America, and the European Alps. In deserts, sandstone cliffs can weather into stupendous arches and shallow caverns, as sand is flaked off the cliff face by wind and chemical erosion.

Since sand can accumulate in so many places, including rivers, beaches, lakes, offshore marine environments, and desert regions such as the Sahara, sandstones are found almost everywhere. Many features of a sandstone outcrop can be used to determine how the original deposition may have occurred. The fossil desert sandhills exposed at Zion Canyon, Utah, USA, for example, still show their original dune shape.

Sandstone is prominent in western USA, notably in Arches National Monument and the natural caves in which the Mesa Verde buildings are ensconced. It may be prominent in roadcuts or on the tops of small hills, where it forms resistant caps. Many vertebrate fossils are preserved in river and lake sandstone. The dinosaur graveyard at Dinosaur National Monument, in Utah, is in sandstone that accumulated during several flood events.

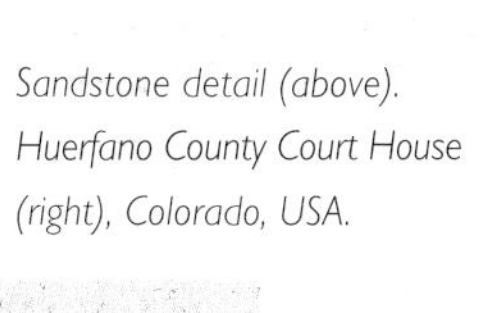

Sandstone detail (above). Huerfano County Court House (right), Colorado, USA.

FIELD NOTES

Even, medium-sized grains. Often forms in layers. Great variation of color.

Forms extensively in marine and land environments, in deserts, rivers, shorelines, and barrier islands.

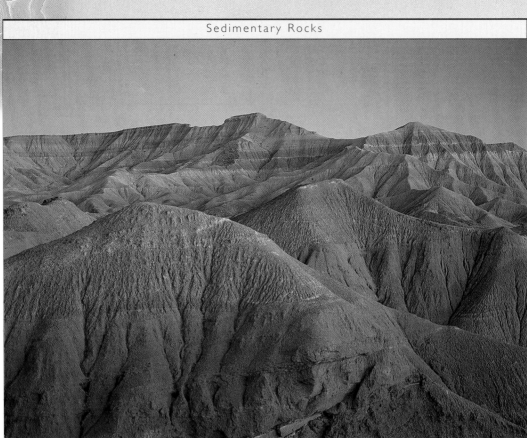

Painted Desert, Arizona, USA

Mudstone

Mudstone is formed from tiny, clay-size particles. Like sandstone, mudstone is found just about anywhere on continents where still water once existed. Most mudstone collects in oceans where the water is calm enough for fine particles to settle. Thick deposits of mudstone are present in most deltas, where rivers enter still water. Layering occurs in thick mud deposits because the clay flakes, being flat, align themselves horizontally.

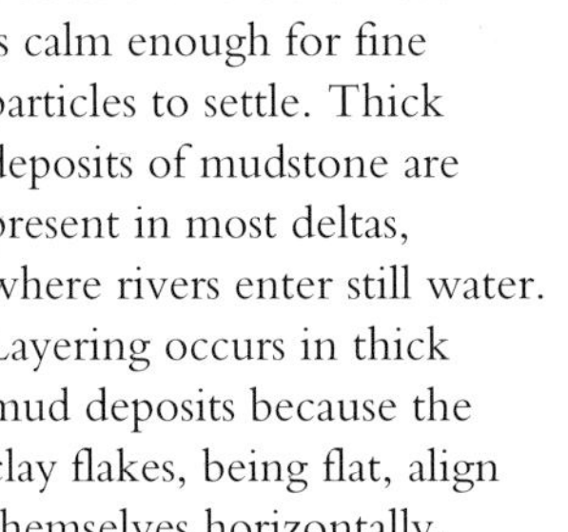

Mudstones are used in the manufacture of bricks and ceramics. Because they weather so easily, the best place to see them is in roadcuts or in desert areas where vegetation is sparse. Multicolored mudstones, called paleosols, represent ancient stacked soil horizons. They are found in desert areas and are easily noticeable by their alternating subdued reds, mauves, grays, and greens. An excellent example can be seen in the Painted Desert of Arizona, USA.

While mudstones may weather to subdued landscapes, turbidites, which contain some sand, may resist erosion and form more hilly topography. They are made of alternating layers of shale and sandstone. Turbidites form mainly beyond the edges of continental shelves when loosened material on the shelf edge slides down the continental slope into the ocean abyss. The downslope flows that form turbidites may attain speeds of more than 100 miles (160 km) per hour, and have been known to cut submarine communication cables. Turbidites may be squeezed up to become continental rocks but rarely form on continents. Minor turbidites are also found in deep lake deposits.

Turbidite detail (above). Banded mudstone (left), Oregon, USA.

FIELD NOTES

Composed of tiny clay particles. Gray to shades of red and distinctly layered.

Sedimentary rocks formed in undisturbed environments. Turbidites found on ancient ocean floors and in lakes.

Karst scenery, Halong Bay, Vietnam

Limestone

Limestone is formed from calcium carbonate. Lime precipitates in warm tropical waters on shallow sea floors, where almost no other sediments cloud the water. Limestone rocks formed in the broad centers of flooded continents and along the edge of continental shelves where reefs were built. As a result, many limestones are composed of the skeletons of reef-forming animals. Freshwater limestone forms in tropical lakes or where hot solutions of calcium carbonate rise to the surface, such as in Yellowstone National Park, Wyoming, USA. As this water cools, the precipitates form flowstones.

The closely related dolomites are calcium magnesium carbonates, whose presence may indicate more arid conditions during deposition. Since such rocks are susceptible to solution, caves, fissures, and sinkholes are a frequent form of weathering. Limestone stores carbon dioxide in a solid state, so the amount of limestone deposited may bear a close relationship to the volume of this greenhouse gas in the atmosphere and its solution in water.

Limestone terrains frequently show the effects of chemical solution. Visually dramatic karst formations, named after the Karst region of the former Yugoslavia, are produced by limestone dissolution. The tower karsts of Asia are particularly striking examples of this type of scenery. Limestones are also prominent at the White Cliffs of Dover, England, and throughout central Texas, USA. Because of limestone's extreme solubility, it often contains underground streams and caves, in which stalactites and stalagmites form. Fossils are extremely abundant and easy to find in limestone.

Limestone detail (above). Pyramid of Kukulkan (left), Chichen Itza, Yucatan, Mexico.

FIELD NOTES

Dense, massive rock. Usually light colored, but may be darkened by organic material. Fizzes in dilute hydrochloric acid.

Sedimentary rock forming in tropical marine waters and warm freshwater lakes.

Banded jasper at Marble Bar, Australia.

Chert and Flint

Like quartz, chert and flint are composed of silicon dioxide, but because they form in the sedimentary environment, they are more likely to contain traces of other elements. All rivers entering the ocean carry dissolved silica. Oceans, already rich in silica from organic sources, consequently become super-saturated with silica, and an ultra-fine silica ooze precipitates in deep water. If no other sediment blankets the ooze, it consolidates to chert, which forms continuously in deep oceans. Where iron is present, red jasper forms. The term flint was used to describe workable nodules of chert that were found in the chalk of western Europe.

Because chert and flint are resistant to weathering, they may be left behind as nodules that accumulate on the surface of slowly dissolving limestone landscapes. Chert layers often outcrop as raised, resistant ridges. Both chert and flint are easy to find in stream channels, where they outlast most other pebbles. Chert pebbles are extremely compact and have no visible crystals. They bounce quite high when dropped on a hard surface and when two pebbles are knocked together, they make a high-pitched sound.

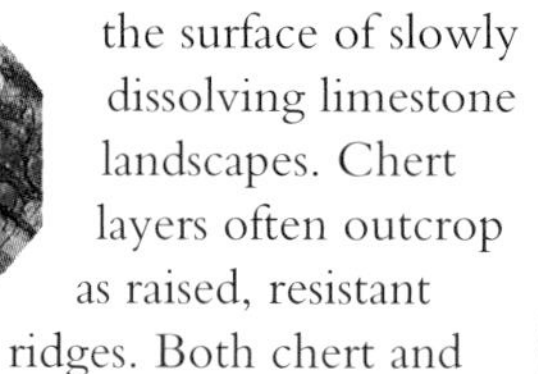

Chert and flint are the materials that formed the technological basis of some of humanity's earliest weapons and tools. These include delicate, razor-sharp, finely worked knives and heads for arrows and spears. Flint was also used to strike sparks to ignite gunpowder in early firearms.

Chert occurs widely in western Europe and North America. Marble Bar in Australia is a classic locality.

Ribbon chert detail (center). Flint spear heads (left and above right).

FIELD NOTES

Extremely tough rock. May be plain or colorfully banded. Dense with no visible crystals.

Associated with ancient ocean floors. Resists erosion, so often found as resistant ridges or massive outcrops.

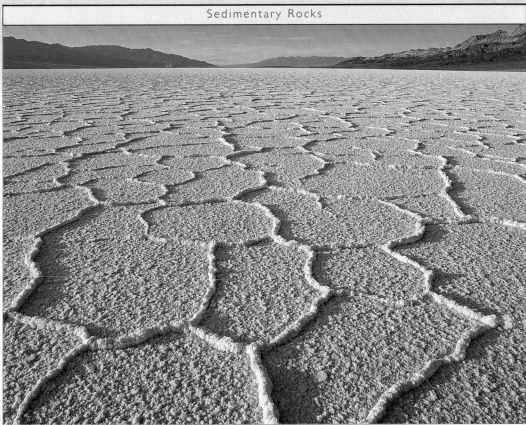

Badwater salt flats, Death Valley, California, USA

Evaporites

An evaporite is the result of mineral-bearing water evaporating and precipitating crystals of its dissolved material. The common evaporites are gypsum and rock salt—gypsum frequently forms beautiful "desert roses". Evaporites are susceptible to later erosion and can dissolve under the surface, causing collapse. They form in terrestrial lakes and river environments, on the shores of arid seas, such as the Gulf of Aden, and in deep, closed basins. The entire Mediterranean Sea was a site of evaporite precipitation during the Miocene. Evaporites typically form in continental rifts when seawater pours in from one end. They are abundant in many Permian rocks around the world.

Being readily resoluble, evaporites are best preserved in deserts. Once dissolved, they precipitate as a white residue on the desert surface, which indicates that there may be crystals nearby. In general, evaporites are not very common. Rock salt is usually found only in desert lakes, such as Lake Eyre, Australia, or Death Valley, California, USA. Gypsum may crystallize in dry lakes and appear as translucent, soft, waxy crystals, gray to pink in color.

In several events during the Tertiary, the Mediterranean Sea was closed at the Strait of Gibraltar. Each time, the entire sea evaporated, leaving behind thick salt deposits on the ocean floor. Being low in density, salt formations may become mobile when buried and rise toward the surface as giant domed structures called diapirs. These are major sites of oil accumulation.

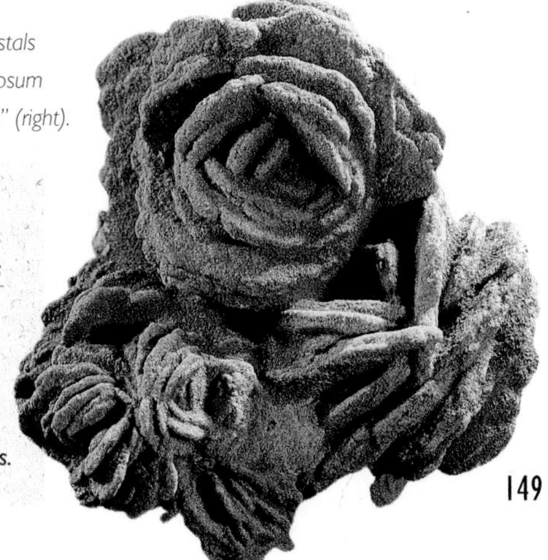

Gypsum crystals (above). Gypsum "desert rose" (right).

FIELD NOTES

Tend to be extremely pure and soft. May occur as large and beautiful crystals.

Deserts and ancient rift basins with high rates of ocean water evaporation. Salt may form underground diapirs.

Apuane Alps, Tuscany, Italy

Marble

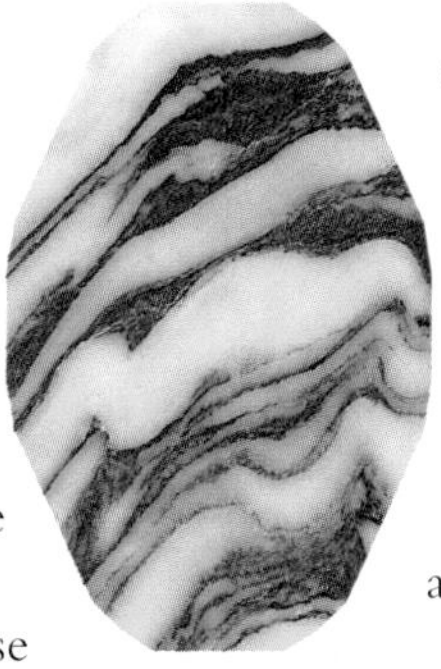

Marble forms from metamorphosed carbonate rock, most usually limestone. It may be found in regionally metamorphosed areas along continent–continent collision zones and in the roots of folded mountain chains, in areas that were once shallow marine shelves where abundant coral reefs accumulated. Granite intrusions can also metamorphose limestone to marble. Examples of marble can be seen in areas such as the Alps in Europe, and Oregon Caves National Monument, USA.

Pure marble, largely calcite with minor amounts of impurities, is white, but depending on the level of metamorphosis and chemical impurities in the original limestone, different colors and crystal sizes may be present. Because it is soft and beautifully colored, it has long been valued as a stone for sculpting. Statues by such famous artists as Michelangelo and Bernini were worked in Carrara marble quarried in the Italian Appenines. Unusual textures and colors also make this a valuable facing stone for buildings and tables. Some of the great buildings of antiquity, including the Parthenon in Greece and the Taj Mahal in India, are made of marble. Sadly, this material suffers greatly from acid rain and industrial pollution and many famous marble statues and buildings are now rapidly deteriorating.

If present in sufficient quantity, marble may weather into the same karst formations as limestone. The rock is easily weathered, so look for it in old quarries and along the sides of valleys that are cut into marine sedimentary rocks.

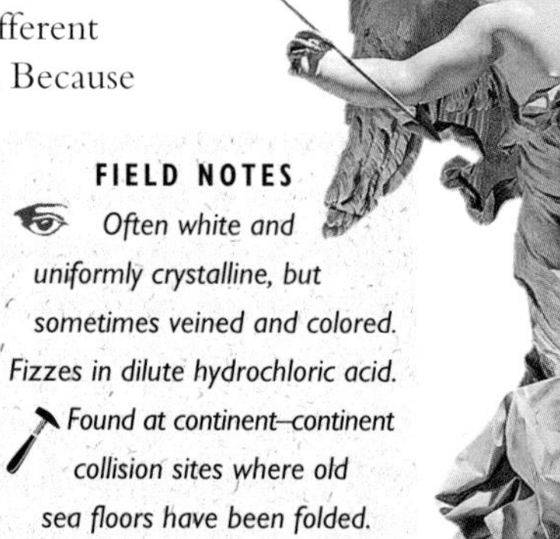

Bernini's "Ecstacy of St Theresa".

FIELD NOTES

Often white and uniformly crystalline, but sometimes veined and colored. Fizzes in dilute hydrochloric acid.

Found at continent–continent collision sites where old sea floors have been folded.

Detail of polished marble slab (above center).

Mt Arapiles, Victoria, Australia

Quartzites

When quartz sandstone with a silica content of 95 percent or greater is metamorphosed, it becomes quartzite. Since sandstones form in so many lowland and marine sedimentary environments, quartzites are found in many metamorphic settings. Look for them in the roots of ancient folded mountain chains at continent–continent collision boundaries. Contact metamorphism also produces quartzite, so it can be found around granite intrusions. Quartz is a very stable mineral, so quartzite will persist through nearly any degree of metamorphism.

Because quartzite is very resistant to erosion and rarely supports vegetation, it forms exposed, rocky landscapes and rugged ridges. It can be easily seen in road cuts, stream channels, and on hillslopes, and usually stands out from the intervening schists. Sandstone crossbeds and bedding will often be preserved in quartzite, providing valuable clues about where the original sandstone came from. A famous quartzite deposit is Mt Arapiles, in Australia.

FIELD NOTES

Commonly light colored. Massive with visible grains. Original bedding may be preserved.

Found at continent–continent collision zones where sandstones or cherts have been metamorphosed.

Quartzite becomes harder the more it is compressed. When it is heated and squeezed, quartz at the contact points between the grains dissolves and then precipitates in the spaces among the grains, increasing the density of the rock. As a result, it is tough and very resistant to cutting. It is therefore rarely used as a building stone. Although most quartzite is white, minor amounts of elements such as iron and manganese may produce red, green, or gray coloring.

Some quartzite in Western Australia is believed to be 3,500 million years old, making it some of the oldest land crust on Earth.

Quartzite outcrop, Mt Narryer, Western Australia. Quartzite detail (above center).

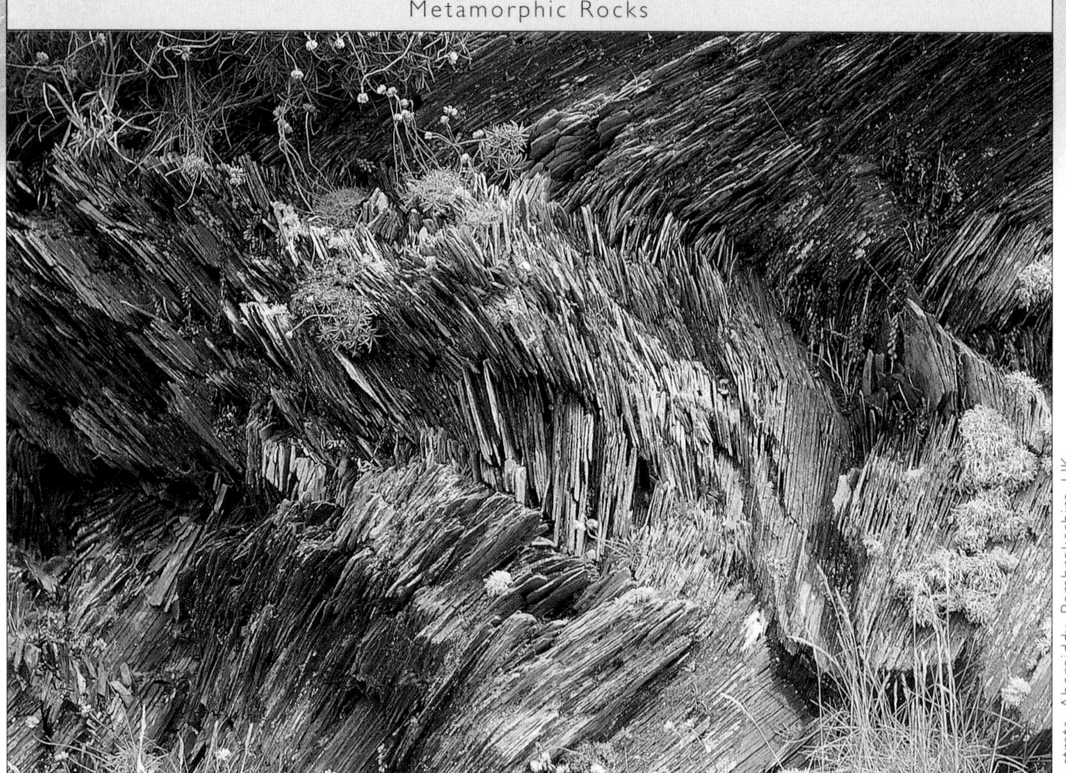
■ Slate strata, Abereiddy, Pembrokeshire, UK

Slate

Slate is metamorphosed mudstone that has been very strongly compressed. It is black to gray and mostly occurs at continent–continent collision boundaries. It is commonly found in the roots of old folded mountain chains, such as the Appalachians in the USA, and the Alps in Europe. It cleaves into sheets because all of its mica minerals are perfectly aligned at right angles to the direction of compression.

Slate detail (above). Slate tombstone (left) in Boston, USA.

Because it cleaves so easily, it sometimes produces sheets of enormous size.

Being very resistant to weathering, slate tends to be exposed in rough and craggy hills and breaks as brittle splinters along its cleavage planes. For centuries, the ability of slate to resist weathering, and even attack by acid rain in industrialized coal-burning England, made slate the preferred roofing. Tiles can be quarried, split, and installed on roofs easily, so slate was an inexpensive building material. Another popular use in former times was for writing slates and blackboards, which took their name from the black color of the slate. A major current use is for the tops of billiards tables, where weight and flatness are essential. The few remaining working slate quarries produce both tabletops and floor tiles. In a few locations, colored slate occurs in red, brown, green, and yellow, often with attractive streaking and texture.

Slate samples are easy to obtain from quarries and road-cuts in regional metamorphic areas throughout Europe.

FIELD NOTES

Mainly black to gray. Cleaves into sheets. Resists erosion.

Forms at continent–continent collision zones, often at the roots of old, folded mountains. May be exposed as rough, craggy outcrops in hills.

Schist eroded by a glacier, Waiho Valley, New Zealand

Schist and Gneiss

Schist and gneiss are regional metamorphic rocks formed at depth in continent–continent collision boundaries. They are characterized by coarse-grained minerals, such as micas, visible to the naked eye. Schist often has a flaky, plate-like appearance, while gneiss shows alternating bands of dark ferromagnesian minerals and light minerals, quartz and feldspar.

Schist represents metamorphosed shale or basaltic rock and is largely formed from minerals that grow during metamorphism, such as muscovite mica and the semi-precious mineral garnet. Where sandstones and mudstones were present in layers, metamorphism turns the sandstone to quartzite and the mudstone to schist. Gneiss forms from either granitic igneous rock or sedimentary rock.

Because of its high mica content, schist may weather to subdued landscapes. Gneiss, which is more resistant to weathering, usually forms ridges and beds that appear as a craggy, rough landscape.

These rocks are among the most common metamorphic rocks because they can form from most other rocks and are produced from deep burial when large volumes of rock are subducted. Some beautiful minerals grow in schist, notably garnet, which is red or brown and forms 12-sided crystals up to the size of tennis balls.

Talc, a common lubricating and dusting powder, is derived from schist that forms from basalt.

Both schist and gneiss exhibit parallel alignment of the mineral grains that grow during metamorphism. This gives them a layered appearance called foliation. When schist with a sedimentary component is exposed to increased temperature and pressure, it will metamorphose to gneiss. Both schist and gneiss are found around the Idaho Batholith, USA, and similar locations.

FIELD NOTES

Coarse-grained rocks. Schist is flaky. Gneiss shows alternate banding.

Found in continent–continent collision zones. Schist erodes easily, but gneiss is more resistant, producing craggy landforms.

Gneiss detail (above center) and schist studded with garnet crystals (above right).

Diamond gem surrounded by uncut diamond crystals

Diamond

Diamond, the hardest natural substance, is highly valued as a rare and precious gem. Surprisingly, it is nothing more than crystalline carbon. It tends to be colorless or pale, although deep shades, known as "fancy colors", occasionally occur. The most sought-after of these colored diamonds is blood red. Less attractive industrial diamonds, known as "bort" and "carbonado", are used in drill bits and saws, and as abrasives and polishes.

Diamonds form under extremely high pressure deep at the base of old continents (known as cratons), such as occur in Africa, Siberia, India, and Australia. They may remain there for many millions of years until caught up as accidental inclusions in volcanic eruptions, which can bring them to the surface in a matter of hours.

Miners mine diamonds from volcanic pipes or recover them from surrounding rivers.

Diamond crystals can be very large. The biggest crystal yet discovered is "The Cullinan", found in 1905 at the Premier Mine near Pretoria in South Africa. It weighed 3,106 carats and was cut into nine large and 96 smaller stones. The largest of all these stones is mounted in the British royal scepter.

As you are unlikely to find your own diamonds, the best places to see them are museums. The National Museum of Natural History in Washington, DC, USA, the Tower of London in England, and the Louvre Museum in Paris, France, all house spectacular diamonds.

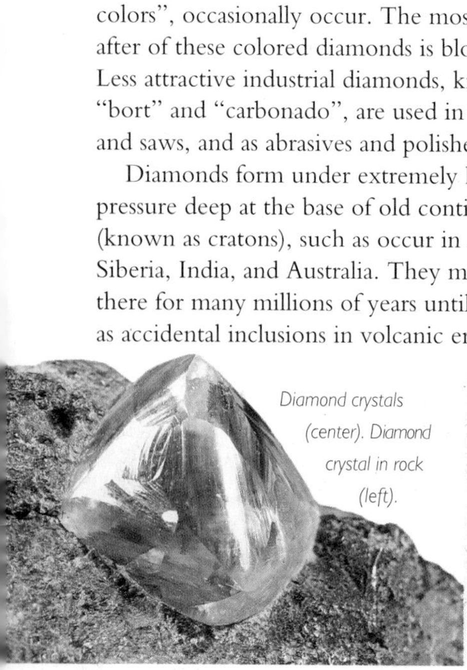

Diamond crystals (center). Diamond crystal in rock (left).

FIELD NOTES

Extreme hardness (10) and adamantine luster.

Australia and South Africa are the main producers. Other sources include Brazil, Russia, Venezuela, India, Namibia, Angola, Ghana, and Borneo.

Identification Diamond (C) has a hardness of 10, a specific gravity of 3.52. It is transparent to opaque, has an adamantine luster and leaves a white streak. It has a perfect octahedral cleavage, breaks with a conchoidal fracture, and belongs to the cubic crystal system. Its crystals are mostly octahedral.

Ruby crystal and gems

Ruby and Sapphire

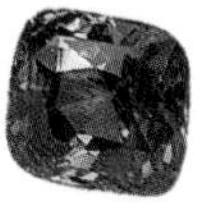

Rubies and sapphires are colored forms of corundum and the most popular colored gemstones used in jewelry today. In its pure form, corundum is colorless and somewhat rare in nature. Ruby is the red form of corundum, and its color is due to traces of chromium. All other forms of corundum are known as sapphire. These come in a range of colors, including blues, greens, and yellows, due to varying combinations of iron and titanium.

Corundum is the hardest gemstone after diamond and a high-quality ruby is more valuable than a diamond of similar quality and size. However, the value drops rapidly with the decline of color, size, and quality.

The color of corundum often occurs in patches or bands and gemstones must be cut carefully for an even appearance. Some display a star-like sheen known as asterism, due to the reflection of light from tiny, needle-like inclusions of rutile in the stone. These are known as star rubies or sapphires.

Corundum grows in aluminum-rich, silica-poor pegmatites and is carried to the Earth's surface by basalts. It may also form in aluminum-rich metamorphic rocks. Resistant to weathering, corundum is often found in rivers.

Identification Corundum (Al_2O_3) has a hardness of 9 and a specific gravity of 3.9 to 4.1. It is transparent to translucent, has a vitreous luster and leaves a white streak. It has no cleavage, breaks with an uneven, conchoidal fracture, and belongs to the hexagonal crystal system. Ruby crystals tend to be tabular to prismatic, while sapphire crystals are often barrel-shaped or pyramidal.

Sapphire crystal and gemstones (above). Ruby crystal growing in rock (right).

FIELD NOTES

Hardness (9) and horizontal striations on crystal faces.

Best locations for rubies are Burma, Thailand, and India. Sapphires are most abundant in Australia and Sri Lanka.

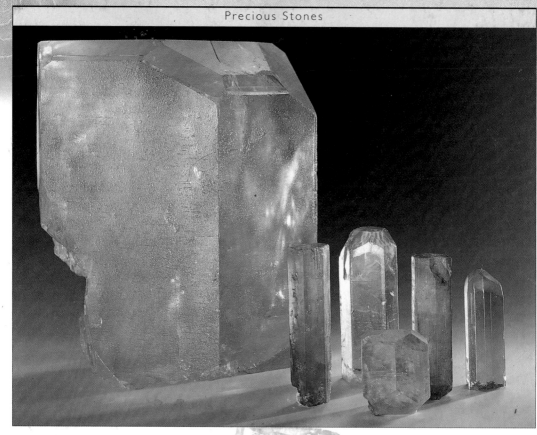

A selection of beryl crystals

Emerald and other Beryls

Emerald is the name given to the highly prized green variety of the mineral beryl. The color is the result of chromium present during crystal growth. The other beryl gemstones are also given special names according to their color: aquamarine (blue-green and light blue), morganite (rose pink), golden beryl (golden yellow), heliodor (yellow to light yellowish-green), goshenite (transparent, colorless), bixbite (red), and bazzite (blue).

Beryl crystals grow, sometimes to gigantic size, in granite pegmatites, together with topaz, quartz, tourmaline, fluorite, and cassiterite. They also form in some regionally metamorphosed rocks. The largest single crystal of gem-quality emerald, weighing 7,025 carats, was found in Colombia in 1961. Massive, poor-quality beryl crystals up to 60 feet (18 m) long have been found.

Emeralds were mined in Egypt around 1500 BC and Cleopatra is said to have possessed an emerald engraved with her portrait. Today, the finest-quality emeralds come from Colombia.

Identification Beryl ($Be_3Al_2Si_6O_{18}$) has a hardness of 7.5 to 8 and a specific gravity of 2.6 to 2.9. It is transparent to translucent, has a vitreous luster, and leaves a white streak. It has an indistinct cleavage and breaks with an uneven to conchoidal fracture. Beryl crystals belong to the hexagonal system and are usually prismatic with flat terminations, although they occasionally form as plates.

FIELD NOTES

Six-sided crystals and hardness (7.5 to 8).

Colombia is the best source of emerald. Other locations for beryls include Austria, Brazil, India, Australia, and the USA.

Uncut and cut crystals of aquamarine (center). Emerald crystals in rock with cut gem (left).

Boulder opal

Opal

Opal is different from most precious gems because of its non-crystalline, or amorphous, nature. Opal comes in a variety of colors, the most valued being black. Other forms include white (milk opal), red (fire opal), or colorless (water opal).

The most striking feature of precious opal is the sheen effect it exhibits when turned in the light, known as "play of color". This is caused by minute spheres of silica that reflect and diffract white light. Smaller spheres produce only blues and greens, whereas larger ones produce the whole spectrum. Black opals highlight play of color most dramatically.

An opal's value depends on its body color, play of color, and soundness. A black opal, free from flaws and with a uniform pattern made up of bright, clear colors, including red, may be worth more per carat than diamond. Common opals do not exhibit play of color and, as a result, are valueless.

Opal forms in surface sedimentary environments when silica-laden waters pass through cracks, joints, and cavities precipitating silica gel. It occurs as thin veins, sheets, or nodular masses that sometimes display a botryoidal habit. Opal may also be found replacing shells, bones, plants or other minerals.

Identification Opal ($SiO_2 \cdot nH_2O$) has a hardness of 5.5 to 6.5 and a specific gravity of 2.1 to 2.2. It is transparent to opaque, has a vitreous, greasy, dull, or waxy luster, and leaves a white streak. Opal is allochromatic and often shows a brilliant play of color. It has no cleavage and breaks with a conchoidal fracture. Opal is non-crystalline and comes in a variety of habits, including massive, stalactitic, and botryoidal.

FIELD NOTES

Play of color and slippery (greasy) feel.

Australia is the largest producer, with Lightning Ridge famous for black opals. Minor sources include Mexico, USA, and Brazil.

Black and white opals (center).
A polished piece of black opal (left).
A cut and polished fire opal (above).

Chrysoberyl crystal and cut stone

Chrysoberyl

Chrysoberyl, once mistakenly considered a form of beryl, comes in various shades of yellowish- and brownish-green, due to traces of iron, titanium and chromium present during formation. It is the third hardest gemstone after diamond and sapphire.

The most prized variety of chrysoberyl is alexandrite, first discovered in the Urals, Russia, in 1830, on the 21st birthday of the Russian Czar, Alexander II. It is notable for its color change, appearing green in daylight but red under incandescent light. It is now considered to be one of the most valuable gemstones. Another notable variety is golden-yellow cymophane, also known as "cat's-eye" chrysoberyl. When cut as a cabochon, light is reflected as a silvery-white chatoyant line from the many fine, oriented inclusions.

Chrysoberyl grows as tabular crystals in granite pegmatites, together with beryl, tourmaline, garnet, and spinel. It also forms in metamorphic gneiss and mica schists. Because of its hardness, it resists weathering and is often found in alluvial deposits. The largest stone found (in Sri Lanka) weighed 0.56 ounce (16 g). Alexandrite localities in the Urals have long since been depleted.

Identification Chrysoberyl ($BeAl_2O_4$) has a hardness of 8.5 and a specific gravity of 3.7. It is transparent to translucent with a vitreous luster and leaves a white streak. It has good basal cleavage and breaks with an uneven to conchoidal fracture. It belongs to the orthorhombic crystal system, with tabular or prismatic crystals, often twinned.

FIELD NOTES

Hardness (8.5). Alexandrite changes from green to red under incandescent light.

Alluvial deposits in Sri Lanka. Mine sources include Brazil, Russia, Italy, Africa, Burma, and the USA.

Cat's eyes (center) and a chrysoberyl crystal (above).

Water-worn topaz crystals and cut stones

Topaz

Topaz occurs in a wide variety of colors, depending on the amount of flourine present during formation. It ranges from colorless, to pale green, light blue, pink, red, purplish, honey-yellow and brown. Dark orange topaz is known as "hyacinth". Some varieties of topaz, particularly the purplish ones, have been heat-treated to enhance their color.

Topaz was a rare and highly valued gemstone until the middle of the nineteenth century, when rich fields were discovered in Brazil. Today, it is very affordable and is extremely popular, particularly in its blue, pink, and honey-colored varieties.

Topaz forms at high temperature in igneous rocks such as granite pegmatites, quartz porphyry, or in veins and cavities in granitic rocks. It is associated with fluorite, quartz, tourmaline, beryl, apatite, and cassiterite. It often occurs as well-formed prismatic crystals that may weigh more than 220 pounds (100 kg). Topaz is also found as an alluvial concentrate.

One of the most distinctive features of topaz is its perfect, easy cleavage. Minute planar cleavage cracks, or incipient cleavage lines parallel to the base of the crystal, distinguish topaz from other colorless stones. Because of its cleavage, topaz requires careful handling when being cut and polished.

Identification Topaz ($Al_2SiO_4F_2$) has a hardness of 8 and a specific gravity of 3.5 to 3.6. It is transparent to translucent with a vitreous luster and leaves a white streak. It shows gold-yellow, cream, or green fluorescence. It has perfect basal cleavage and breaks with an uneven to subconchoidal fracture. It belongs to the orthorhombic crystal system with crystals forming vertically striated prisms.

FIELD NOTES

Hardness (8) and perfect basal cleavage.

Brazil is the world's best source of topaz. Other sources include Sri Lanka, Mexico, Japan, and Russia.

Topaz displaying typical crystal form (center and below).

Orange grossular garnet crystals

Garnet

Garnets are a related group of minerals whose chemical compositions can vary continuously from one to the other. Key members of this group, often cut as gemstones, include almandine (red to violet red), spessartite (yellow, rose, or orange to reddish-brown), pyrope (deep red), grossular (white, yellow, yellowish-green, brownish-red, orange, or black), and andradite (colorless, yellow-green, or brown to black). This last group includes the most prized garnet—the emerald-green variety known as demantoid. The name garnet probably derives from the Latin *granatum malum*, for pomegranate. The inside of this fruit resembles red almandine and pyrope garnets.

Pyrope ranks among the most popular of the garnets, and fine crystals have been mined and used in Czechoslovakia since the Middle Ages. Many forms of garnet do not occur as gem-quality crystals and are used industrially as abrasives.

Garnets are common worldwide, particularly in metamorphic rocks. Pyrope garnet forms in many magnesium-rich, ultrabasic, igneous rocks, including peridotites and serpentinites. Grossular and andradite garnets grow in calcium-rich metamorphic rocks, such as marble. Almandine garnet forms in iron-bearing, regionally metamorphosed rocks, such as schist.

Identification Minerals in the garnet group have a hardness of 6.5 to 7.5 and a specific gravity of 3.4 to 4.6. They are translucent to opaque, have a vitreous luster, and leave a white streak. They have very indistinct cleavage and break with a conchoidal fracture. Garnets belong to the cubic crystal system and display a variety of equant habits, particularly rhombododecahedrons.

A variety of cut garnets (above). Andradite garnet crystals in marble (below).

FIELD NOTES

Crystals form distinctive soccerball shapes.

All varieties of garnet are abundant in the USA. Other sources include Australia, Brazil, Sri Lanka, and the Czech Republic.

Green elbaite crystals

Tourmaline

Tourmaline is a complex borosilicate of aluminum and other elements, noted for its beautiful, multi-colored, prismatic crystals. In fact, tourmaline exhibits a greater range of color than any other gem, with the spectacular watermelon tourmaline changing color from green to red along the length of the crystal.

The tourmaline family is made up of seven distinct groups: schorl (black), elbaite (many colors), dravite and buergerite (brown), rubellite (pink), chromdravite (green), and uvite (black, brown, yellow-green).

Tourmaline crystals display a number of very interesting properties. They are pleochroic, meaning that they appear differently colored when viewed through different axes. Also, if both terminations of the crystal are present, they are morphologically different. If the crystal is heated, the ends become positively and negatively charged, attracting dust particles.

Tourmaline is mostly found in granites and pegmatites but also in some metamorphic rocks. Crystals may grow to spectacular sizes and are found with a wide range of minerals, including quartz, feldspar, beryl, and topaz. Being resistant to erosion, tourmaline often occurs in alluvial deposits.

Identification Tourmaline minerals have a hardness of 7 to 7.5 and a specific gravity of 3 to 3.3. They are transparent to opaque, have a vitreous luster and leave a white streak. They have poor cleavage and break with an uneven to conchoidal fracture. They belong to the hexagonal crystal system, displaying a prismatic habit with vertical striations.

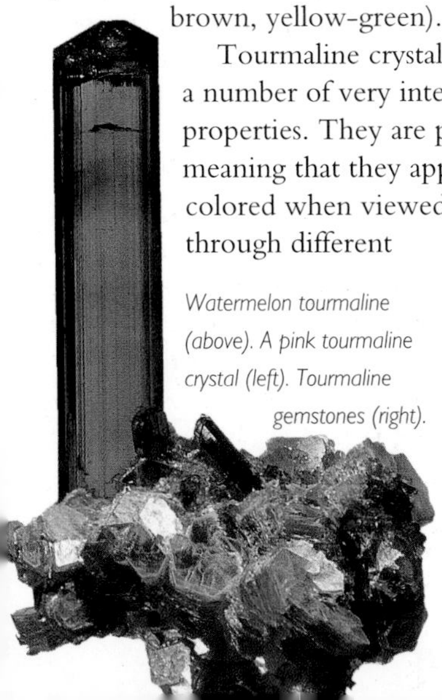

Watermelon tourmaline (above). A pink tourmaline crystal (left). Tourmaline gemstones (right).

FIELD NOTES

Vertically striated, columnar crystals may be multi-colored or show different colors through different axes. Has electrical properties when heated or rubbed.

Sources include Afghanistan, Brazil, Russia, and the USA.

An peridotite nodule from basaltic lava

Peridot (Olivine)

Peridot is the name given to the mineral olivine when it is of gem quality. Olivine is an idiochromatic mineral, always occurring in a shade of green. The ratio of iron to manganese varies in this mineral, with the maganese end member known as forsterite and the iron end member known as fayalite.

Olivine is an essential rock-forming mineral, making up much of the Earth's mantle, where it combines with pyroxene to make up the heavy, coarse-grained, green rock known as peridotite. These rocks are brought to the surface by basaltic volcanoes and are occasionally blown out as volcanic bombs. Olivine also occurs in meteorites and was found in basalts recovered from the moon by *Apollo* astronauts.

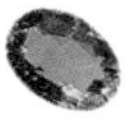

Olivine is most often found in tiny granular masses, intergrown with other minerals. Large, gem-quality olivine crystals are extremely rare, although small crystals are fairly common. These occur as short, columnar, or thick tabular forms as well as wedge-shaped terminations in basic and ultrabasic igneous rocks, such as basalts. They also form in some marbles.

Identification Olivine $(Mg,Fe)_2SiO_4$ has a hardness of 6.5 to 7, a specific gravity of 3.3 (forsterite) to 4.2 (fayalite), and leaves a white streak. It is transparent to translucent and has a vitreous luster. Olivine has an indistinct cleavage and breaks with a conchoidal fracture. It belongs to the orthorhombic crystal system, forming short, prismatic crystals, although it is usually found in granular masses.

FIELD NOTES

Olivine is always a shade of green and often contains black chromite inclusions.

The major sources are Burma, and Arizona in the USA. Other sources include Brazil, Sri Lanka, and Russia.

A selection of cut peridot stones (far left). Detail of an olivine crystal (center). A peridot talisman (left) from ancient Egypt.

Rock crystal

Quartz

One of the most common minerals on Earth, quartz, comes in a wide variety of colors and forms. The purest variety is the colorless form known as rock crystal. One of the most popular quartzes is the violet form, amethyst. Other varieties include citrine (yellow), smoky (brown to black), milky (white), rose (pink), sapphire quartz (blue), and morion (black). Rutile inclusions are common.

Quartz is an important rock-forming mineral and an essential component of acidic igneous rocks, together with feldspars, micas, amphiboles and pyroxenes. Quartz forms hexagonal prisms that are often horizontally striated and terminated by rhombohedra or pyramidal shapes. Crystals can grow to enormous sizes in pegmatites. A single crystal of approximately 40 tons was found in Brazil, and one of 70 tons was found in Kazakhstan.

In ancient Rome, large rock crystals were cut and polished into balls, which rich citizens used to cool their hands in summer. A rock crystal ball weighing 106 pounds (48 kg) can be seen in the National Museum in Washington, DC, USA.

As a key rock-forming mineral, quartz is common in all geological environments and it makes up much of the dust in the air. As a result, any gem with a hardness less than quartz (7) will be worn down by exposure to the elements and is not generally considered to be truly precious.

Identification Quartz (SiO_2) has a hardness of 7, a specific gravity of 2.65 when pure, and leaves a white streak. It is transparent to opaque and has a vitreous luster. Quartz has no cleavage and breaks with a conchoidal fracture. It belongs to the hexagonal crystal system, usually forming prismatic, horizontally striated crystals.

Quartz stained by hematite (below). Detail of rutilated quartz (center). A cut amethyst gemstone (right).

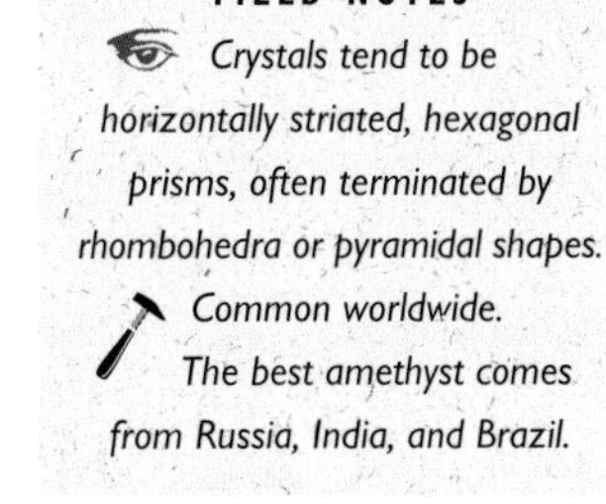

FIELD NOTES

Crystals tend to be horizontally striated, hexagonal prisms, often terminated by rhombohedra or pyramidal shapes.

Common worldwide.

The best amethyst comes from Russia, India, and Brazil.

Twinned spinel crystals in quartz

Spinel

The term spinel refers to a group of oxide minerals including magnetite, chromite, and the gem member also known as spinel. The name derives from the Latin *spina*, meaning spine or thorn, in reference to its sharp, octahedral crystals. Gem spinels occur in a wide range of colors, including shades of red, blue, bluish-green, green, and violet. Spinel has long been valued as a gemstone and, in the past, was often mistaken for ruby or sapphire. Although large, gem-quality crystals of spinel are quite rare, they tend to have fewer flaws than ruby and thus often came to be used by royalty. The best-known ruby-red gem spinel is the Timur-Rubin, weighing 352 carats and included in the British Crown Jewels. Another famous spinel is the 2½ inch (5 cm) oval Black Prince's Ruby, which adorns the British Imperial State Crown.

Spinel most commonly grows as well-formed octahedral crystals or twinned octahedra. It forms in basic igneous rocks, and in a wide variety of metamorphic rocks, including serpentinite, gneiss, schist, and marble. It also weathers into secondary alluvial deposits where it is often associated with rubies and sapphires.

Identification Spinel ($MgAl_2O_4$) has a hardness of 8, a specific gravity of 3.6, and leaves a white streak. It is transparent to translucent and has a vitreous luster. Spinel has imperfect cleavage and breaks with a conchoidal fracture. It belongs to the cubic crystal system, forming octahedral crystals that are frequently twinned.

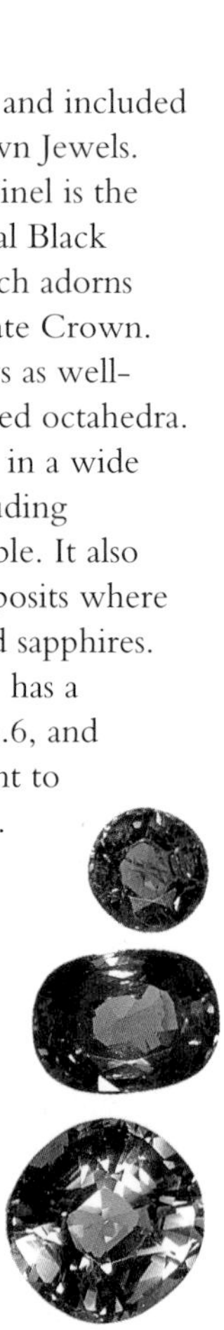

Detail of a ruby-red spinel crystal (center). A variety of cut spinel gemstones (left and right).

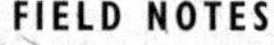

FIELD NOTES

Sharp, octahedral crystals, often twinned. Hardness (8).

Best sources are Sri Lanka and Burma. Minor sources include Italy, Germany, Russia, Thailand, and the USA.

Albite crystals

Feldspars

Feldspars are aluminum silicates and make up more than 50 percent of the Earth's crust. They are the most common rock-forming minerals, being five times more common than quartz. They occur widely in all rock types and giant crystals are common in pegmatites. They form two major subgroups: the potash (potassium) feldspars, including orthoclase, microline, and adularia; and the plagioclase (calcium and sodium) feldspars, including oligoclase, albite, and labradorite.

A number of gemstones belong to the feldspar family. Moonstone is a transparent orthoclase which diffracts reflected light, creating a silvery-blue, cloudlike sheen called schiller. Amazonstone is a bright green variety of microcline with a silvery sheen. Sunstone is an oligoclase that reflects light from tiny, oriented, flake-like inclusions of goethite or hematite. Labradorite is a grey plagioclase with a beautiful iridescent sheen. Perfectly transparent, golden-yellow orthoclase may be faceted for jewelry, as may the pure colorless variety, adularia. With a hardness of 6 to 6.5, feldspar is not particularly durable, and tends to scratch and chip easily.

Identification Potash feldspars ($KAlSi_3O_8$) and plagioclase feldspars ($NaAlSi_3O_8$ to $CaAl_2Si_2O_8$) have a hardness of 6 to 6.5, a specific gravity of 2.5 to 2.7, and leave a white streak. They are transparent to translucent and have vitreous lusters. Feldspars have two good-to-perfect cleavages and break with an uneven fracture. The plagioclase feldspars belong to the triclinic crystal system and the potash feldspars to the monoclinic crystal system (with the exception of microcline, which is triclinic).

Moonstone cabochons (center).
An uncut amazonstone crystal (above).

FIELD NOTES

Various members show distinctive sheens. Two flat cleavage faces often apparent.

Worldwide. The best source of amazonstone is Russia. Moonstone common in gem gravels in India and Sri Lanka.

Zircon crystal in carbonatite

Zircon

Zircon is a zirconium silicate that is colorless in its pure form, but comes in a variety of yellows, browns, and reds. It displays a subadamantine luster and, like diamond, has the ability to break up white light into its spectral colors, particularly in cut stones. Consequently, colorless zircons have long been used as diamond simulants. However, the birefringence, or double refraction, of zircon is so strong that when a cut stone is viewed with a hand lens, the edges of the rear facets appear double. Although it has a hardness of 7.5, zircon is extremely brittle, and the edges of cut stones chip easily.

Zircon crystals grow in many types of igneous and metamorphic rocks, and are also found in sedimentary rocks and alluvial deposits weathered from the primary zircon-bearing sources. Most gem-quality zircon comes from alluvial deposits. Although most gem-quality crystals are small, low-quality crystals of up to 16 pounds (7 kg) have been found in Canada.

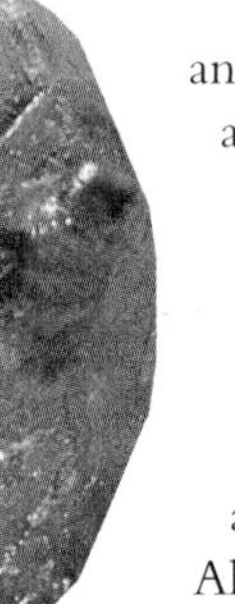

Zircons contain traces of radioactive elements that decay at a known rate and can be used to determine the mineral's age.

Identification Zircon ($ZrSiO_4$) has a hardness of 7.5, a specific gravity of 4.6 to 4.8, and leaves a white streak. It is transparent to translucent and has a sub-adamantine luster. Zircon has imperfect cleavage parallel to the prism face and breaks with an uneven fracture. It belongs to the tetragonal crystal system, forming prismatic and bipyramidal crystals, which are frequently twinned.

A variety of zircon crystals (left and center). Cut zircon gemstones (right).

FIELD NOTES

Subadamantine luster, strong dispersion, double refraction, and brittleness.

The best sources of gem-quality zircon are alluvial deposits in Sri Lanka, Cambodia, and Thailand.

Rough reniform hematite

Hematite

Hematite is an iron oxide varying in color from dull brown to brownish-red, bright red, grey, or a lustrous black. Its name derives from the Greek, *haima*, meaning blood, because of its extremely diagnostic cherry-red streak.

Hematite is an important ore of iron. The earthy-red ochre varieties are used as a polishing powder known as rouge. They have also been used as a raw material for red paint since ancient times. Compact crystalline varieties, called specular iron, were once ground into highly polished flat pieces displaying a black metallic luster and used by the Aztecs as mirrors. Hematite is common, inexpensive, and is popular for jewelry. Its only drawback as an ornamental is its weight.

Common in many countries, hematite occurs in sedimentary, metamorphic, and igneous rocks. It forms as a hydrothermal or secondary mineral, as concretions, or by filling cavities and fractures.

Hematite may grow as tabular or rhombohedral crystals. The tabular crystals sometimes form rose-shaped aggregates, known as iron roses, which are highly sought after by collectors. Other habits are massive, crusty, granular, earthy, radiating-fibrous, and reniform (known as kidney ore). The hardness of hematite varies from 6.5 for the crystalline massive varieties, to as low as 1 for the fibrous and earthy varieties.

Identification Hematite (Fe_2O_3) has a hardness of 6.5 (in crystalline form), a specific gravity of 5.0 to 5.3, and leaves a cherry-red streak. It is opaque, with a dull, metallic luster and breaks with a conchoidal fracture. Its crystals are hexagonal (trigonal) and it occurs in a variety of habits.

Polished hematite stones (center). A necklace made from hematite beads (below).

FIELD NOTES

Distinctive cherry-red streak. High specific gravity (5.3). Compact varieties may be attracted to a hand magnet.

Common worldwide, including sites in North America, Europe, and Australia.

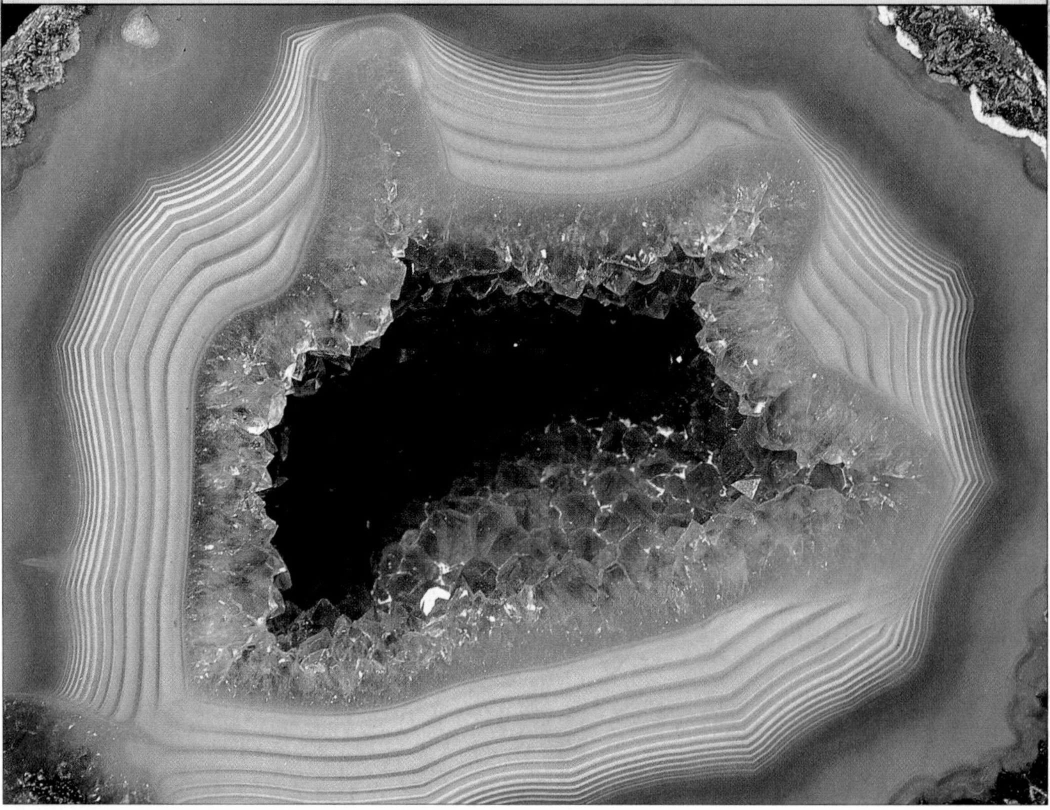

A polished geode with bands of agate and amethyst crystals

Agate

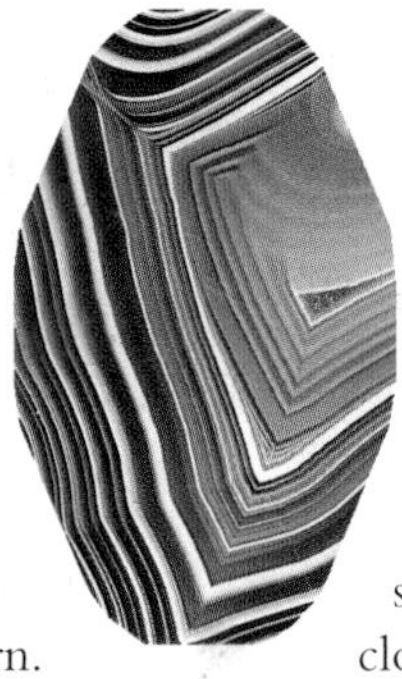

The cryptocrystalline quartzes, or chalcedony, comprise a wide group of minerals made up of tiny, interlocking microcrystals of quartz. Its members are the most visually diverse of any group, in terms of both color and habit. Collectors and jewelers have given them a wide variety of names, depending on their color and pattern.

Chalcedony precipitates from silica-bearing groundwater in rock cracks and cavities. Agates consist of brightly colored, thin, parallel bands that follow the contour of the cavity where the silica gel solidified, and this concentric arrangement is best seen in sawn and polished slabs.

Agate and quartz crystals formed in a rhyolite geode cavity (above). Agate detail (center).

Sometimes, if the cavities are not completely filled, the centers may contain crystals of amethyst, rock crystal, or smoky quartz.

Agates are named according to their appearance. Blue lace agate has bands of delicate translucent, mauve-blue and opaque white. Cloud agate has several dark, translucent centers set in a cloudy white background. Ruin agate is an interesting rare form where agate-filled cavities have been shattered by earth movement and then the broken pieces have been recemented by chalcedony of a different color. Moss agate is a transparent chalcedony with green, moss-like inclusions of chlorite. Onyx is a black-and-white-banded agate, and sardonyx has reddish-brown and white bands. Other varieties include star agate and ogle-eyed agate.

Agate was prized by the peoples of ancient Sumeria and Egypt, who used it for ornaments, receptacles, amulets, and charms. Artists in ancient Greece and Rome carved agate cameos and intaglios, using the different-colored layers to accentuate aspects of the design.

FIELD NOTES

Agates are distinguished by characteristic colors and banding.

Forms in cracks and cavities, often in volcanic rocks. Like all cryptocrystalline quartz, they are found worldwide.

Chrysoprase

Other Chalcedony

There are many unbanded varieties of chalcedony that are popular with collectors and jewelers. Some of these include pale blue or yellow chalcedony; a beautiful orange-red translucent variety known as carnelian; a reddish-brown variety called sard; and plasma, a dark green chalcedony. Heliotrope is an interesting form of plasma, with numerous red spots that give it the common name bloodstone. Chrysoprase, a bright apple green due to the presence of nickel, is the most highly sought after variety. It is sometimes referred to as "Australian jade", because of its appearance and the fact that its most important sources are mines in Queensland, Australia.

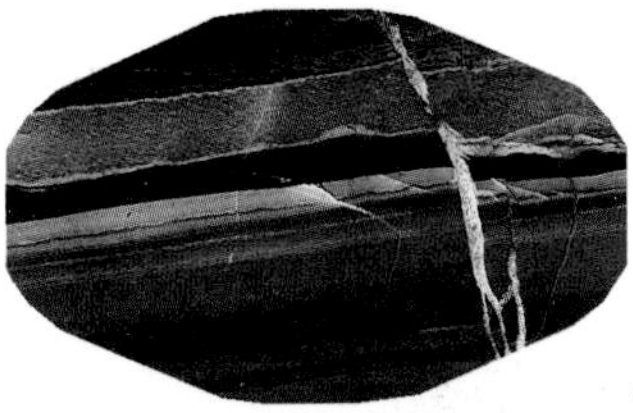

Jasper is a mixture of chalcedony, quartz, and opal. It is opaque and comes in a variety of colors, including red, green, and brown, or in interesting combinations of these.

The use of colored jasper, chalcedony and agates dates back to Greek and Roman times when they were drilled and carved into numerous ornamental objects. In the Urals, in Russia and the Ukraine, boulders of jasper weighing hundreds of pounds have been sculpted into beautiful objects. The popularity of chalcedony for carving and sculpture is largely due to its toughness and ready availability in large, uncracked lumps.

Identification Cryptocrystalline quartz (SiO_2) has a hardness of 7 and a specific gravity of 2.6 to 2.9. It is transparent to translucent, has a waxy, vitreous, or dull luster, and leaves a white streak. It has no cleavage and breaks with a conchoidal fracture. It most often occurs as botryoidal crusts and in geodes.

Detail of banded jasper (center). Cut and uncut specimens of carnelian (below), a variety of chalcedony.

FIELD NOTES

Under a microscope, chalcedony reveals itself to be either a fine crystalline aggregate, or made up of layers with a reniform surface.

Forms in cracks and cavities. Found worldwide.

Rhodochrosite on quartz

Rhodochrosite

Rhodochrosite is a manganese carbonate that comes in various shades of pink and red. It may be found as crystals but it more commonly occurs in a massive, granular, stalactitic, globular, nodular, or botryoidal habit. In these forms, rhodochrosite is often finely banded in hues of translucent to transparent pink and opaque white. These beautifully colored forms of rhodochrosite are often carved into decorative objects. Only a few stones are known to have been cut from extremely clear crystals.

Large rhodochrosite crystals (center) are rare. A polished slice of rhodochrosite (below).

Rhodochrosite is found in medium-temperature hydrothermal veins associated with copper, lead, and silver sulfides, and in altered manganese deposits. It is fairly common as a secondary mineral in the oxide zone of sulfide minerals, and may be used as an ore of manganese when available in large enough masses. When exposed to the air, rhodochrosite will develop a thin, dark oxidation crust.

The main source of rhodochrosite is San Luis in Argentina, where the local silver mines were worked by the ancient Incas. It is also mined at Capillitas near Andalgala and Catamarca east of Tucumán in Argentina.

Identification Rhodochrosite ($MnCO_3$) has a hardness of 4 and a specific gravity of 3.4 to 3.7. It is translucent, has a vitreous luster and leaves a white streak. It has one perfect cleavage and breaks with an uneven to conchoidal fracture. Rhodochrosite belongs to the hexagonal (trigonal) crystal system, but is more often found in granular, massive, reniform, or stalactitic forms.

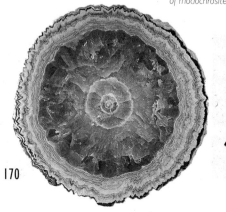

FIELD NOTES

Pink or red. In nodular form it shows distinctive pink and white banding. Soluble in warm hydrochloric acid.

Main source is San Luis in Argentina. Other sources include Romania and USA.

Uncut rhodonite with cut and polished stones

Rhodonite

Rhodonite is a manganese silicate that, like rhodochrosite, derives its name from the Greek *rhodos*, meaning rose. As the name suggests, its color varies from pink to various shades of red, with cardinal red being the most prized. It has been used for decorative purposes since the nineteenth century and has recently become fashionable as a gemstone.

Large, transparent rhodonite crystals suitable for cutting into gemstones are extremely rare. Generally, rhodonite grows in massive cryptocrystalline aggregates, often containing black spots, bands, or fine veinlets of manganese oxide. Such inclusions stand out strikingly against the mineral's pink hues. This opaque, compact form of rhodonite has the greatest use in jewelry, being ideal for cabochons, necklace beads, and ornate carvings.

Rhodonite occurs in ore veins and forms through metamorphism of rocks rich in manganese, such as deep marine sediments and impure limestones. Although it is often associated with the oxide ores of manganese, rhodonite itself is not considered an economical manganese ore.

Major deposits of rhodonite are located in the Urals, at Sverdlovsk, Ukraine, where the mineral is extremely popular. Rich deposits also occur in Australia, particularly at Broken Hill, where large red crystals are found associated with the lead–zinc lode.

Identification Rhodonite ($MnSiO_3$) has a hardness of 5.5 to 6 and a specific gravity of 3.3 to 3.7. It is translucent, has a vitreous or pearly luster and leaves a white streak. It sometimes shows dark red fluorescence. Rhodonite has two perfect cleavages and breaks with an uneven fracture. It belongs to the triclinic crystal system, but generally occurs in massive and granular aggregates, or as stalactitic encrustations.

FIELD NOTES

Pink or red coloring with characteristic black veining.

Major deposits are located at Sverdlovsk, in the Urals in Ukraine. Crystals are found at Broken Hill in New South Wales, Australia.

Detail of polished rhodonite (center). Naturally occurring rhodonite (right).

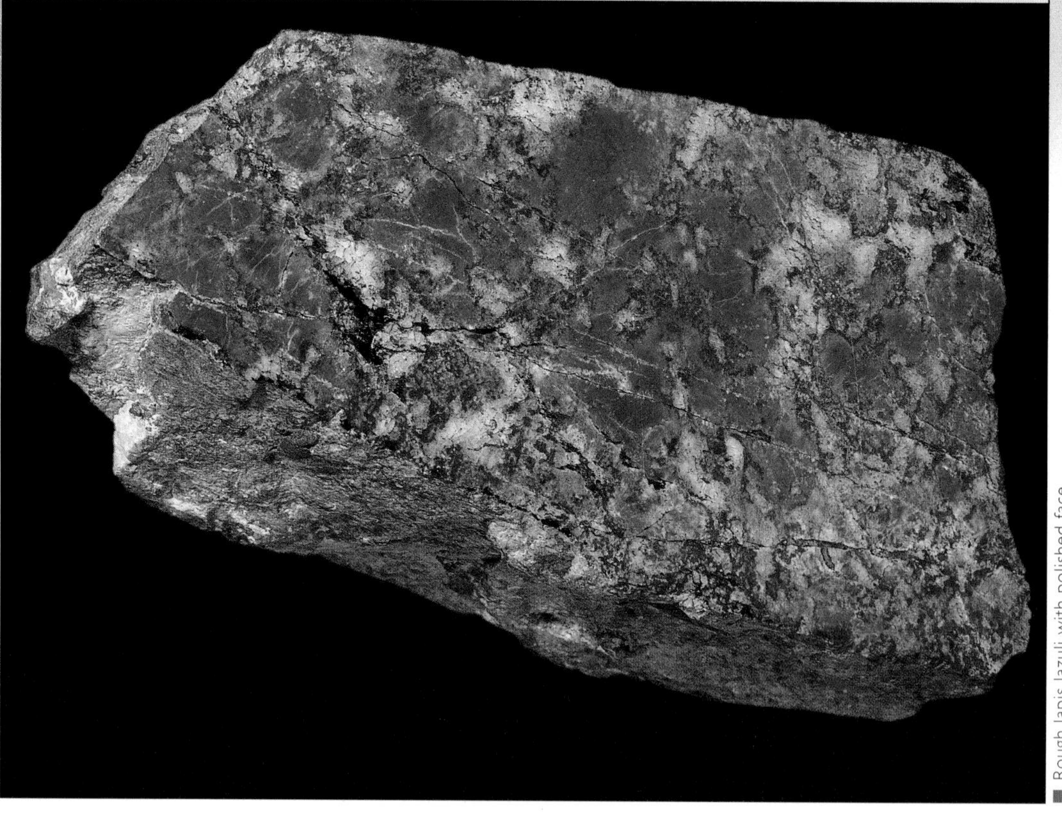

Rough lapis lazuli with polished face

Lapis Lazuli

Lapis lazuli is a rock containing a number of minerals. Its quality and value is determined by the color and abundance of the blue mineral lazurite. This mineral is a feldspathoid of the sodalite family and varies in color from deep blue to azure or greenish-blue. Other minerals occurring in lapis lazuli include hauyne, sodalite, wollastonite, pyroxenes, amphiboles, and calcite. Traces of pyrite pepper the rock with an unmistakable, golden-yellow sparkle. Lapis lazuli forms in regional and contact metamorphic rocks associated with calcite and pyrite.

References to lapis lazuli occur as long ago as 2650 BC, in the Sumerian epic, "Poem of Gilgamesh". In ancient Egypt, it was used by the Pharaohs and fabulous examples were found in the tomb of Tutankhamen. The oldest and most famous lapis lazuli deposits in the world are the mines of Sar-e-Sang in the mountains of Badakhshan Province in Afghanistan. These deposits have been worked for some 6,000 years and still produce the best lapis lazuli.

Identification The primary mineral in lapis lazuli, lazurite $(Na,Ca)_8(Al,Si)_{12}O_{24}(S,SO)_4$, has a hardness of 5.5, a specific gravity of 2.4 to 2.5, and leaves a light blue streak. It is opaque and has a greasy, dull luster. Lazurite has one imperfect cleavage and breaks with a conchoidal fracture. It belongs to the cubic crystal system but crystals are rare. Lazurite generally occurs in massive and compact forms, or as fine, granular aggregates.

Lapis lazuli detail (center). Sumerian panel (left) made of lapis and shell.

FIELD NOTES

Blue rock peppered with golden-yellow pyrite crystals.

The best lapis lazuli comes from Afghanistan. Other sources include Chile, Tadjikstan, Colorado, USA, and the Pamirs region of Central Asia.

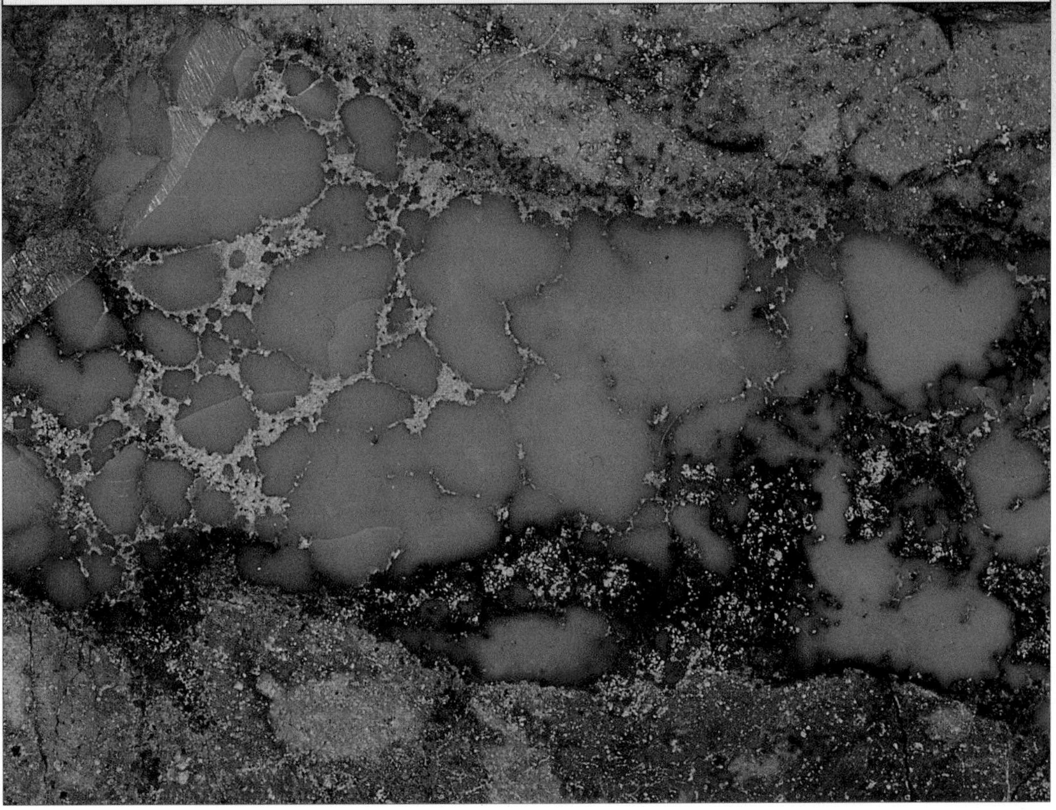

Turquoise mass in rock

Turquoise

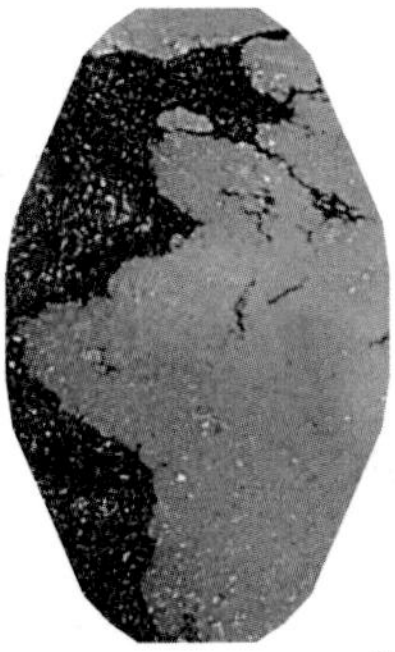

Turquoise is a hydrated phosphate of copper and aluminum. Its color varies from bright blue, colored by copper, to blue–green, colored by traces of iron.

Turquoise is opaque and often mottled with brown veinlets of limonite or black manganese oxide. It rarely forms crystals, more often occurring in massive, granular, cryptocrystalline, stalactitic and concretionary habits. It may be confused with the other aluminum phosphates—variscite, wardite and lazulite.

Turquoise forms in arid areas as a secondary mineral in the weathered zone of aluminum-rich igneous and sedimentary rocks. It forms crusts on the weathered rocks or fills cavities.

Turquoise has been used as a gemstone for thousands of years, dating back to ancient Egypt. It is usually cut as cabochons or used for carving. The ancient Aztecs used small tiles of turquoise to decorate elaborate masks. Turquoise is sensitive to heat so great care must be taken with cutting or polishing. If it is heated to 480°F (250°C), it will turn an unattractive green.

Identification Turquoise ($CuAl_6(PO_4)_4(OH)_8{\cdot}5H_2O$) has a hardness of 5 to 6, a specific gravity of 2.6 to 2.9, and leaves a white streak. It is opaque and has a greasy, waxy luster. Turquoise has no cleavage and breaks with a conchoidal fracture. It belongs to the triclinic crystal system but crystals are rare. It generally occurs in a massive form, as compact veins and crusts.

FIELD NOTES

Bright blue-green with veinlets. Greasy luster.

Best source is the Ali-Mirsa-Kuh Mountains in Iran. Another good source is the USA, particularly Arizona, Colorado, Nevada, New Mexico, and Utah.

An Aztec turquoise mask (left). Turquoise detail (center). A turquoise earring (above).

Malachite and azurite

Malachite and Azurite

Malachite and azurite are carbonates of copper that generally form together. Malachite is a banded, rich green mineral, while azurite is deep azure blue. They were originally used as copper ores before their ornamental applications were realized.

Azurite was important in the ancient Orient as a blue pigment in mural painting, and today it remains important in paint production. Its use in jewelry is limited because, with a hardness of 3.5 to 4, it is too soft.

Malachite was popular with the Greeks and Romans, and was often worn in the form of amulets to ward off evil spirits. It was a popular decorative stone at the court of the Russian Czars. The majestic malachite columns of St. Isaac's in Leningrad are from this period.

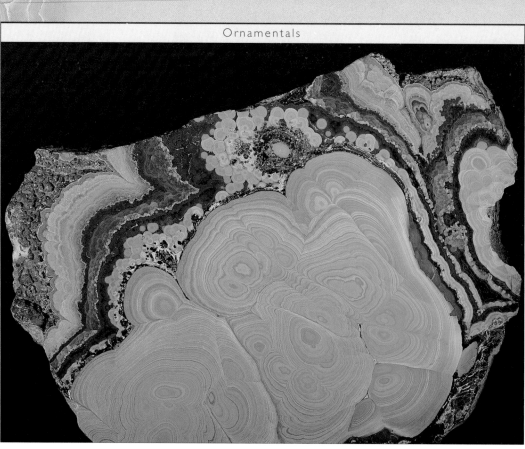

Malachite and azurite both occur commonly as stalactitic, nodular, or botryoidal masses, although azurite may be found as tabular crystals. Both are very common in the oxidized zone of copper deposits, together with iron oxides and other secondary copper minerals, such as cuprite, chalcocite, and chrysocolla.

Identification Malachite $Cu_2CO_3(OH)_2$ has a hardness of 4, a specific gravity of 3.3 to 4.1, and leaves a pale green streak. Azurite $Cu_3(CO_3)_2(OH)_2$ has a hardness of 3.5 to 4, a specific gravity of 3.7 to 3.9, and leaves a light blue streak. Both are translucent to opaque, with a vitreous to dull luster. They break with an uneven conchoidal fracture and have a monoclinic crystal structure. Malachite crystals are prismatic while azurite crystals are tabular or columnar. Both occur more commonly in granular and massive forms.

Polished malachite (above) from Zaire, Africa. Detail of unpolished blue azurite (center).

FIELD NOTES

Malachite is an opaque, banded green, while azurite is a deep azure blue. They are usually found together. Both react to hydrochloric acid.

Sources include Australia, Namibia, Russia, and USA

Cut and uncut specimens of jadeite

Jade

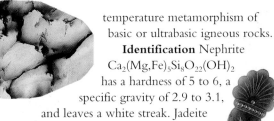

The term jade refers to two separate minerals: nephrite (a member of the amphibole group) and jadeite (a pyroxene). Because of their incredible toughness, these minerals have developed great cultural and historical significance, being used for stone axes, hammers, fish hooks and other implements.

Nephrite varies in color from black through dark green, brown, yellow, and white. Jadeite may be green, yellow, white, pink, purple, orange, bluish, brown, or gray. The emerald-green variety of jadeite, known as imperial jade, is highly sought after. Ornamental nephrite and jadeite occur in massive habit and are difficult to distinguish from each other visually—in fact, they were considered the same mineral until the nineteenth century.

Both nephrite and jadeite are made up of microscopic, interlocking crystals making them the toughest naturally occurring minerals. They can withstand enormous pressure and are more elastic than steel.

Nephrite and jadeite form during the high-pressure, low-temperature metamorphism of basic or ultrabasic igneous rocks.

Identification Nephrite $Ca_2(Mg,Fe)_5Si_8O_{22}(OH)_2$ has a hardness of 5 to 6, a specific gravity of 2.9 to 3.1, and leaves a white streak. Jadeite ($NaAlSi_2O_6$) has a hardness of 6.5, a specific gravity of 3.3 to 3.5, and leaves a white or gray streak. Both are translucent to semitransparent. They lack cleavage and break with a hackly fracture. They belong to the monoclinic system but generally occur in massive form or as granular aggregates.

FIELD NOTES

Jadeite and nephrite are best distinguished from one another by their specific gravity.

Best sources of jadeite are China, South America, and Mexico. Best sources of nephrite are Canada and New Zealand.

Carved nephrite knives (above) from New Zealand. Detail of polished nephrite (center) from New South Wales, Australia.

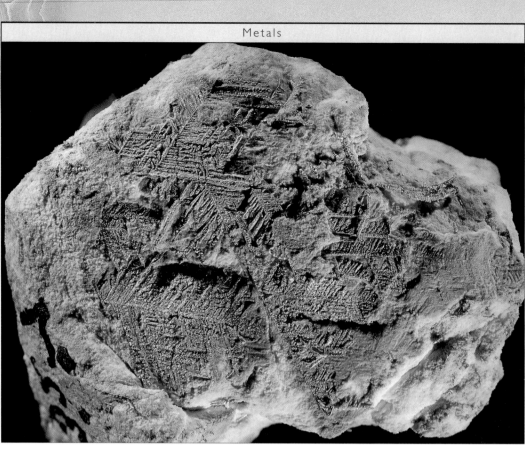

Dendritic gold from Western Australia

Veined Gold

Gold is a precious metal that resists forming compounds with water or oxygen. As a result, it usually occurs in an uncombined metallic form in ore-bodies. Like silver, it is referred to as an elemental, or native, metal.

Native gold is a bright, rich yellow and is resistant to tarnish. It is sometimes confused with pyrite or chalcopyrite (known as "fool's gold"), but can be easily distinguished by its golden-yellow streak. Pyrite and chalcopryite leave greenish black streaks.

Gold will sometimes mix with other metals such as silver, copper, rhodium, iridium, and platinum, which may change its color. For example, when gold contains silver, it is much paler, or if it contains copper it is redder.

Gold jug (left) from the tenth century. Veined gold in quartz (center).

Gold forms as a native metal in hydrothermal igneous veins. It may be associated with quartz, and sulfide minerals such as stibnite, chalcopyrite, arsenopyrite, pyrite, pyrrhotite, tellurides, and silver minerals. Gold sometimes crystallizes as cubes or octahedra, but more usually as irregular or dendritic masses. It has also recently been discovered that gold may be precipitated in soils by bacteria and other organisms.

Occasionally, gold appears as thick encrustations within the quartz veins. Generally, however, the gold occurs only as small specks and, in many cases, the gold particles are too small to be seen, remaining locked up within the sulfide minerals. During processing, these rocks have to be crushed and ground to expose the gold, which can then be dissolved using mercury or cyanide. With mercury, gold forms a heavy amalgam, that can then be separated from the treated ore. In the cyanidation process, gold reacts with sodium cyanide to form sodium cyano-aurite, which can then be treated to extract the gold.

FIELD NOTES

Distinguishable by high specific gravity, softness, malleability, and yellow streak.

The Witwatersrand mines of South Africa produce a third of the world's gold. Other sources are Australia, Russia, and the USA.

Gold nuggets

Alluvial Gold

Because of its high specific gravity and resistance to erosion, gold will often weather from vein deposits. The grains and flakes of gold will then find their way downhill, eventually ending up in river deposits or sedimentary rocks formed from these deposits. While these grains are usually tiny, larger lumps, known as nuggets, are sometimes found. Because gold is so malleable, the nuggets are often rolled or worn into rounded shapes.

Gold nuggets are rarely larger than pebbles, but there are some famous exceptions. The largest pure nugget, weighing 156 pounds (70.9 kg) was the "Welcome Stranger" found at Moliagul, Victoria, in Australia in 1869. The "Holtermann Nugget", found in 1872 at Hill End, Australia, and weighing 630 pounds (286 kg), is often named as the largest gold nugget. However, this is not technically correct. It was actually a slab of slate containing 220 pounds (100 kg) of veined gold.

Gold nugget (above) from Mt Kari, Papua New Guinea. Gold nuggets (center). An 1869 sketch (right) showing gold panning in California, USA.

FIELD NOTES

Grains, flakes, or nuggets of soft, golden-yellow material that will tend to concentrate in the bottom of the pan.

Gold can be panned wherever streams or rivers drain away from gold deposits.

Panning is the age-old method of searching for alluvial gold, although extraction processes have become more sophisticated. Whatever the method of extraction, the principle is much the same—dense gold particles are separated from the lighter sand and gravel particles.

Identification Gold (Au) has a hardness of 2.5 to 3, and a specific gravity of 19.28 (when pure). It is opaque, with a metallic luster, and leaves a golden-yellow streak. It has no cleavage and breaks with a hackly fracture. Gold belongs to the cubic crystal system and grows as crystals or dendritic masses in veins.

Silver in a sawn slab of white barite

Silver

Silver, like gold, occurs as a native metal, although it is more often found combined with the sulfides of lead, zinc, and copper (galena, sphalerite, and chalcopyrite). In its native form, silver usually occurs as star-shaped or tree-like aggregates, as dendrites and wire-like forms, or as compact masses. It occasionally crystallizes as cubes or octahedra.

Silver forms as a primary mineral in hydrothermal, supergene deposits and in veins, usually associated with lead, zinc, or copper. It is very soft and is usually alloyed to make it stronger. The best known alloy is sterling, which is 92.5 percent silver and 7.5 percent copper, and has long been valued for its workability and resistance to corrosion. The main applications for silver are for coinage, jewelry, medicine, chemistry, and photography. Because of its thermal and electrical conductivity, it is also important in electrotechnology.

The first significant silver mines were those of the pre-Hittites of Cappadocia in eastern Anatolia, and by 2000 BC, lead ores were being smelted to obtain silver. Today, the largest producer of silver is the Guanajuato Mine, in Mexico, where silver has been extracted since AD 1500. More than half the world's silver reserves are held in Mexico, USA, Canada, Peru, Kazakhstan, and Russia.

Identification Silver (Ag) has a hardness of 2.5 to 3 and a specific gravity of 9.6 to 12 (when pure). It is opaque, has a metallic luster, and leaves a glossy white streak. It has no cleavage and breaks with a hackly fracture. Silver belongs to the cubic crystal system, but crystals are rare, forming as octahedrons or hexahedrons. Silver more often occurs as wires, dendrites, or plates.

FIELD NOTES

High specific gravity. Softness, malleability, and color.

Guanajuato Mine in Mexico is the world's best source. Also North America and Russia. Fine dendritic and wire-like crystals are found at Köngsberg in Norway.

Viking amulet made from silver (above). Silver crystals (center).

A sawn stony iron meteorite from the Atacama desert, Chile

Rocks from Outer Space

Meteorites are bodies of material that, while traveling through space, are trapped by the Earth's gravitational field. If they survive their journey through the atmosphere, they strike the Earth's surface. Of thousands of meteorites that strike the Earth each year, fewer than 10 are generally recorded. The rest land in oceans or deserts.

As meteorites enter Earth's atmosphere they partly melt. As they slow down, their surface solidifies again, forming a dark, glassy fusion crust. Four distinct types of meteorite are found. The first, composed completely of nickel and iron, are known as irons or siderites. If cut and polished, the nickel and iron appear as light and dark parallel bands. A gigantic, 36.5-ton siderite is housed in the American Museum of Natural History.

The second type are stones or aerolites. Composed of silicate material, these have an oxidized crust and consist largely of pyroxene and olivine. The third group are composed of both metal and silicate and are known as stony irons or siderolites.

FIELD NOTES

Look for the presence of a fusion crust, and the presence of nickel-iron alloy.

If you recover an object that fell from the sky, submit it to a museum or university for expert examination.

The fourth group are tektites—splashes of molten meteoritic and crustal material from large impacts. Because they are predominately crustal in composition, however, they are not always considered to be true meteorites.

Meteorites are important because they are likely to represent samples of planetary material and their compositions are used to estimate the average composition of other planets as a whole. The irons and stony irons may contain iron, cobalt and nickel sulfides, carbides, phosphides, and sometimes graphite. Crater structures so large that they are visible only from a satellite tell us that the Earth has been struck by giant meteorites in the past.

An australite (center), a tektite from Australia. A meteorite (above) recovered from the Arizona crater, Flagstaff, USA.

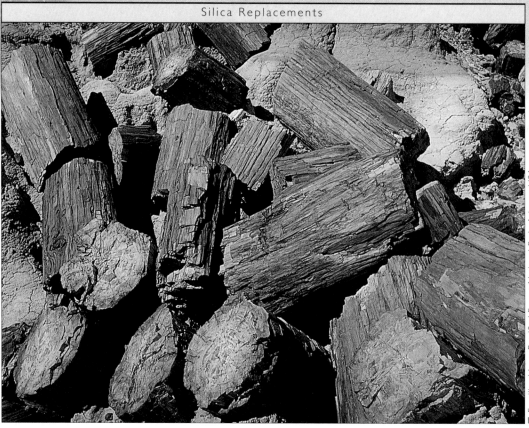

Fossilized logs, Petrified Forest National Park, Arizona, USA

Silica Replacements

Wood, bone, shell and other material may become fossilized or "petrified" (literally meaning "turned to stone") by silicates, such as jasper, cryptocrystalline quartz, and even opal. The material retains its original structure as it is gradually petrified, but may take on a variety of different colors.

This replacement occurs near the ground surface where silica is abundant in the groundwater. It is still not fully understood what geological factors determine whether the replacing material will be cryptocrystalline or opaline.

Tiger's eye is a different type of silica replacement where fibers of golden or blue asbestos have been replaced by cryptocrystalline quartz, creating a gemstone that polishes up to a high luster but retains its fibrous sheen. This results in a brilliant, thin band of chatoyant light rolling across the stone at right angles to the parallel fibers as the stone is turned. Tiger iron is the name given to the beautiful banded ornamental where yellow tiger's eye is interbedded with red jasper and dark grey hematite.

Fossil forests can be found throughout North America, Europe, and Africa. The most famous site for fossilized wood is the Petrified Forest National Park near Holbrook in Arizona, USA. Here a tree measuring 220 feet (65 meters) and 10 feet (3 meters) in diameter has been found. At White Cliffs, in Queensland, Australia, whole animal skeletons have been found replaced by opal (see p. 252).

Identification Silica replacements have the physical properties of cryptocrystalline quartz or opal. The delicate structure of the original material will often be preserved in minute detail.

FIELD NOTES

Silica replacements have physical properties of chalcedony or opal. Original structure is preserved in minute detail.

Petrified Forest, Arizona, USA and White Cliffs, Australia, are notable sites.

Quartz replacement in a piece of fossilized wood (left). A detail of tiger iron (center).

Jet showing conchoidal fracture

Jet

Jet is attractive, fine-grained, black rock, often classified as a type of coal because of its high carbon content. It is lightweight, strong, and is hard enough to take a good polish. As a result, it has been used for centuries for jewelry and ornaments.

Unlike coal, which is associated with land rocks, jet is found in sedimentary marine rocks. It is generally believed that jet formed from floating logs and the remains of other woody plant material that became waterlogged and sank into the mud on the sea floor. In this oxygen-deprived environment, they were preserved and became compressed like coal.

Jet mining in England started as long ago as 1500–1400 BC, near what is now Yorkshire, and reached its peak toward the end of the nineteenth century. Being black and easy to carve, jet became popular for highly intricate ecclesiastical and mourning jewelry during the Victorian era. Today, there is little demand for it in jewelry, but it is still carved into decorative pieces and figurines.

FIELD NOTES

Fine-grained and black. Light and warm to the touch. Burns easily with smell of burning coal. Insoluble in toluene

England is historically the most famous source of jet.

Cannel and anthracite, both high-carbon coals, are sometimes used as substitutes for jet, and are beautiful in their own right. They have a vitreous luster even higher than that of jet. Albertite, a form of asphalt, has been used to imitate jet, but when rubbed with the solvent, toluene, albertite leaves a brown stain as it is slightly soluble.

Identification Jet (C) has a hardness of 2.5 to 4, a specific gravity of 1.3 to 1.4, and leaves a black streak. It is opaque, with a resinous to vitreous luster, has no cleavage, and breaks with a conchoidal fracture.

A vase made from cannel coal (above). A detail of jet (center).

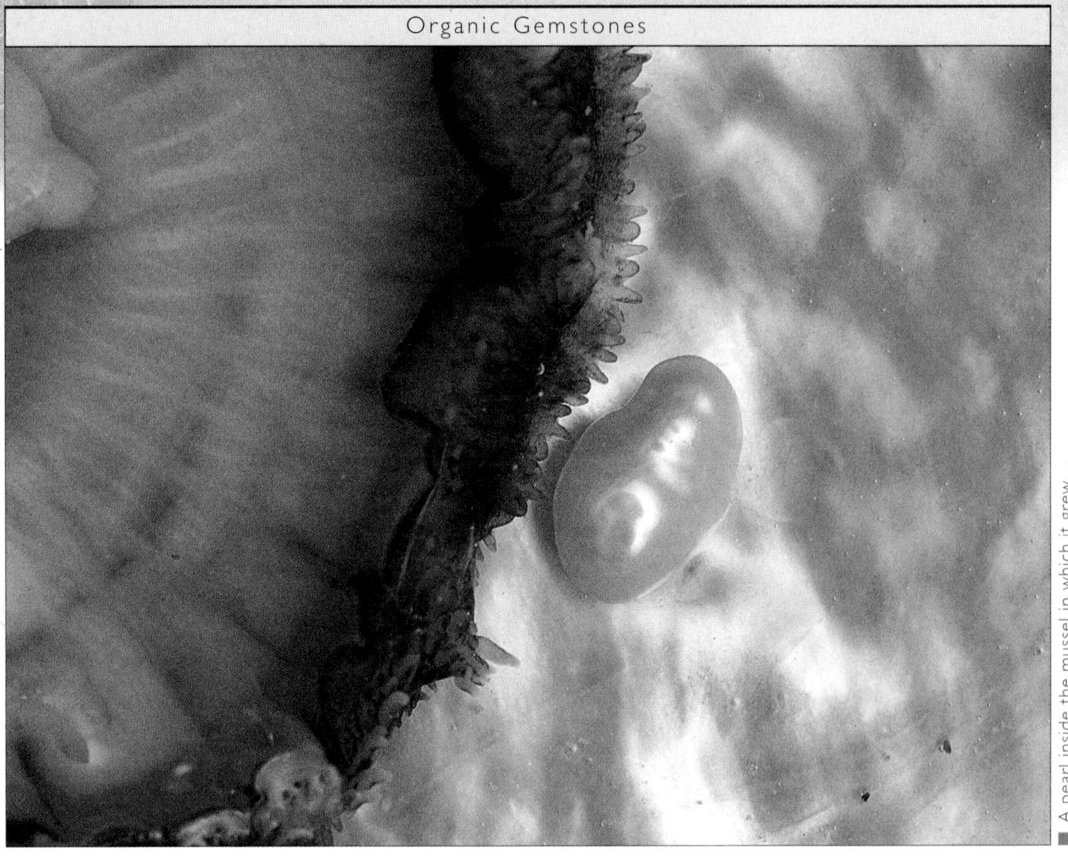

A pearl inside the mussel in which it grew

Pearls

Pearls form inside living mollusks in subtropical and tropical waters. They most commonly grow in oysters but do also occur in conches, and some freshwater clams and mussels. Pearls range in color from white to silver-white, cream, yellow, pink, greenish, bluish, and black. Their size may vary from less than 1⁄24 inch to 2 inches (1 mm to 5 cm).

Pearls develop when a foreign particle becomes trapped within the shell, irritating the oyster. The tissues of the mantle react by secreting layer upon layer of nacre (thin, platy crystals of aragonite bonded with conchiolin) around the irritant. This pearl-forming process takes about seven years and the concentric layers of varying thickness can be seen easily when a pearl is cut in half or X-rayed.

Cultured pearls are produced by placing a small bead, ground from mother-of-pearl shell and wrapped in a piece of mantle, into the body of a three-year-old oyster. Implanted oysters are left for up to four years in special cages suspended from floats in sea farms. Cultured pearls have about one fifth of the value of natural pearls.

The largest known pearl is the Pearl of Laotze (Pearl of Allah) weighing 14 lb 1 oz (6.37 kg). It was found at Palawan in the Philippines in 1934 in the shell of a giant clam. In 1984, the San Francisco Gem Laboratory valued the pearl at between US$40–42 million.

A collection of pearls (above), including pink conch pearls, black clam pearls, and cultured pearls. Oyster pearls (center inset).

FIELD NOTES

Precious pearls have a pearly luster with a weak, rainbow iridescent sheen.

Form in mollusks in tropical and subtropical waters, particularly around Australia and South East Asia.

Identification Pearl has a hardness of 3.5 to 4, a specific gravity of 2.6 to 2.9, and leaves a white streak. It is translucent, has a pearly luster, no cleavage, and breaks with an uneven fracture. In natural pearl, the fluorescence is sky-blue and variable; in cultured pearl, it is greenish-yellow and variable.

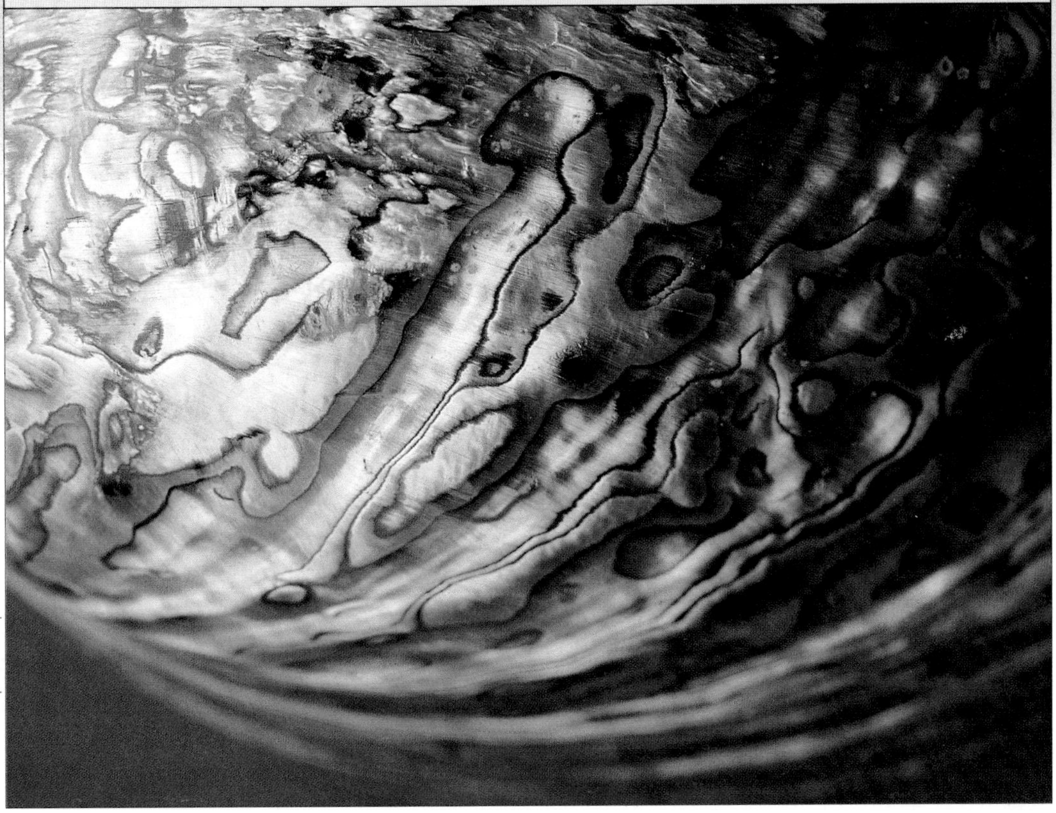

Detail of a polished paua shell from New Zealand

Shells

Shells are made up of calcium carbonate and grow in the ocean, forming part of a living ecosystem. As a result, they should not be collected. Despite this, three types of shell have been used as gems for centuries: those made up of pearly or nacreous layers, such as mother-of-pearl, abalone, and New Zealand paua; those with layers of different colors; and the operculum, or cat's-eye.

Pearly shells are ground back to the colorful, iridescent, nacreous layers and polished, or pieces are fashioned into shapes that are mounted in jewelry. Shells with layers of two or more colors are often carved so that a figure in one color stands out against a background of another color. These low-relief carvings are known as cameos, a term first applied to carved layered stone, such as onyx or sardonyx. Being easier to carve, shell cameos quickly became very popular. Helmet and queen conch were the main shells used for cameos.

The third form of shell used in jewelry is the operculum, or cat's-eye. This is the cover for the shell opening of the large sea snail, *Turbo petholatus*, which is found in the tropical waters of Australia, Fiji, Tahiti and Samoa. The operculum may vary in diameter from ¼ inch to 3 inches (5 to 75 mm) and is called a cat's-eye because of its domed shape and circular green, brown, and white markings.

Identification Pearly shells are recognized by a nacreous interior with a pearly luster. Shell cameos are all curved, reflecting the shape of the original shell. An operculum shows a spiral growth pattern on its underside. Shell will react in acid, although this will damage the surface.

Detail of an abalone shell (center). Manufacturing buttons from mother-of-pearl in New South Wales, Australia, 1933.

FIELD NOTES

Nacreous interior and pearly luster. Will react with weak hydrochloric acid.

Shells grow in the ocean, forming part of a living ecosystem. Consequently, they should not be collected.

Rock Sites

Mt Augustine, Alaska, USA

Explosive Volcanoes

Volcanoes vary in shape according to their magma composition. The more silica the magma contains, the steeper the volcano will be. So the high-silica andesitic volcanoes of Kamchatka (Russia), Japan, Alaska, Indonesia, and the western side of the Americas are all steep. These volcanoes occur above subducting slabs where silica collects and mixes with basalt magma. Most high-silica volcanoes produce ash and form soft, conical mountains, such as occur in the Mojave Desert, California, and Sunset Crater, Arizona, USA. Typically, these have radial drainage—many gullies running straight down all around the mountain, as seen at Mt Yoti, Japan.

All high-silica volcanoes are explosive, and their perfect cone shape changes dramatically upon explosion. Some of the mountain may be left, as when Mt St Helens exploded in Washington State, USA, or only the perimeter of the cone may be left, as when Mt Hakone erupted in Japan.

After an explosion, a lake may form in the resulting crater. Subsequently, a small cone forms in the center of the lake, as in Lake Toya, Japan, and Crater Lake, Oregon, USA. Eventually, the small cone grows to the size of the original volcano, and the next explosion occurs. This pattern is recorded around Mt Fuji, Japan. The growth and explosion of andesitic volcanoes continues until subduction stops feeding them magma.

An 1831 sketch (right) of a volcano, Julia Island, Sicily. Sunset Crater (below) Arizona, USA.

FIELD NOTES

Volcanoes along subduction lines, as in Russia, Japan, Alaska, Indonesia, New Zealand, North and South America. Crater lakes can be seen at Lake Toba, Indonesia; Crater Lake, USA; Lake Toya, Japan.

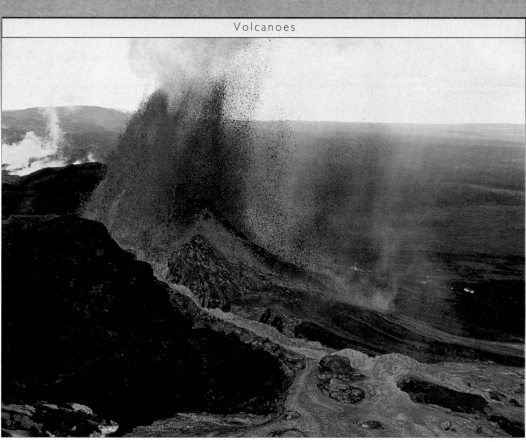

Mt Kilauea, Hawaii, USA

Shield Volcanoes

Hot-spot volcanoes form from fast-running, low-silica basalt magma. On land, they have low-angle cones and are often referred to as shield volcanoes, after ancient Roman shields. When they form underwater, they start with a steeper shape because the lava freezes much faster and does not travel as far. The volcanic pedestal is formed from fractured pieces of pillow basalt (see p. 189). The shape flattens to the shield form as the cone builds above sea level, as in the Hawaiian Islands.

A hot-spot volcano has a limited life, forming as a plate passes over a hot spot deep in the mantle (see p. 43). The volcano builds in size as the plate moves steadily on. Once the plate has moved sufficiently, a new volcano will appear over the stationary hot spot and the old one will become inactive and eventually erode. Continental hot-spot volcanoes erode to subdued hills, as can be seen all along the eastern coast of Australia. Oceanic hot-spot volcanoes erode to sea level to produce flat-topped pedestals called guyots. These subside below sea level as their load depresses the ocean floor. This series of events is clearly recorded in the Hawaiian Island chain and surrounding reefs.

An eroded oceanic hot-spot volcano in Tahiti, Polynesia.

FIELD NOTES

Hawaiian Islands, USA
Yellowstone National Park, USA
Mt Erebus, Antarctica
Mt Kilimanjaro, Tanzania
Tristan da Cunha, Atlantic Ocean
Réunion Island, Mauritius
Lord Howe Island, Australia
Barrington Tops, NSW, Australia

Giant's Causeway, Antrim County, Northern Ireland

Columnar Jointing

Columnar jointing is one of the best known geological structures. Columns, usually with five or six sides, can measure from 2 inches (5 cm) to 10 feet (3 m) across. Columnar jointing occurs when hot lava or welded volcanic ash cools and contracts.

To release the contraction stresses, triple cracks develop from points of chilling on the top surface. The angles between the cracks are about 120 degrees and they influence each other, joining into polygons. The cracks penetrate into the rock, eventually meeting similar cracks initiated by cooling at the bottom.

The resulting basalt or ignimbrite columns are usually oriented at right angles to the cooling surfaces. Horizontal surfaces tend to have vertical columns, but when lava flows cool against sloping creek banks, the columns will be inclined, indicating the topography beneath. When the cooling pattern is distorted by nearby lava tubes or steam vents, some columns develop in radiating shapes that suggest flowers.

Thick ignimbrite can develop columns up to 1,000 feet (300 m) tall. Basalt lava thins as it flows, so columns up to 50 feet (15 m) are more usual.

Teleki's East African expedition encounters a columnar basalt ravine (left). Devil's Tower (right).

FIELD NOTES

Devil's Tower, Wyoming, the Palisades, New York, and along the full length of the Columbia River Gorge, Washington and Oregon, USA. The Giant's Causeway, Ireland. Cape Raoul, Tasmania, Australia.

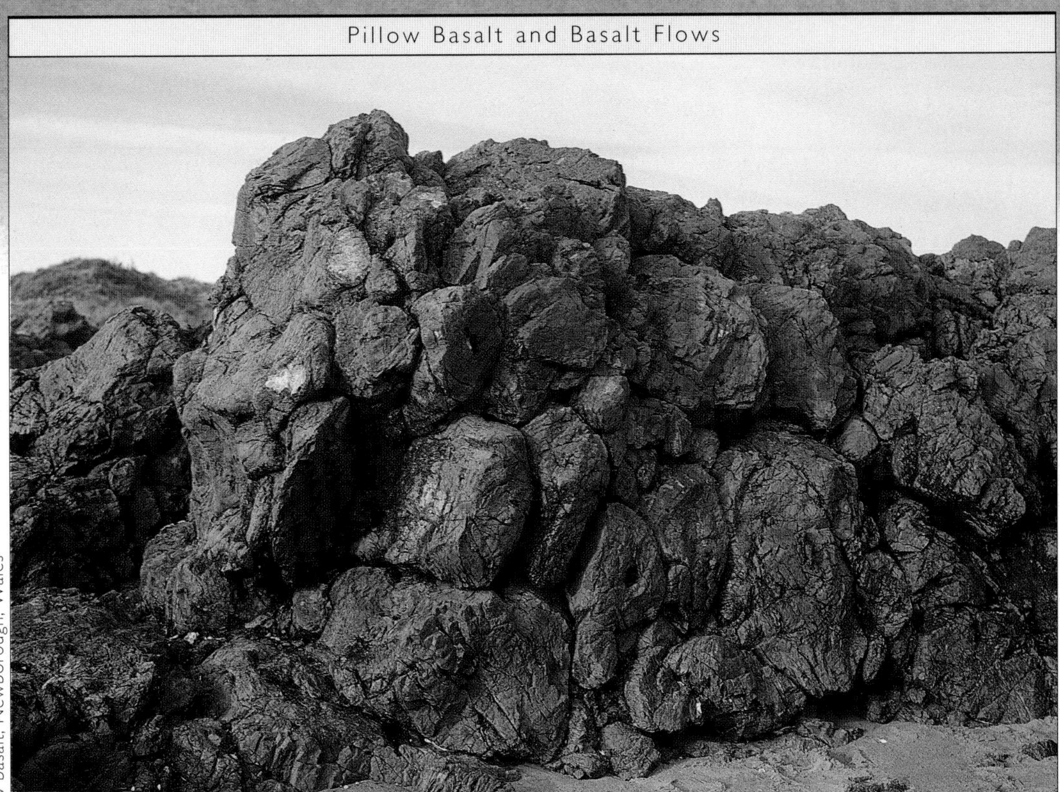

Pillow basalt, Newborough, Wales

Pillow Basalt and Basalt Flows

When basalt magma flows into water, the outer surface chills, becoming solid, while the inside remains hot and continues to flow. As the front of the flow freezes, magma bursts out of its new basalt tube and surges forward before chilling again. The top shape of the flow is rounded, much like a long pillow, and the spaces between the tops are V-shaped. As new pillows flow over layers of earlier pillows, they conform to the shape between the old tops. These new pillows will then have V-shaped bottoms and rounded tops.

Pillows form where oceans spread, so pillow basalt is a very common rock form, although usually hidden beneath the oceans. Pillow basalt along the Columbia River Gorge, Washington State, USA, was formed by a flow of basalt that interrupted rivers, causing lakes, which then filled with basalt in this pillow form. Most of the underwater flows near volcanic islands are pillow-shaped, so it is common to see them around such coastlines.

Basalt flows that chill in air have two typical forms. Slow air-cooling causes a thin skin, which is then stretched and wrinkled as the flow continues. The resulting ropy texture is called *pahoehoe* lava. Further cooling thickens the skin, so cracks develop, and the lava surface becomes blocky, and is called *a-a* lava. With cooling downslope, there is a transition from ropy lava to fractured blocky lava. This is visible in all surface basalt flows and around any active volcano.

Pillow basalt (right) forming on the slopes of an underwater volcano in Indonesia.

FIELD NOTES

Craters of the Moon lava field, Idaho, and Columbia River Gorge, Washington, USA.
Volcanoes National Park, Hawaii.
Around the coasts of Japan, New Zealand, Tahiti, Hawaii, Italy, and the west coast of South America.

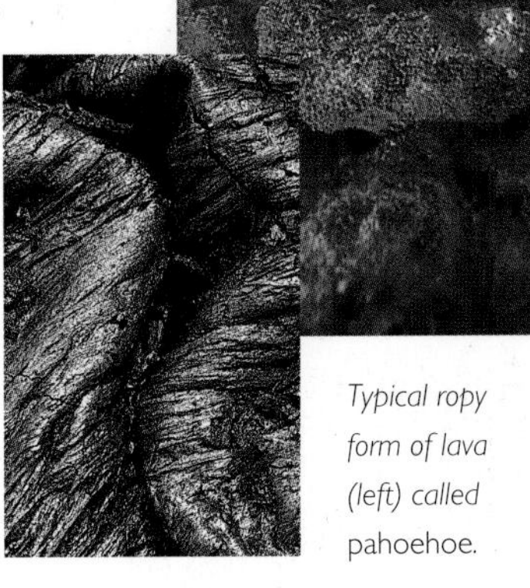

Typical ropy form of lava (left) called pahoehoe.

The Twelve Apostles, Victoria, Australia

Sedimentary Remnants

Erosion is Nature's sculptor, carving some of her best art in flat-lying layers of sedimentary rock by isolating small parts of once-continuous beds. Water enters vertical cracks that are generated when the layers of sediment flex, and weathering widens them until most of the layer is removed. This process creates mesas and buttes that erode further to natural bridges, arches, pinnacles, and pillars, some capped by balanced rocks. Seaside cliffs may also erode in this way, creating such formations as the Twelve Apostles, in Victoria, Australia.

An 1869 engraving of buttes in Colorado (center) by John Wesley Powell. A natural arch (below left), Utah, USA.

Mesas and buttes are common in central Australia, and North and South America. Some of the most impressive eroded landforms can be seen in Utah and Colorado, USA. In Arches National Park, Utah, many closely spaced, intersecting vertical fractures divide the thick Entrada Sandstone into thin vertical slabs. The fracturing results from the presence of a layer of salt beneath the sandstone. The soft, buoyant salt flows with pressure, and in doing so, removes support from beneath sections of the slabs. Unsupported sections fall down, forming natural bridges, caves, and arches.

At Bryce Canyon, Utah, USA, quite different patterns form when silty and limey lake deposits are eroded by rain. The result is myriad banded, pastel-colored spires that form a spectacular natural amphitheater.

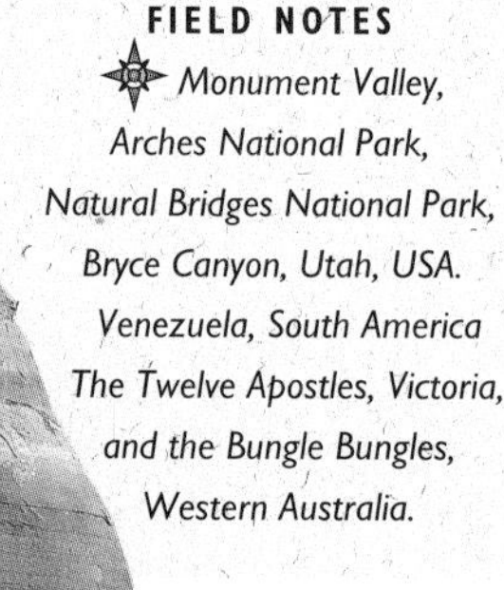

FIELD NOTES

Monument Valley, Arches National Park, Natural Bridges National Park, Bryce Canyon, Utah, USA. Venezuela, South America The Twelve Apostles, Victoria, and the Bungle Bungles, Western Australia.

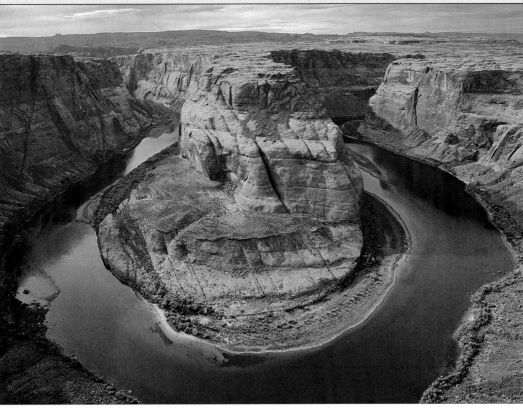

Horsehoe Bend, the Colorado River, Arizona, USA

River Bends

When water flows across a plain with almost no gradient, small local rises can deflect the flow from the average downhill direction. This leads to meandering streams. In low-gradient areas, these streams will not cut very deeply into the land surface.

Most meandering streams flow in loose sediment, where the position of the stream channel is constantly changing. Erosion on the outside of bends is accompanied by deposition on the inside of bends, and the bends creep both sideways and downstream. This occurs mostly in mid-plate regions, affecting rivers such as the Amazon, in South America, and the Darling, in Australia. The Alatna River, in Alaska, USA, has spectacular, tight bends that increase the actual length of the river dramatically.

In areas of strong tectonic uplift, such as the Rocky Mountains, in Colorado, USA, streams can become deeply incised in the bedrock. These have very abrasive bedloads, allowing rapid down-cutting into the relatively soft sediment of the surrounding area. Deep entrenchment depends on uplift, but also requires soft bedrock, or the process ceases as soon as a hard barrier occurs in the river, slowing erosion to less than the uplift rate.

Entrenched meandering rivers change their position little, as is evident at the Goosenecks on the San Juan River, Utah, USA. The pattern of bends has not changed since the river first began.

FIELD NOTES

Meandering rivers are common in mid-plate regions and areas of strong tectonic uplift.
Amazon River, South America
Mississippi River, USA.
Alatna River, Alaska, USA.
Colorado and Green Rivers,
Goosenecks, San Juan River, USA.

East Alligator River (above left), Australia.
Goosenecks (above), San Juan River, USA.

Carlsbad Caverns, New Mexico, USA

Caves

Limestone caves form when calcium carbonate dissolves out of limestone. Water entering narrow cracks in the limestone enlarges them to passageways, and sometimes to ballroom-size caverns. Many levels may develop in the one cave system, leaving upper caves dry. Some caves are formed by wave action around the perimeters of lakes, and may have been inhabited by animals and humans through time.

Water leaking into dry caves evaporates, leaving calcite on the cave surface. Beautiful shapes form as ornamentation on the cave walls, floors, and ceilings. Cave decorations include stalactites, stalagmites, columns, draperies, flowstones, oolites (cave pearls), and helictites (or mysteries, the formations that project sideways from walls and other formations). Coloring in caves is the result of impurities in the groundwater, with iron oxide producing most of the yellows, browns, and reds.

Drapery formations in Carlsbad Caverns, New Mexico, USA.

Cave systems can be vast. The Mammoth–Flint Ridge cave system in Kentucky, USA, consists of 190 miles (306 km) of connected passages and chambers. One of the world's deepest caves, extending 1 mile (1.6 km) underground, is the Gouffré de la Pierre St Martin, on the border of France and Spain. Papua New Guinea also has enormous caves, formed in thick, inclined limestone layers. One of these, Atea Kananda, is just under a mile (1.5 km) deep.

Stalagmites and stalactites in Jenolan Caves, NSW, Australia.

FIELD NOTES

Carlsbad Caverns, New Mexico, USA
Mammoth–Flint Ridge cave system, Kentucky, USA
Jenolan Caves, NSW, Australia
Gouffré de la Pierre St Martin, French/Spanish border
Atea Kananda, Papua New Guinea

Undara lava tube, Queensland, Australia

Lava Tubes

Basalt flows can become hollow inside. When flowing basalt begins to cool, the interior cools more slowly than the outer skin. This results in the top of the flow becoming hard while the inside is still able to move. Should the lava supply dwindle, the interior of the lava tube may drain and what remains is an empty tunnel.

Lava tubes are relatively rare, but many stretch for several miles. They can have the dimensions of a large railway tunnel, or be quite small, just a few feet (a meter or so) across. Sometimes, the lava has flowed along a creek bed, so that the lava tube is an isolated tunnel, such as the Undara lava tubes in Queensland, Australia. Ornamentation is common in lava tubes, with solidified dribbles of lava and spatter from popping bubbles decorating the walls.

Lava stream in tube, Surtsey Island, Iceland.

FIELD NOTES

Undara lava tubes, Queensland, Australia
Thurston Lava Tube, Big Island Volcanoes NP, Hawaii, USA
The Shoshone Ice Caves, Idaho, USA
Family caves of Rapanui people, Easter Island

Sections of the roof may collapse giving access to the tube, and eventually, the entire roof may collapse, forming a long depression. Lava tubes often have flat floors, created when later flows drained through the cooling tube. Sometimes, several flows will fill the tube completely, leaving horizontal layers. Occasionally, benches are seen along the sides of the tubes, revealing past "flood" levels. In cold areas, ice may form in lava tubes and persist all year round.

Lava tubes may also be found on any recently formed islands, such as Hawaii and Tahiti. On Easter Island, the family caves in which the Rapanui people stored relics, are, in fact, lava tubes. In Hawaii, the Thurston Lava Tube, complete with stairs and interior lighting, has been opened to the public.

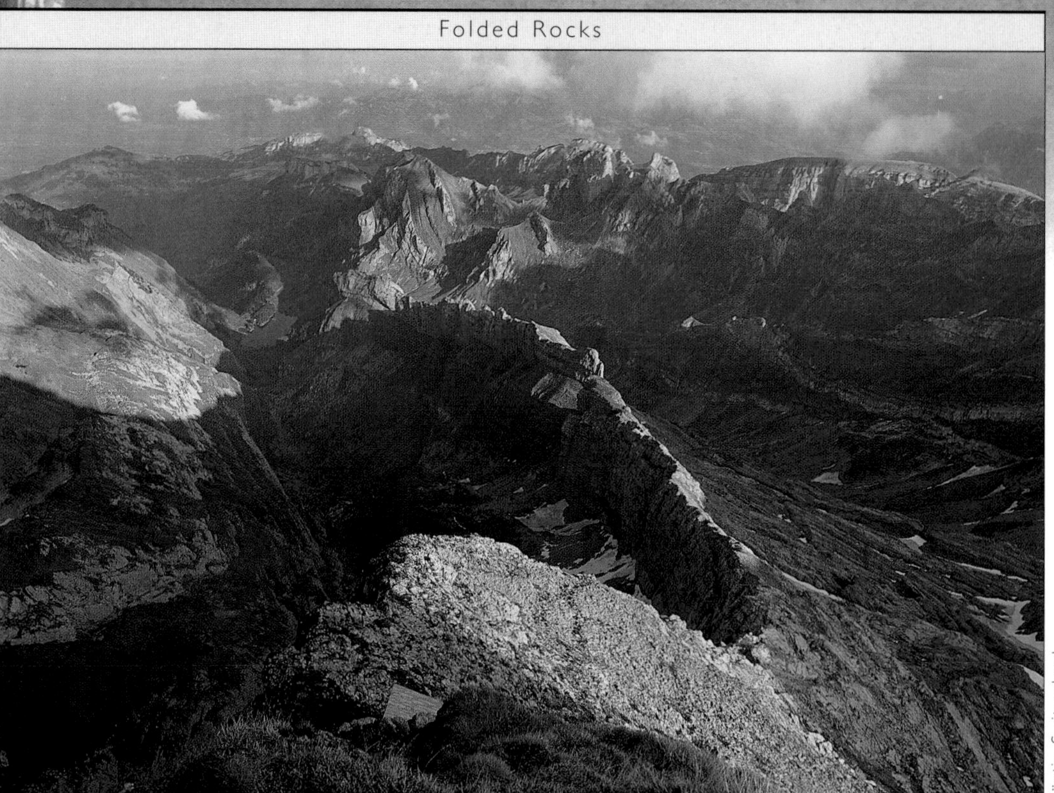

Mt Säntis, Switzerland

Folded Rocks

Sediment deposited in water forms rocks with uniform flat, horizontal layers. Subsequent geological events may fold these layers, changing their angle. Erosion working on inclined sediments makes different scenery from that formed in unfolded sediments. Compare the scenery produced by the horizontal layers of the Grand Canyon, Colorado, USA, with that produced by the sloping layers of the European Alps, or the Andes, in South America.

When continents collide, horizontal compression can incline rock layers at angles up to 90 degrees. Folding on a continental scale forms fold mountains. The European Alps and the Himalayas are examples of currently active continental collision folding, while the Appalachian Mountains of the USA are the eroded remnants of an ancient continental collision.

Localized folding of rocks may occur when buoyant masses of magma and salt rise through overlying sediment layers, forming circular domes. Rocks also fold when dragged along a fault. These folds often reveal the movement of the fault, as seen along the San Andreas Fault, in California, USA.

Folds can be open, such as the gently sloping folds near Barstow, in California, USA. They can also be more tightly folded, such as seen in the dramatic folds exposed in the Jurassic limestone of Stair Hole, at Lulworth Cove, in Dorset, UK. Here, the once-horizontal beds have been folded into spectacular S-shapes. Very small, tight folds develop in deeply buried rock layers when they become warm and plastic.

FIELD NOTES

Andes, South America
Zagros Mountains, Iran
European Alps
Rocky Mountains, North America
Ural Mountains, Russia and Ukraine
Stair Hole, Dorset, UK
Himalayas, Nepal

Tight folds in Hammersley Gorge (center), Western Australia. Open folded hills in north-west Argentina (below).

Canyonlands National Park, Utah, USA

Gorges and Canyons

Gorges are deep, steep-sided slots, cut below the surrounding countryside by water. Special conditions are required to produce the deep floor and steep sides, as valleys usually widen faster than they deepen. The existence of a gorge indicates either active uplift, or some recent event, such as a fault or a volcanic eruption. Canyons cut into hard metamorphic rocks maintain their shape well.

Cheddar Gorge (right), UK. The Colorado River, USA (below).

The Grand Canyon in Arizona, USA, developed from the slow uplift of the Colorado Plateau. The Colorado River kept pace with the uplift and still maintains its base level. The Grand Canyon keeps its remarkable shape because it is located in arid country, where erosion of the gorge sides is slow. Waimea Canyon on Kauai Island, Hawaii, is a volcanic canyon, cut into soft ash and occasional lava layers by heavy rainfall.

FIELD NOTES

Waimea Canyon, Kauai Island, Hawaii, USA
Grand Canyon, Arizona, USA
Verdon Gorge, Provence, France
Cheddar Gorge, UK
Katherine Gorge National Park, Northern Territory, Australia
Yangtze Gorges, China

Canyons can form from the widening of fractures in sandstone, as occurred in the Blue Mountains, in NSW, Australia. Joints formed some 80 million years ago, and were widened into gorges by recent uplift.

■ Wind River Range, Wyoming, USA

Glacial Valleys

As glaciers grind their way through valleys, rocks become embedded in their bases and scrape a characteristic U-shaped path. The width of a valley glacier does not change downvalley, as do rivers, so glacial valleys are uniformly wide between side tributaries. Loose rock fragments are continuously falling into the valley, mostly down into the gap between the ice and the valley sides, ensuring that erosion of the valley sides keeps pace with the valley bottom.

Smaller tributary valleys carry smaller glaciers, and so wear down more slowly. As a result, they often end abruptly, well up the main wall of the valley. These are called "hanging valleys" and the glaciers that created them are known as "hanging glaciers". The ice and rock debris of a hanging glacier avalanches onto the main glacier, depositing rocks along its edge, and sending shattered ice across its surface.

Retreating glaciers melt back up their valleys, past hanging glaciers. Hanging glaciers persist longer, having greater elevation, and so continue to enter the main valley after the main glacier has melted past. Their terminal moraines, or rock piles, collect on the floor of the main valley, forming rock fans.

FIELD NOTES

Glacial activity occurs at high altitudes and high latitudes.
Glacier National Park, British Columbia, Canada
The fiords of Norway
The Andes, Chile and Argentina
The Himalayas, India and Nepal
The European Alps

A precariously balanced rock, deposited by a glacier (left), Mer du Glace, French Alps. Briksdalsbreen Glacier (centre), Norway.

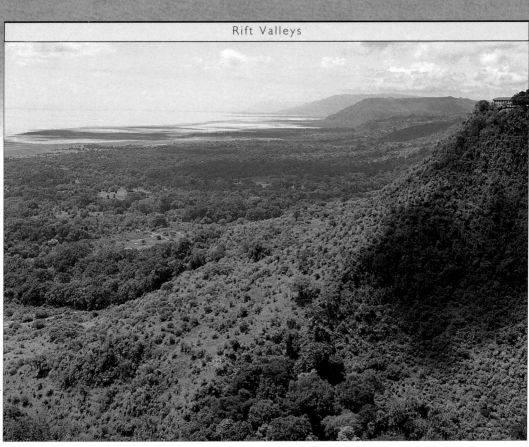

Lake Manyara, Tanzania, Africa

Rift Valleys

Rift valleys form mid continent, directly above divergent convection zones. They are the first step in continental break-up. Continental rifting begins when a dome develops and the surface cracks. Three splits form, radiating from the center at roughly 120 degrees to each other. These intersecting rift valleys extend in length and gradually widen further.

Fissures open as the rift valleys form and basalt lava rises to create a floor in the rift, well below sea level. High scarps are common features of rift margins and drainage from surrounding areas creates lakes on the rift floor, such as Lake Victoria and the several other lakes of the East African Rift. Volcanoes develop on rift floors, creating beautiful and unusual volcanic wetland scenery.

A rift valley at Almannagjá, Thingvellir, in Iceland.

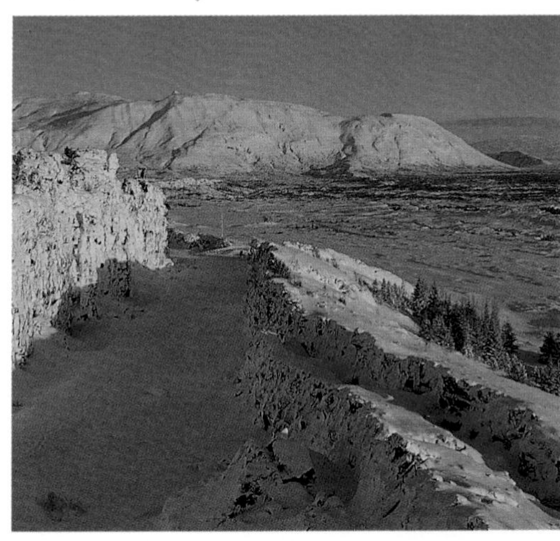

The East African Rift, the Red Sea Rift, and the Gulf of Aden Rift are three arms that originally formed over a dome at their point of intersection. When one rift reaches the ocean, sea water enters and evaporation causes salt to be precipitated. This process results in the buildup of evaporite deposits. In the East African Rift, major volcanic activity blocked off the East African arm of the rift from the ocean, making the salt deposits accessible.

The East African Rift valley contains some of the earliest traces of human development. It has been suggested that rifting in Africa created a series of rapidly changing environments which, in turn, provided the conditions for humans to evolve.

FIELD NOTES

East African Rift, Africa
Red Sea Rift, Africa
Gulf of Aden Rift, Africa
Baikal Rift, Russia
Rhine Graben, Germany
Oslo Rift, Norway
Rio Grande Rift, New Mexico, USA
Iceland Rift

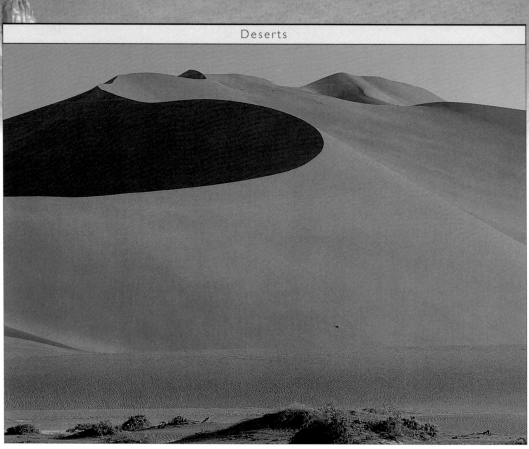

Namib Desert, Namibia

Deserts

A desert is a region where there is very little available water. This includes areas such as Antarctica, as well as the dry, sandy regions people more often recognize as deserts. Low rainfall in deserts often results in sparse vegetation, with landforms standing out in sharp relief, giving a clear picture of the geology of the region.

Many deserts occur mid-continent because the winds have released all their moisture by the time they reach so far inland. These dry winds erode desert landscapes into spectacular shapes and build vast sand dunes. Variable wind directions create arcuate, curved, sand dunes, but if wind direction is consistent, straight sandhills up to 75 miles (120 km) long form, as in central Australia. Strong winds strip sand from among pebbles, leaving a layer of pebbles exposed on the surface. These are often coated in iron oxide and have a red-brown polish called desert varnish.

"A Sand Wind on the Desert" (below), 1821, in northern Africa. Gibson Desert wildflowers (below left), Western Australia.

Some deserts, such as the Atacama Desert in Chile, are entirely natural. Here, dry offshore winds from the Andes Mountains blow across the region and out to sea. They are so constant that rain has not fallen there in living memory. In other cases, the activities of human habitation have contributed to desertification.

FIELD NOTES

Kalahari Desert, Botswana
Sahara Desert, North Africa
Atacama Desert, Chile
Central Australian deserts
Sonoran, Great Basin, Mojave, and Chihuahuan Deserts, USA
Gobi Desert, Mongolia and China
Thar Desert, India and Pakistan

Colorado Plateau, Utah, USA

Crossbeds

Crossbeds are inclined layers seen within otherwise horizontally deposited beds of sandstone. When bedding is horizontal, the differences in color and composition between consecutive layers of sandstone and mudstone are obvious. In the case of crossbed layers, however, the compositional difference between the layers is not nearly so apparent.

Crossbeds are formed by the movement of sand in dunes by wind or water. Sand grains roll or bounce up a dune, then collapse down its lee side as an avalanche, adding a discrete layer—a crossbed. Between the deposition of each sand layer, fine dust or mud is deposited in the lee of the dune. Later dunes may then cover the earlier dunes.

When the grains cement into rock, the thin, fine layer allows the sandstone to break along the crossbed surfaces, particularly when compression has turned the fine sediment to mica. Erosion works at different rates on layers of various hardness, revealing the crossbeds.

The distinctive pattern of crossbedding in Navajo Sandstone (left and below).

Wind-deposited dunes have undulating tops, and sandstone formed in this way reveals the wind-blown shape of desert dunes. River-deposited beds have flat tops, and the crossbeds slope from top to bottom of each bed. During a flood, the river picks up sediment, but after the peak level, this is dropped. Large volumes of sand are then deposited quickly with the rapid fall in stream velocity. Sand is dumped on the river bottom, slumps down dune faces, and one sand wave covers another, forming flat crossbedded layers.

FIELD NOTES

Crossbeds are found on all sedimentary basins
Zion Canyon, Utah, USA
Blue Mountains, NSW, Australia
The Old Red Sandstone, Devon, UK
The Karoo Sandstone, South Africa

Whakarewarewa Thermal Reserve, Rotorua, New Zealand

Geysers

The term geyser comes from the Icelandic word *geysir*, meaning spouter or gusher. A geyser is the spectacular exit of water and steam from a subterranean cavity through a vent in the ground. The continuous discharge of water from geysers is probably not possible, although steam jets issuing from hot ground areas are related and some in Iceland and Hawaii are harnessed to generate geothermal power.

Geysers occur only in volcanic areas with hot rock close to the surface. They are most common above subducting zones and hot spots, where temperatures near the surface exceed boiling point. Water discharge is cyclic, and is produced by a sequence of events unique to each geyser.

Tall geysers occur only under special conditions when water steadily fills a cool cavity near the surface. Beneath this cavity, and connected to it, is a hot-walled cavity that receives the overflow from the upper cavity. The walls of the lower cavity are well above boiling point and they turn the overflowing water to steam, which generates a vast and sudden increase in volume.

The water in the upper cavity is then forcefully ejected by the steam. Once the water is gone, the steam is free to escape, no more steam is made, pressure falls, and the show is over. Before the next performance, the system must rebuild to flash point.

Explorer John Wesley Powell (left) at Yellowstone in 1869. Beehive Geyser (below), Yellowstone, USA.

FIELD NOTES

Yellowstone National Park, Wyoming, USA (300 geysers, including Old Faithful, which shoots every 90 minutes)
Hawaii, USA
Rotorua, New Zealand
Iceland

Wolfe Creek meteorite crater, Western Australia

Meteorite Craters

A meteor describes any material traveling through space. When such an object strikes the surface of a planet, it is known as a meteorite. There are craters made by meteor impacts all over the Moon, Mars, and Mercury, and on all other dry planets and satellites of the solar system. The Earth has had its share of these, but plate tectonics and erosion continuously erase the record. As a result, major craters still visible tend to be in old, dry areas.

Gosse Bluff, Central Australia, a 130-million-year-old crater with a giant "shattercone" in the foreground.

Because meteors strike the Earth at such high velocity, their craters are much larger than the meteors that made them. Meteor Crater in Arizona, USA, is ¾ mile (1.2 km) in diameter and 600 feet (180 m) deep, but is thought to have been made by a meteorite only 100 feet (30 m) in diameter traveling tens of thousands of miles per hour on impact. This event took place 25,000 years ago, and since that time, most of the rim of the original crater, now about 100 feet (30 m) high, has been eroded, even in such dry conditions as prevail in Arizona.

The high compressional shock of a meteor impact creates radiating supersonic waves. The heat of impact fuses rock, forming glass. Deeper rock is shocked, producing an unusual cone-in-cone structure, characteristic of high-impact sites. Much material is ejected on impact, including chemicals vaporized from the meteor. Some material from the Acraman Crater in Australia has been found more than 160 miles (250 km) from the impact site.

FIELD NOTES

Meteor Crater, Arizona, USA
Wolfe Creek, Henbury, and Gosse Bluff, NT, Australia
The Acraman Craters, South Australia
Chicxulub Crater, Yucatán, Mexico

Every truth has two sides; it is well to look at both, before we commit ourselves to either.

"The Mule", *Fables*
AESOP (c. 620–560 BC), Greek fabulist

CHAPTER SEVEN

FOSSILS FIELD GUIDE

USING *the* FOSSILS GUIDE

This field guide introduces some common fossil types as well as special sites where finds of great significance have been made.

Whether you want to get out into the field and feel the thrill of finding your own specimens, or just understand more about fossils, this guide provides the amateur with a useful introduction.

This panel *refers to the biological classification of the fossil group.*

Mollusca: Cephalopoda

Dactylioceras (ammonite), Jurassic

Mollusca: Cephalopoda

Ammonites: Asteroceras (large shells) and Promicroceras (small shells)

Nautiloids and Ammonites

Cephalopods are the most active and highly developed marine carnivores. Although many forms had coiled or straight shells, some, such as octopus and squid, appear never to have developed major skeletons. These animals have complex eyes and brains, tentacles with sucker pads, and parrot-like beaks with which they consume their prey. Cephalopods use jet propulsion to move through water. The oldest known octopus is from the late Cretaceous of Palestine.

Ammonites, which are now extinct, and nautiloids were perhaps the most agile and intelligent invertebrate carnivores, living in ocean waters of all depths. Since these animals lived in the water column rather than on the sea floor, the rocks in which their shells are found may not actually reflect the environment in which they lived. Ammonites and nautiloids had straight or coiled chambered shells, in which the old living chambers were either empty or lined with mineral (cameral) deposits that acted as ballast.

FIELD NOTES

Cretaceous limestones, Texas and Kansas, USA, and Europe
Triassic rocks, California, USA
Devonian rocks, New York, USA
Permian limestone, central Europe
Cretaceous lagoonal shales (dwarf forms)
Jurassic rocks of the Himalayas

€ O S D C P Tr J K T

Older chambers were separated by walls called septae that left distinct suture marks where they joined the outer shell. In nautiloids, these are seen as a simple sigmoid curve, but in ammonites, there are many types of complicated, jagged suture lines. Detailed lineages of ammonite evolution have been arrived at by studying the suture patterns.

The abundance of ammonites and their presence in many types of sedimentary rock makes them one of the most useful biostratigraphic indicators of all. Coiled cephalopods are very common in the middle and late Paleozoic and the Mesozoic. Ammonites became extinct at the end of the Cretaceous. The small numbers of nautiloid surviving after this time make them relatively rare in Cenozoic rocks.

Nautiloids and ammonites were creatures of open tropical seas, and are abundant in limestones from tropical reefs. They may also be found in shales both from deeper water and from restricted lagoons. Some of the more interesting forms are dwarf species that presumably lived in the hyper-saline waters of such lagoons, but most are found in limestone.

Like the bivalve mollusks, the shells were made of the pearlescent calcareous mineral aragonite. In most fossil shells, the aragonite has been converted to calcite and the pearly luster has been lost, but in some forms, usually preserved in shales, the shells retain their beautiful luster. Unfortunately, the shells of most ammonites were so thin that they are

228

Color diagrams *supplement the text by showing a cutaway of a basic fossil structure. These may include a depiction of the living animal.*

Field Notes panels

Collecting Sites: *A list of a few choice locations around the world where these fossils can be collected, giving information on the geological nature of the area where most relevant. For more detailed site localities, contact your nearest club or society (see p. 276).*

€ O S D C P Tr J K T

Timeline: *From Cambrian to Tertiary. A bar across the timeline indicates when these fossils were living—darker shading equals greatest abundance. Letters refer to Geological Periods outlined on p. 91.*

The text *provides information on the biological evolution of various fossil forms and what to look out for in the field.*

***This panel** indicates the Geological Period and environment of the site/s.*

*The **illustrated banding** at the top of the page identifies either general categories of fossils or sites of particular importance—see the key below right.*

Jurassic–Cretaceous Continents

Apatosaurus and Diplodocus bones, Colorado, USA

Jurassic–Cretaceous Continents

Centrosaurus bone bed in Alberta, Canada

Dinosaur Sites

years, from the late Triassic
e Cretaceous, dinosaurs
Their fossilized remains
ery continent, including
tribution is skewed to
ns. One of the most
rt of North America.
usually a poor place for
g the Mesozoic Era,
much of western
ed by the flood plains
hat drained into an
ns were home to
d other animals.
nosaur Triangle,
Utah
do,

Centrosaurus bone bed excavation.

FIELD NOTES

Dinosaur National Monument, Utah, USA
Cleveland-Lloyd Quarry, Utah, USA
American Museum of Natural History, New York, USA
British Museum (Natural History), London, UK
Royal Tyrrell Museum, Drumheller, Alberta, Canada

€ O S D C P Tr J K T

such as *Diplodocus*, *Apatosaurus*, *Camarasaurus*, and *Barosaurus*, but there are other plant-eaters, such as *Stegosaurus*, *Dryosaurus*, and *Camptosaurus*, as well as meat-eaters, such as *Allosaurus* and *Ceratosaurus*. This unusual abundance is a result of bodies collecting in a sandbar as they were washed downstream in a huge river. Dinosaur National Monument is open to the public and many fine specimens can be seen, still partly embedded in the quarry face. Other sites in the Dinosaur Triangle include the Cleveland-Lloyd Quarry south of Price.

Younger, Cretaceous-age deposits to the west and north of the Dinosaur Triangle contain a different set of dinosaurs. Dinosaur Provincial Park in Alberta, Canada, has been set aside to protect its vast fossil deposits. Coupled with the Royal Tyrrell Museum near Drumheller, this area is a mecca for dinosaur enthusiasts because of the beautiful specimens that are on display in the museum or have eroded out of the ground.

The Alberta specimens come from toward the end of the dinosaurs' reign and include a variety of strange duck-billed dinosaurs and various types of horned dinosaur.

In one bone bed, lie the remains of thousands of individuals of a horned dinosaur similar to *Centrosaurus*. This species appears to have lived in huge herds, perhaps comprising tens of thousands of individuals, that migrated across the continent in search of food. These migrations entailed dangerous river crossings during which many animals died, their remains swept away by the river. The carcasses collected where the flow of water slowed, producing the bone beds we find today.

The skull of Camarasaurus (top), and a Triceratops skeleton (right).

255

...rved as fossils.
...ey are found as
...olds, preserving the
...ature pattern so
...an identification.
...straight nautiloids,
...at a time lying in the
...are found in the Atlas
...occo. These are often
...ck shops and are
...any remaining shell is
...mmonites occur in the
...ay be only a fraction of
...ss, although some

monite ctylioc

***Color field sketches** supplement the text by illustrating some basic fossil types.*

Field Notes panels

*✦ **Location:** The site or sites where these significant fossils have been found. These are usually inaccessible to amateur collectors.*

*🏛 **Collections:** Institutions such as museums and universities where significant finds can be viewed.*

€ O S D C P Tr J K T

***Timeline:** A bar across the geological timeline shows the age range of the fossils. Where this is a specific site, an arrow indicates its age. The letters refer to the Geological Periods* *(see p. 91).*

***Secondary photographs or illustrations** show some of the fossil finds from this particular site or type of site.*

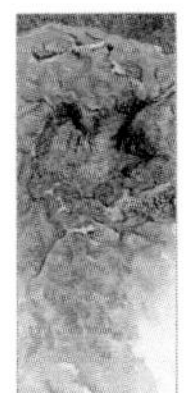

Collecting Fossils

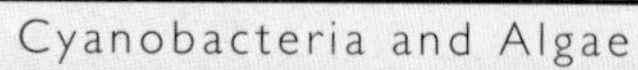

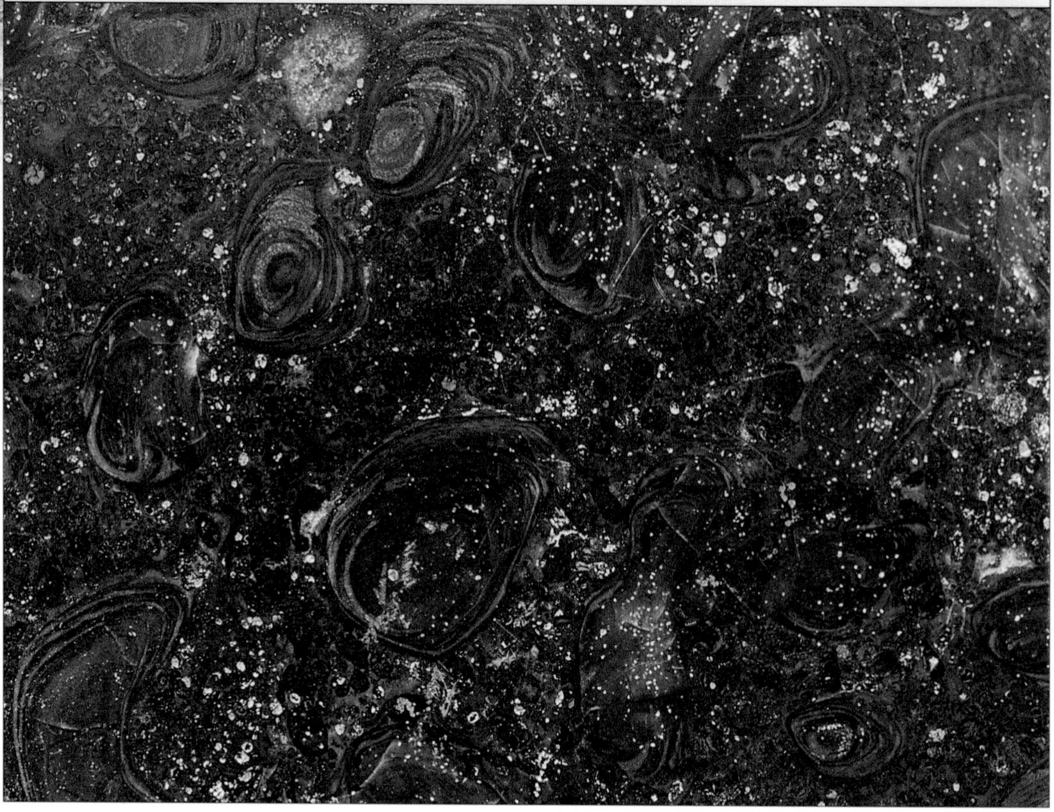

Cut section through a stromatolite

Stromatolites

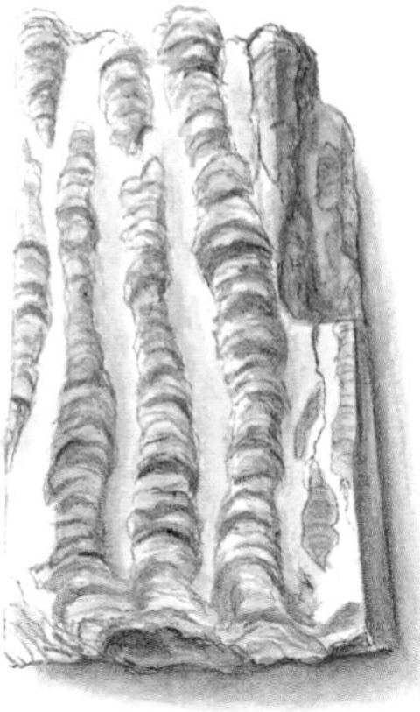

fossil stromatolite

Stromatolites are layered structures produced by the growth of so-called algal mats, primarily of various sorts of algae and cyanobacteria, which, like plants, photosynthesize their food. They are among the oldest fossils, some dating back more than 3,500 million years. Ancient stromatolites lived in waters with a range of salinity. Needing light, they are limited to shallows where sunlight can penetrate.

Stromatolites are prokaryotic (simple-celled) organisms and their evolution has been very slow. Many forms lived during the Precambrian, creating some of the earliest reefs, but today, they are limited geographically and morphologically. Good examples of living stromatolites are found in many regions, such as the hot springs in Yellowstone National Park. Fossil stromatolites may be common in carbonate rocks younger than 3,500 million years, but are not common after the early Paleozoic.

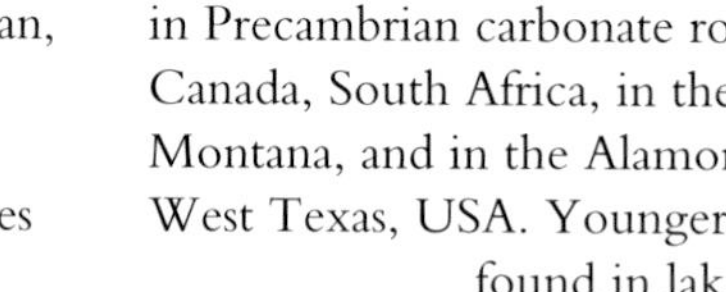

Look for stromatolites in Precambrian carbonate rocks of southern Canada, South Africa, in the Altyn Limestone of Montana, and in the Alamore Formation of West Texas, USA. Younger stromatolites can be found in lake deposits of the Green River Formation in Wyoming, and in the Canning Basin of Australia. Look for road cuts or weathered vertical surfaces where the delicate layers of stromatolites can be seen in cross-section as domes or bulbous forms. They also occur as layered planar structures with interlayered carbonate and silica. Some stromatolites are mineralized and beautifully colored.

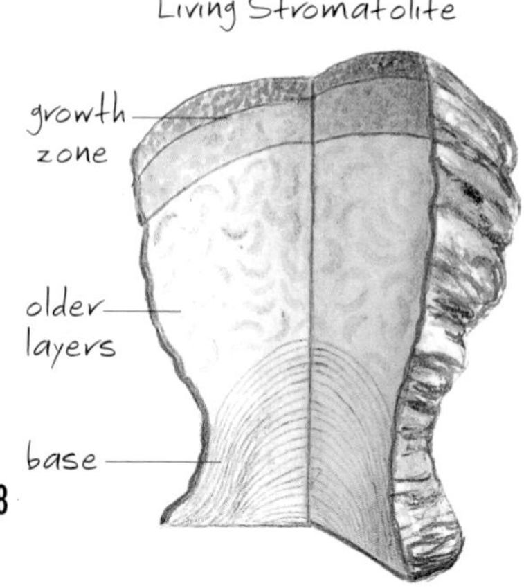

FIELD NOTES

Van Horn, Texas, USA
Waroona Group, North Pole, Western Australia
Manitounuk Island, Hudson Bay, and Beresford Lake, Manitoba, Canada
Living examples: Green Lake, New York, USA
Shark Bay, Western Australia

Є O S D C P Tr J K T

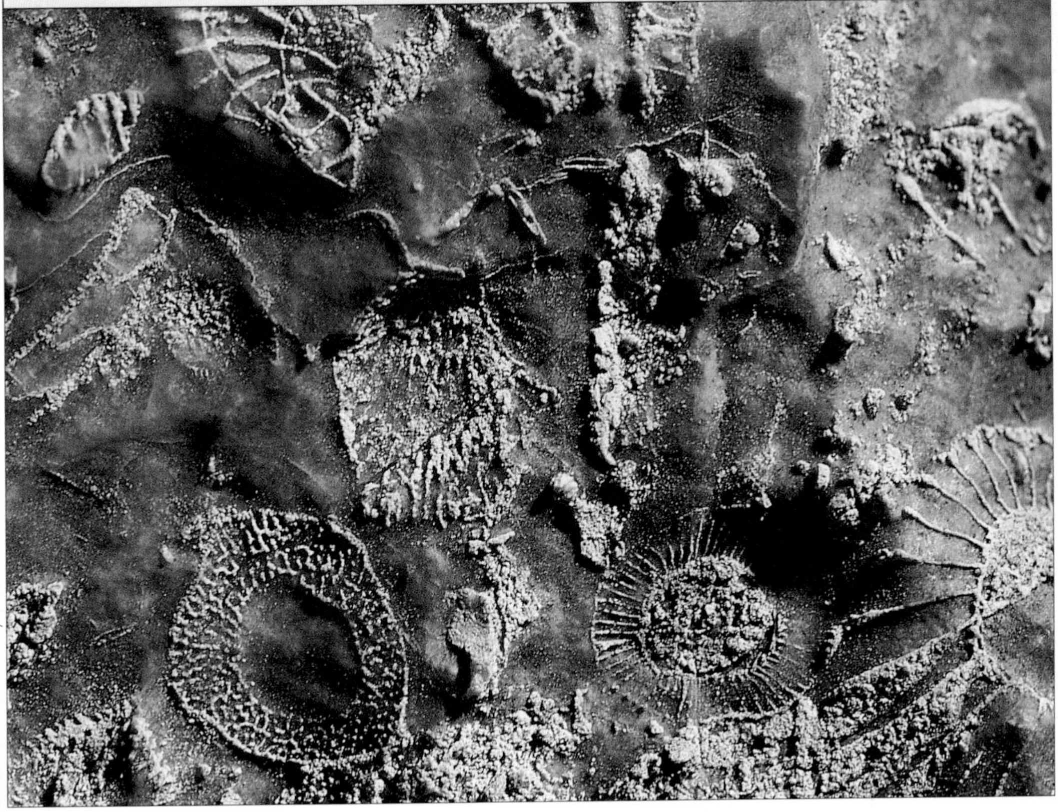

Cambrian archaeocyathids in limestone

Sponges and Relatives

The simplest multicellular organisms are the sponges and related forms, such as stromatoporoids and the early Cambrian archaeocyathids. Sponges, which still thrive today, have a two-layered body with an internal skeleton composed of soft spongin, or of microscopic calcite or silica particles (spicules). Most sponges live in marine waters from shallow to deep. Tall, vase-like, or branching sponges are found in calm water, while shorter, encrusting forms live in areas with fast currents.

Sponges have changed little in the past 600 million years and are found throughout the fossil record. These do not include the familiar bath sponges, as they are too soft to fossilize. Double-walled sponges, such as archaeocyathids, flourished only in the early Cambrian and then became extinct. Stromatoporoids (Cambrian to Cretaceous) formed layered calcareous skeletons that resemble sheets or cabbage heads.

Although relatively rare in the fossil record, sponges may be locally abundant. Whole sponges are found in Paleozoic limestones, but the best place for whole, easily recoverable skeletons is in shale. Siliceous spicules are present in many limestones. Most sponges are preserved only as single spicules—the siliceous types can be extracted from limestone by dissolving the matrix in weak acid. Some siliceous and calcareous sponges are preserved as entire fossils, ranging from peppercorn-size forms up to some the size of clothes baskets.

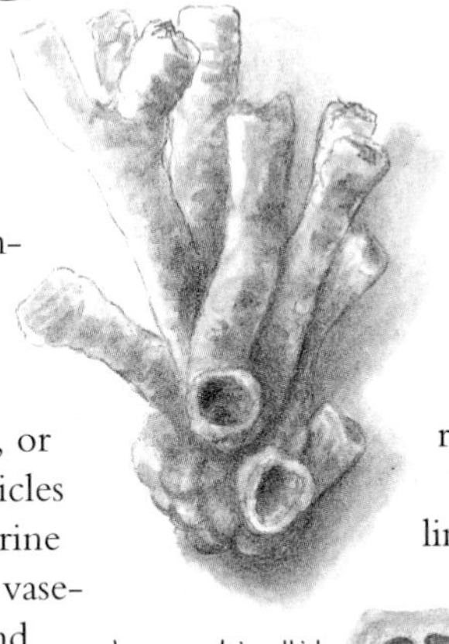

hexactinellid sponge

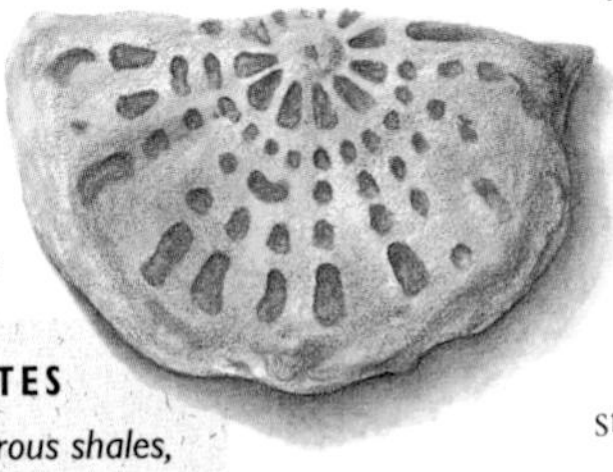

hyalosponge

FIELD NOTES

Late Carboniferous shales, central North America
Lower Cretaceous limestones, UK
Jurassic limestones, Germany
Permian Reef, Texas–New Mexico
Devonian limestones, UK and Pennsylvania, USA
Ordovician limestones Kentucky, USA
Cambrian limestones, Nevada, USA

Є O S D C P Tr J K T

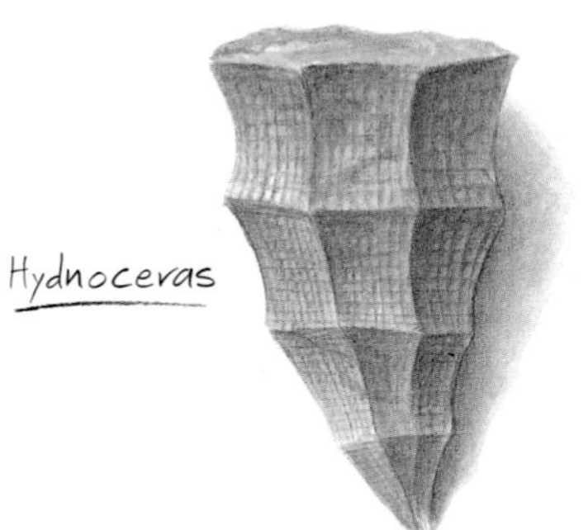

Hydnoceras

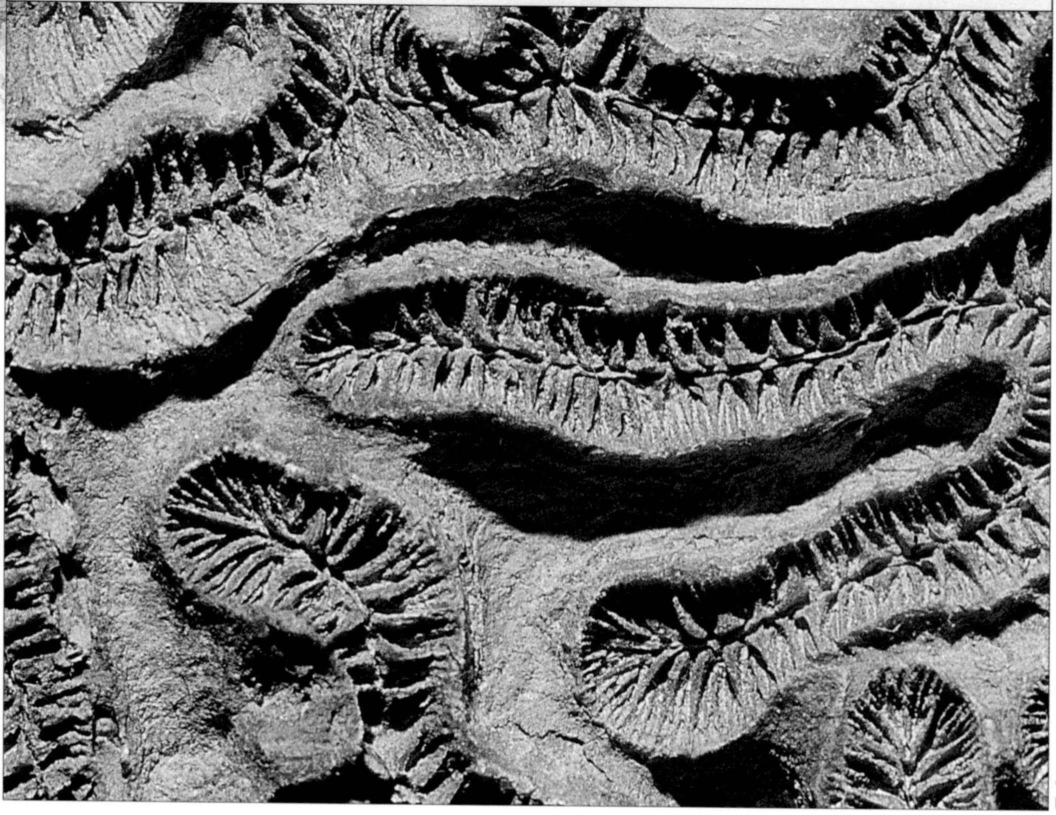

Rhipidogyra, scleractinian coral, Jurassic

Corals

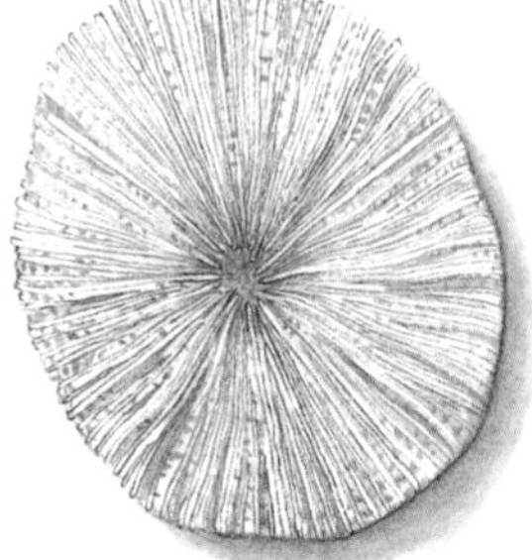

scleractinian *Micrabacia*

The phylum Cnidaria consists of jellyfish, corals, sea anemones, and the Anthozoa or "flower animals". These relatively simple animals have complex folds (mesentaries) in their gut cavities that are frequently reflected in their skeletons. They have a single opening, or mouth, to the gut cavity where food enters and solid wastes are eliminated. Cnidarians are carnivores, and immobilize or kill their prey with specialized stinging cells (nematocysts) on their tentacles.

Although some fossil jellyfish are preserved as impressions from as long ago as the late Precambrian, it is the corals that are most frequently found as fossils. They have a dense, calcium carbonate skeleton and occur either as massive colonies of thousands of cloned individuals, or as single, usually cone-shaped corralites, commonly called horn corals. Modern corals live in relatively clean, warm (usually tropical) sea water of normal salinity, and may form reefs that cover vast areas. Ancient corals also seem to have preferred warm, tropical ocean environments. Unlike their colonial cousins, solitary corals are often found in shales, indicating that they preferred soft, muddy bottoms. But horn corals are also found, along with the colonial forms, in areas that once had hard sea floors. Primitive tabulate and rugose corals lived

FIELD NOTES

Most middle and late Paleozoic limestones
Lower Carboniferous rocks, UK
Late Carboniferous shales of mid North America
Carboniferous rocks, Missouri, USA
Devonian limestones, New York, USA, and Devonian rocks, Kentucky, USA

Є O S D C P Tr J K T

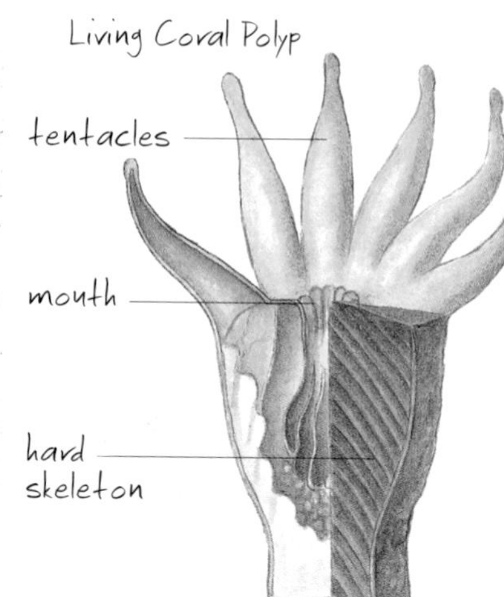

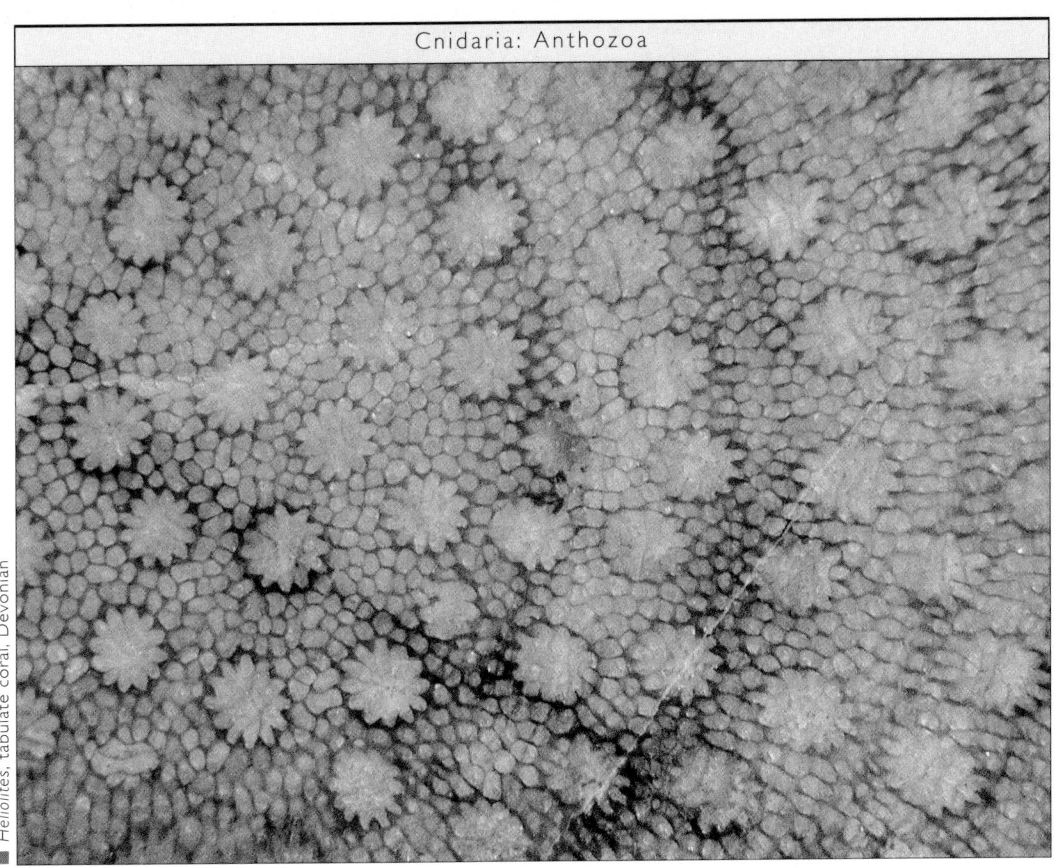

■ *Heliolites*, tabulate coral, Devonian

during the Paleozoic, while only the more advanced scleractinid corals are found in Mesozoic and Cenozoic rocks. The tube-shaped tabulate corals frequently resemble a series of joined organ pipes or pan pipes. Their name comes from the simple horizontal divisions (tabulae) that block off older portions of the skeletal tubes during the growth of the individual coral. Rugose and scleractinian corals have vertical, radially aligned walls (septae) that support the mesentaries and project into the coral living space. These two major groups differ principally in the arrangement and number of their septae. Rugose corals are often found as horn corals, although they also occurred in colonial form, while tabulate corals are always colonial.

rugose coral

Corals are among the most common fossils on Earth. Rugose and tabulate corals are abundant in Paleozoic limestones and shales but are much easier to clean and recover from shales, where they simply weather out. Scleractinid corals are rare in the Mesozoic but are fairly common in Cenozoic limestones, where they are mostly found as massive reefs, although solitary scleractinian corals may be found. Solitary and colonial rugose and scleractinian corals occur as whole body fossils and are relatively easy to recognize because of their numerous radial septae. The most abundant fossil corals are scleractinian corals in the reefs that make up many tropical islands. By standard definition, any corals older than 10,000 years are fossils, so it is not difficult to find fossil scleractinian corals exposed on these islands in great abundance. Living corals have a symbiotic relationship with colonies of algae in their tissues, which help provide them with oxygen.

scleractinian coral

rugose coral *Arachnophyllum*

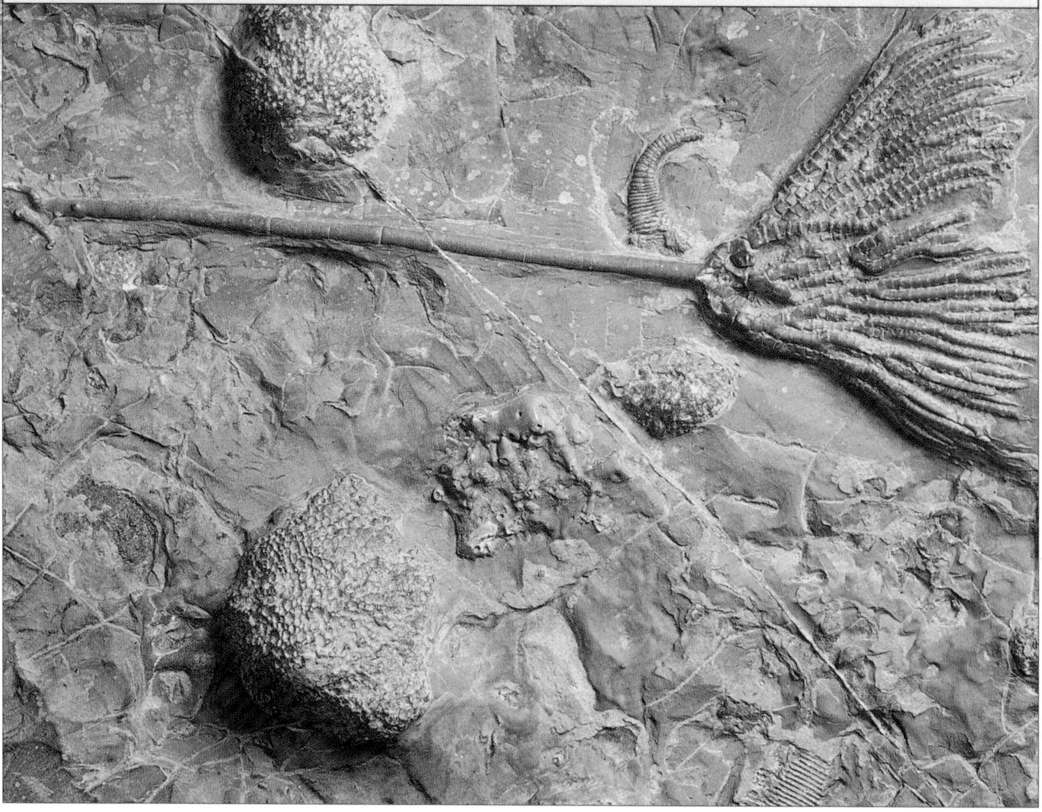

■ *Gissocrinus* crinoid and *Favosites* coral communities

Reef Communities

The massive deposits that were once reefs are an abundant source of fossils for the amateur collector. Large numbers of soft-bodied and skeleton-producing organisms inhabit reefs, although only the skeletal remains may be preserved.

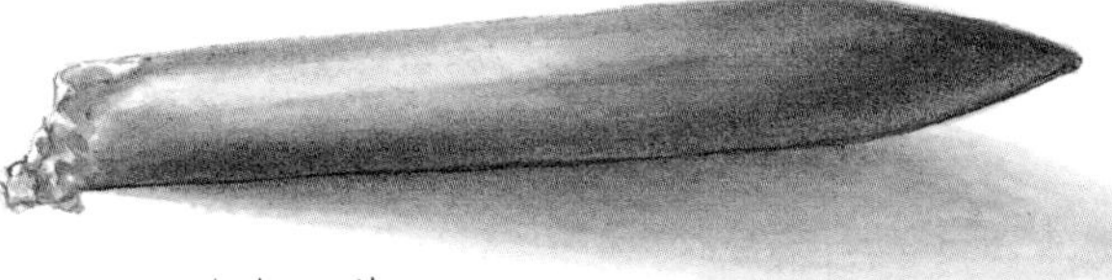

belemnite

Reefs can be up to several thousand miles long, or as small as a few square feet, and stand on hard and stable sea floor. They are usually regions of incredibly rich biodiversity in which organisms are ecologically linked in a complex and intricate food web.

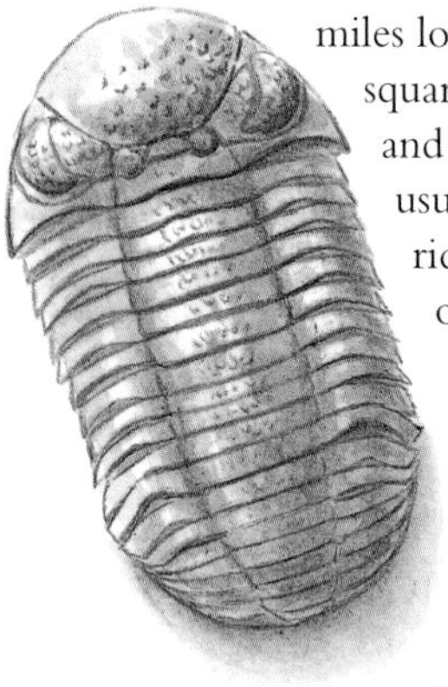

trilobite

The kinds and shapes of animals that live in reef communities may be distributed across the reef, each inhabiting sections at the depth best suited to it.

Although modern reefs are dominated by the more advanced scleractinid corals, the composition of ancient reefs has varied a great deal through time. For example, Cretaceous reefs are commonly formed of a type of bivalve called rudists, and Permian reefs may be mainly composed of brachiopods, sponges, and even foraminifera. The earliest Cambrian reefs were mostly composed of archaeocyathids, but middle and late Cambrian reefs are almost entirely made up of stromatolites.

Although small "patch" reefs, such as those oysters build, may form in temperate waters, large accumulations of calcium carbonate skeletal debris are really only possible in warm, tropical, ocean waters. Most reef organisms live within the photic zone—the top level of the water column where light can penetrate—and require clean, non-stagnant water of normal salinity.

Since reefs are composed of vast numbers of organisms, it

FIELD NOTES

Carboniferous, Illinois, USA
Middle Paleozoic rocks, England and Scandinavia
Middle Paleozoic carbonates from the Great Lakes, USA, to Arctic Islands
Edwards Formation, Texas, USA
Tropical islands
Petroleum Museum, Midland, Texas: reconstruction of a Permian Reef

Є O S D C P Tr J K T

Community of brachiopods, bryozoans, and bivalves

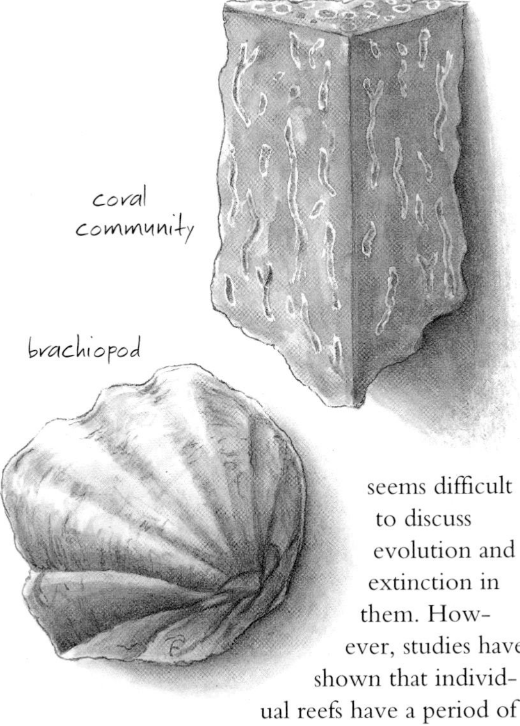

seems difficult to discuss evolution and extinction in them. However, studies have shown that individual reefs have a period of development and growth and a finite lifespan, measured in tens of thousands of years, much like an individual organism.

We also know that the composition of reefs has changed through time, reflecting the tempo and mode of evolution of reef-dwelling organisms. Reefs, especially accumulations of organic debris, are quite common in environments with high concentrations of calcium carbonate. They are particularly good sites for fossil-collecting, because of the high likelihood of finding a wide variety of different fossils.

When collecting from ancient fossilized reefs, it is worth considering the structure of the reef when it was alive. The best fossils of reef-building organisms are usually found at the edge farthest from the seaward side of the reef complex. On the seaward side, the activities of waves, especially during storms, smashed shells or skeletons into small pieces, so complete specimens are rare. The remains here consist of coarse breccia formed mostly of broken shells.

Behind the reef, in the lagoon, if there were one, conditions were quieter, but only a few skeleton-producing organisms lived there. Although excellent specimens are sometimes found in exceptional circumstances, normally there will be little more than a strand of broken shells gently washed from the reef itself.

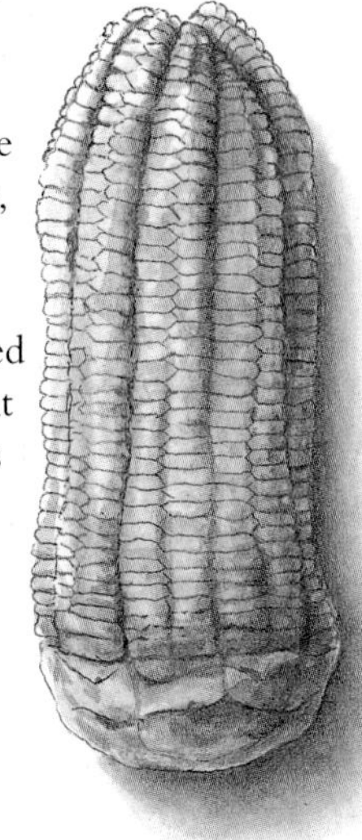

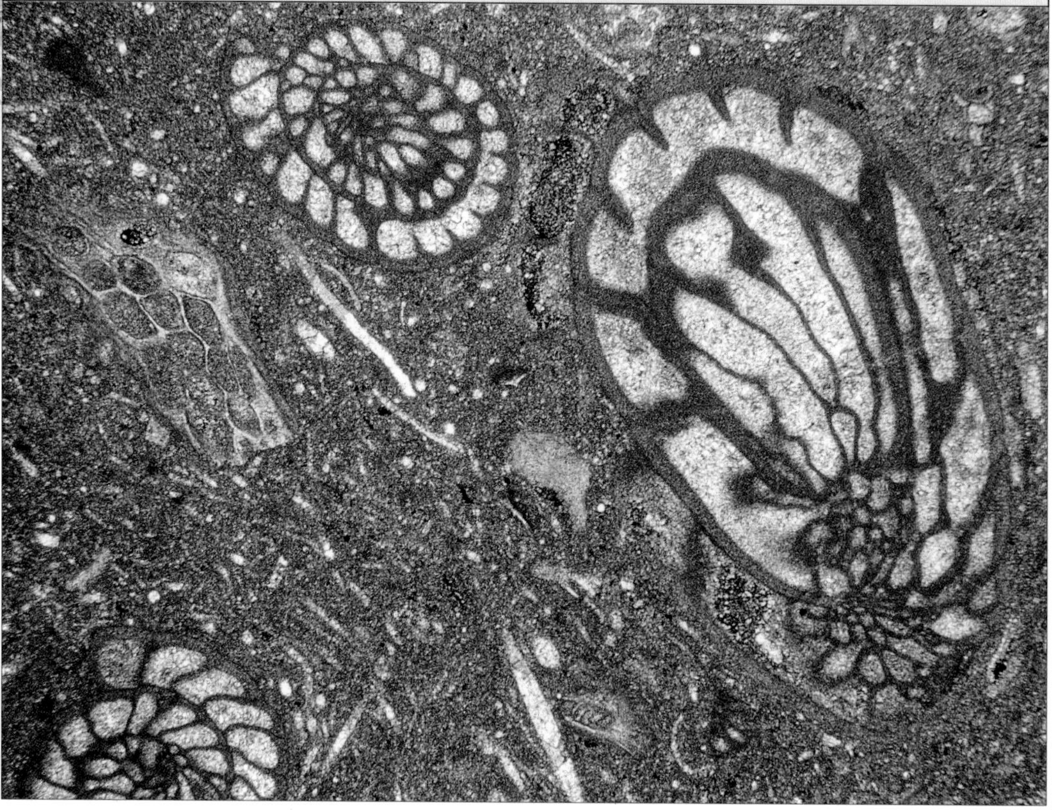

Fossil foraminifera in limestone

Microfossils

Although for the amateur, collecting microfossils is a challenge, many find it a rewarding pastime. Most microfossils are so small that they can be seen only under magnification. There are many types, but most common microfossils are the shells of single-celled organisms, such as foraminiferans and radiolarians. In spite of being the smallest skeletons on Earth, they are among the most intricate and detailed of any organism, often complex assemblies of glassy girders and spikes.

Radiolarians, with siliceous skeletons, and foraminiferans, also known as forams, with calcareous shells, are found in both fresh and sea water. Deep-sea muds are almost entirely composed of radiolarian skeletons. The evolution of many forams is recorded in deep-sea sediments, and they are useful in dating strata for geological purposes, such as oil exploration.

Forams and radiolaria are found in many limestones and are also preserved in deep-water shales. Fusulinid forams the size of rice grains are very common in some later Paleozoic rocks. Many chert formations, such as the Arkansas Novaculite, seem largely made of chert derived from radiolarians. These can be recovered in the residue after dissolving limestone, and forams after washing shales. When present, fusulinids are easy to see because they look like rice grains. Other microfossils may be present in shales and limestones. Some forams are as large as small coins.

globigerinid foram

quinqueloculinid foram

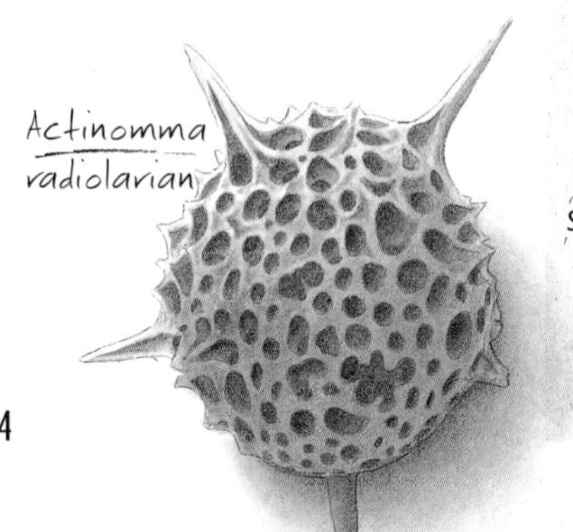

Actinomma radiolarian

FIELD NOTES

Devonian limestones, Iowa and Oklahoma, USA
Jurassic rocks, Montana, USA
Pleistocene rocks, California, USA
Carboniferous shales, Texas, USA
Silurian limestones, Oklahoma, USA
Silurian rocks, Tennessee, USA
Eocene rocks, Alabama, USA
Chalk of the UK coast

Є O S D C P Tr J K T

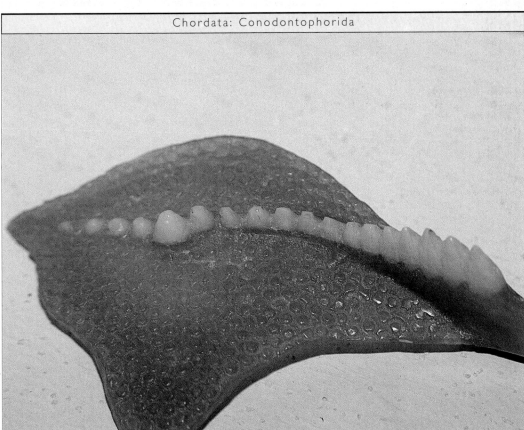

A *Palmatolepis* conodont

Conodonts

Conodonts are microscopic, up to ⁵⁄₁₆ inch (8 mm), phosphatic tooth-like structures found abundantly in Paleozoic sedimentary rocks. After hundreds of years of debate, the discovery of a conodont-bearing animal in Scotland indicates that they are the remains of early fish-like vertebrates. They are found in many different types of sedimentary rocks, representing a wide variety of marine environments, so the original animals must have been tolerant of wide environmental variation. Up until now, all we have known about these structures is that certain types appear in groups known as associations and that these associations change through time; in fact, conodonts make very important index fossils for much of the Paleozoic. Changes in their shape and associations presumably reflect the evolution of the conodont-bearing animals.

Ozarkodina

Conodonts are very common in Paleozoic limestone. If this is dissolved in weak acid, phosphatic conodonts may be left in the residue. Because these are very small, examine the residue with a microscope or strong hand lens. Conodonts are particularly abundant in Ordovician through Carboniferous rocks. They also appear in mudstones and can be removed by soaking these in detergent and water.

Look for shiny, tooth-like structures in the residue. Some are as simple as long cones, but most have elaborate blades and cusps. Extreme heat blackens conodonts, so color indicates degree of heat exposure.

FIELD NOTES

Ordovician and Silurian rocks, Missouri, USA
Carboniferous rocks, Oklahoma, Iowa, and Missouri, USA
Harding Sandstone, Colorado, USA
Ordovician shales, Minnesota, USA
Ordovician through Permian rocks, Arkansas, USA
Devonian rocks, Iowa, USA

Є O S D C P Tr J K T

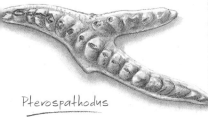

Pterospathodus

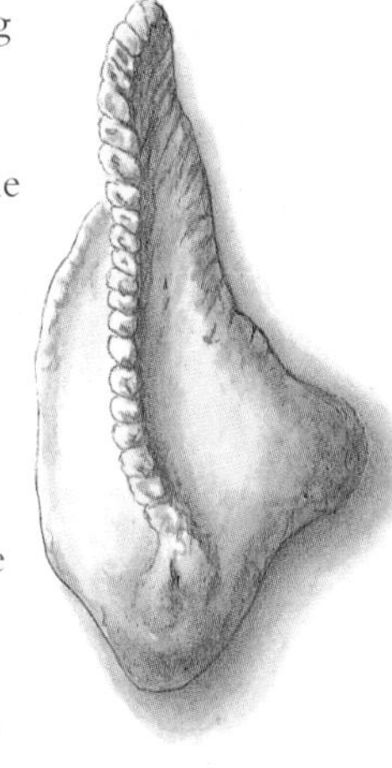

Palmatolepis

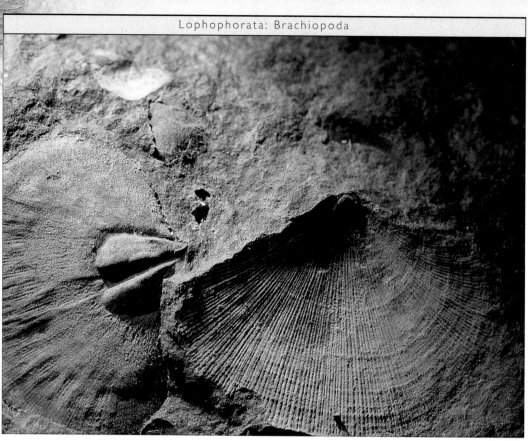

Strophomena, Ordovician

Brachiopods

The bodies of brachiopods are surrounded by two shells, or valves, that are joined along a common hinge, as in bivalve mollusks, such as clams. Unlike bivalve mollusks, brachiopods have dorsal (brachial) and ventral (pedicle) valves, while bivalve mollusks have left and right valves. So for unspecialized brachiopods, the plane of body symmetry passes through the center of the valves, with each side being a mirror image of the other.

Depending on the group, brachiopod shells are either chitinophosphastic (inarticulate), or calcareous (articulate). Inside the shells are complex sets of muscles, arranged like pulleys, that both open and close the shells.

Brachiopods usually grip the substrate with a fleshy structure called a pedicle, but many extinct forms had spiny shells that were cemented to the sea floor. Frequently the shells have a fold on one valve that fits into a groove (sulcus) on the other. As adults, brachiopods are attached, but some inarticulate forms burrow.

Although solely marine, ancient brachiopods occupied a variety of oceanic environments—some Permian reefs are largely formed of interlocking, spiny forms. Brachiopods may be found attached to sea floors or isolated in shales that represent what were originally very muddy environments. Modern inarticulate brachiopod forms burrow in sand or mud, often on mudflats that are exposed at low tide. All brachiopods are filter feeders.

First appearing in the Cambrian, they rapidly diversified into many lineages, with a wide variety of

Rhynchopora

Microspirifer

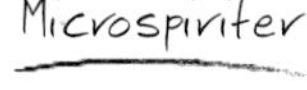

Meristella

FIELD NOTES

Carboniferous and Permian limestones, UK
Paleozoic limestones and shales, central Oklahoma, USA
Devonian rocks, New York, USA
Glass Mountains, Texas, USA
Paleozoic limestones, Appalachian Mountains, USA
Cambrian limestones, Ohio, USA

Є O S D C P Tr J K T

Paraspirifer, Devonian

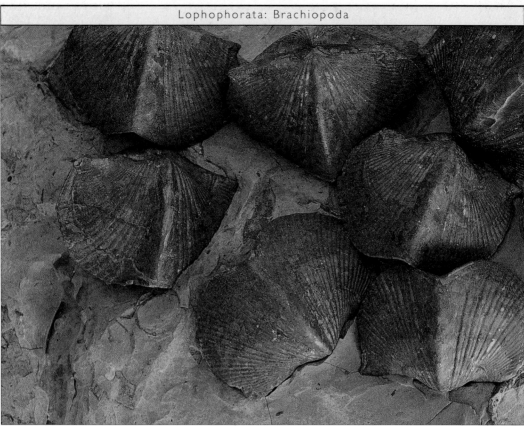

shell shapes. Among the most interesting are the wing-shaped spiriferids. The greatest brachiopod diversity occurred during the Paleozoic. They declined dramatically at the mass extinction on the Permian–Triassic boundary.

Brachiopods survive today, but they are now limited to cool and temperate waters, most commonly around Japan, Australia, New Zealand, and the North Atlantic coastline of North America. They tend to live in deeper waters. Fossil brachiopods range in size from a few millimeters to more than 10 inches (25 cm) in length, but most forms are 1–3 inches (2–7 cm). They are common as fossils in Paleozoic shales, sandstones, and limestones, and are important biostratigraphic index fossils worldwide.

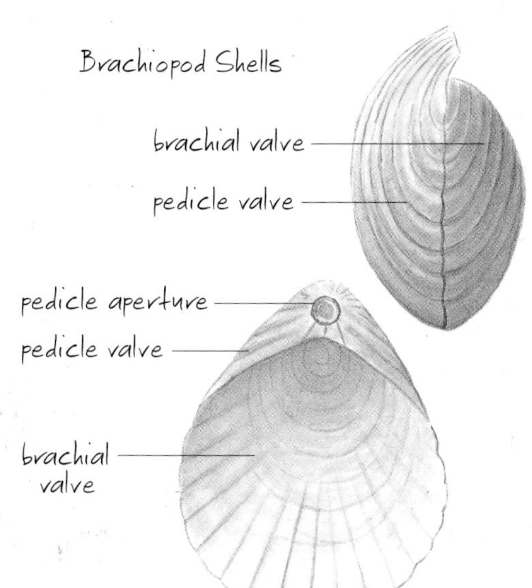

Megastrophia

Although very common in Paleozoic rocks, they can be found in much smaller numbers in Mesozoic and Cenozoic rocks. More often than not, brachiopods are found with both valves articulated, since at death the shells tend to stay closed and together. Those preserved in mudstones may be somewhat flattened by compaction, but are the easiest specimens to extract, since the clay can be washed away.

Many brachiopods had exaggerated surface folds (plications), which probably served to strengthen the shells. When brachiopods are exquisitely preserved, delicate spines can be found on the valves, particularly in the Permian productid brachiopods. In well-preserved specimens, the delicate, calcareous brachidium may be preserved within the shell—this is the structure that supported the spiral, ribbon-like, filter-feeding lophophore during life.

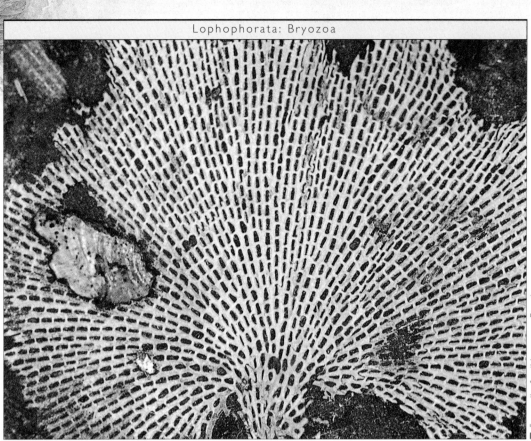

■ *Fenestella*, fossil bryozoan, Permian

Bryozoans

The "moss" animals are unfamiliar to most people, but they are common animals that live in colonies and are found on many seashores. Unlike the corals, they have a true mouth and anus, and a ciliated filter-feeding mechanism called a lophophore. Yet they are tiny animals, only about 1⁄24 inch (1 mm) long. Bryozoans prefer shallow ocean water, but are found at depths of up to 18,000 feet (5,500 m). Most colonies are only about 1¼ inches (3 cm) across, but some fossil forms are up to 2 feet (60 cm) wide.

Most bryozoan colonies are fixed, but some can move slowly around. In the Cambrian, bryozoans quickly diversified into a number of twig-like, fan-like and encrusting forms. They are very common in carbonates and shales. In some shales they are the most abundant fossil—representing what were once vast undersea meadows of the twig- and fan-like forms. These tend to be well preserved in Paleozoic limestones and shales.

The best places to look for bryozoans are where the surface has weathered. The branching twig-like forms are especially common in Ordovician rocks. Lacy or fan-like forms are abundant in the early Carboniferous and are best found on the weathered surface of limestones. On each bryozoan, look for tiny pin-pricks, which are the openings in which the bryozoan animals lived. Modern bryozoans are common in tropical and temperate climates but because of their small size, they can be extremely difficult to see.

Archimedes

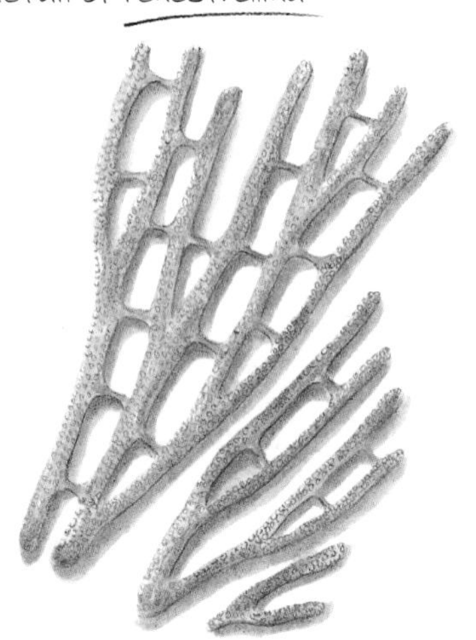

detail of *Fenestrellina*

FIELD NOTES

Silurian through Permian rocks, Kansas, USA
Ordovician rocks, Minnesota and Kentucky, USA
Eocene, Mississippi and North Carolina, USA
Carboniferous shales, Texas, USA
Devonian shales and limestones, UK

Є O S D C P Tr J K T

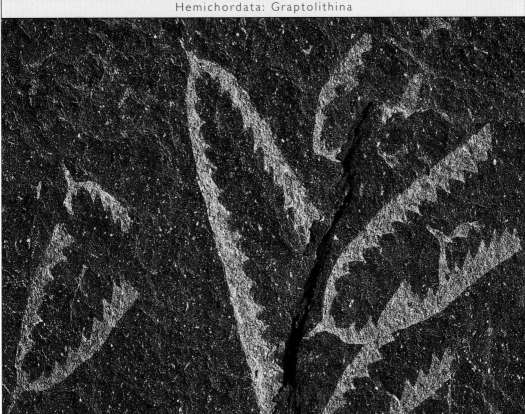

Didymograptus, "tuning fork" graptolites, Ordovician

Graptolites

Graptolites, an extinct group of colonial floating (planktonic) organisms, are quite unlike any living forms. We now place them in the phylum Chordata, with the vertebrates, but they were very unusual. Some may have been attached to the sea floor, but during the early Paleozoic, most floated in vast numbers, perhaps on or close to the ocean surface. They probably filtered planktonic microorganisms.

Climacograptus

Most of what we know about graptolites today comes from thin carbon films left in many Paleozoic shales and limestones, although some have been removed as three-dimensional structures from fine-grained limestones. A colony of graptolites is made up of many branches (stipes) with different arrangements of little cups (theca) in which the animals lived. They are important index fossils and are quite common around the world in early Paleozoic rocks. From the Ordovician through the Silurian, numbers gradually dwindled.

Graptolites are most plentiful and easiest to find in thinly layered black shales that are easy to break apart, and in some thinly stratified deep-water limestones. Open the matrix along the bedding planes and look for thin serrated lines that reflect light well. In limestones, look for serrated black strips of carbon. Dendroid graptolites built complex, branching colonies, while simpler colonies of graptoloid forms were suspended by floats (nema).

FIELD NOTES

Ordovician rocks, British Columbia and Ontario, Canada, and UK
Ordovician shales, Sweden and New York, USA
Silurian shales, Marathon Basin, Texas, USA
Silurian rocks, New York, Oklahoma, and Nevada, USA

Є O S D C P Tr J K T

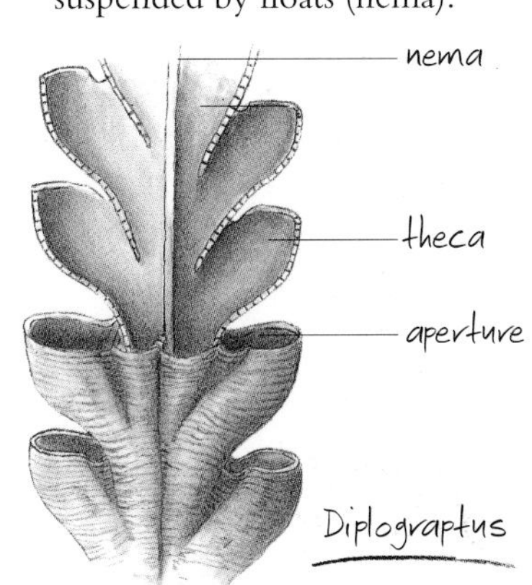

Diplograptus

Trinucleus, Ordovician

Trilobites

Trilobites were the most numerous and successful marine organisms of the early Paleozoic. These segmented "pill-bug"-like arthropods ranged in length from several millimeters to 3 feet (90 cm). They had many legs, each with a set of gills, and most had sophisticated, multifaceted eyes. Like most arthropods, trilobites could grow only by shedding their old skins. Consequently, many of the trilobite fossils that are found are actually shed external skeletons and not the trilobite animals themselves.

Olenoides

The name "trilobite" is derived from the three lobes into which each body segment is divided—two lateral pleural lobes and an axial lobe. They may also be divided into three longitudinal sections, the cephalon (or head), the thorax (made up of variable numbers of hinged segments or somites), and the pygidium (or tail), made up of from one to 30 fused somites. The cephalon is formed from individual segments joined at facial sutures, which split apart during growth. The edges of the shell may be smooth or have a variety of spiny projections.

Most trilobites lived in shallow ocean waters, and on reefs. It has been proposed, however, that some of the spiny trilobites may have been capable of swimming. Trilobite tracks and trails, reflecting a wide variety of their activities, have provided some of the best examples of the daily activities of these extinct organisms. Their skeletons are found in all types of sedimentary rock, from limestones to shales.

Trilobites are relatively common in Cambrian, Ordovician, and Silurian rocks throughout the world. By the Devonian, they had declined to the point of being relatively rare finds.

Calymene (rolled up)

FIELD NOTES

Atlas Mountains, Morocco
Early Paleozoic rocks, northwest Scotland
Ordovician and Silurian rocks, Illinois, USA
Cambrian rocks, Utah and Virginia, USA
Arbuckle Mountains, Oklahoma, USA
Cambrian, British Columbia, Canada

Є O S D C P Tr J K T

Ellipsocephalus, Cambrian

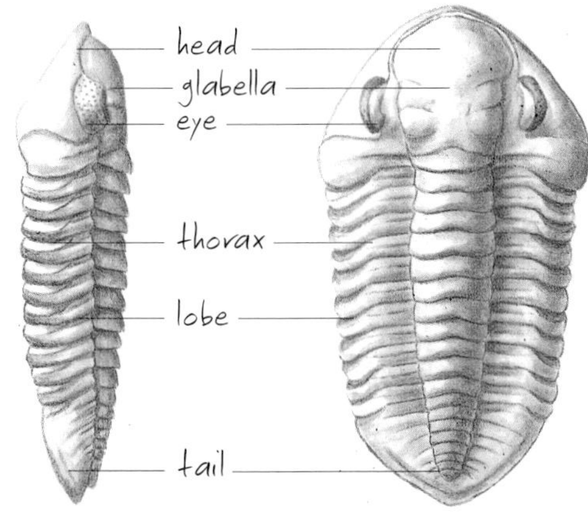

Trilobite exoskeleton

In the late Carboniferous and Permian, they are extremely rare. They are most likely to be found in limestones and shales.

When preserved in shales, trilobites are usually flattened, and often the original shell has been destroyed, leaving only an external cast. In some shales, however, entire skeletons are preserved in excellent detail, allowing X-rays to be taken. Specimens in limestone are frequently preserved in three dimensions, with the skeletal material intact. Trilobites were gregarious animals and are frequently found in large numbers, either in deposits from storm activities or on bedding surfaces. It is not unusual for them to be locally abundant in Cambrian and Ordovician rocks as deposits or debris left by storms. It is also not unusual to find enrolled trilobites, curled up upon themselves, apparently a defensive behavior.

The oldest known published scientific paper on trilobites dates back to 1698. Trilobites are particularly popular with amateur collectors, some of whom collect them exclusively. Good specimens can be found in rock and fossil shops, but inspect carefully before you purchase. In some cases, large, complete specimens have been pieced together from parts collected separately. More than 1500 species of trilobite are known, and many are yet to be described.

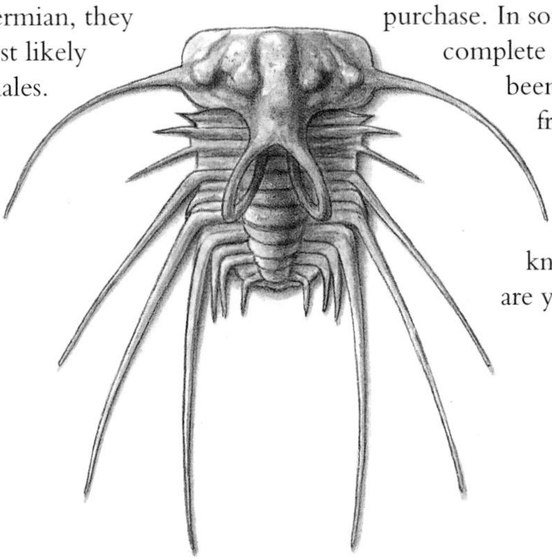

Dicranuras

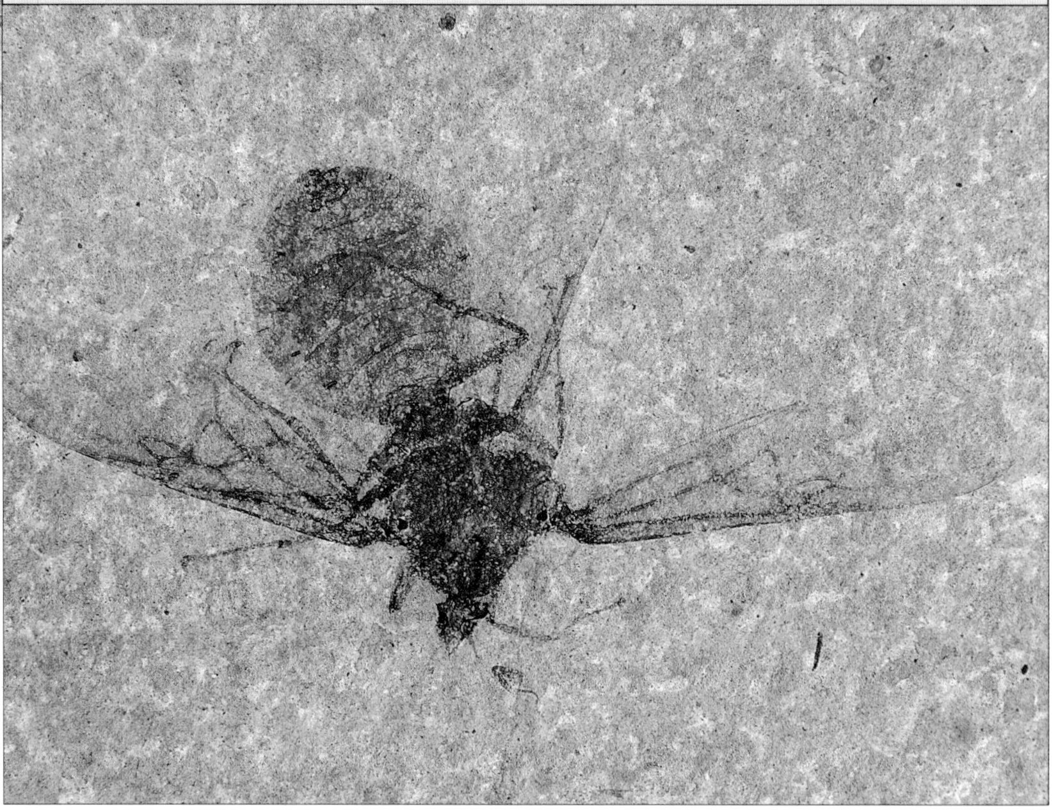

Fossil bee, Eocene

Insects

Insects are small, mobile organisms with jointed external skeletons made of chitin, a complex sugar. They are the most abundant form of multicellular animal life in the terrestrial realm. Insects live in a bewildering array of environments, from the base of an eyelash to Antarctica. The largest number of insects is found in the tropics, and this is the ancient environment best represented by fossil insects. Members of the phylum Arthropoda, insects evolved from centipedes in the Silurian. Their bodies are divided into three segments: head, thorax, and abdomen.

Despite their great abundance, insects rarely fossilize because of their size and the delicacy of their skeletons. Yet for all their rarity in the fossil record, the inclusion of insects in fossil tree sap (amber) has resulted in fossils of the most remarkable preservation. This can be so perfect that muscle fibers and even DNA fragments can be extracted for study. Some sites, such as the Green River Shale in Wyoming, are famous for insects preserved in delicately layered shales.

To find insects in shale, look for small fragments of wings or bodies that have a slight shine. Use a hand lens to make certain. Most amber does not contain insects, so check any pieces carefully. Some of the largest fossil insects were Carboniferous and Permian cockroaches, sometimes more than 20 inches (50 cm) long. Insects are the only invertebrates that developed the ability to fly.

Libellulium

FIELD NOTES

"Baltic Amber", Samland Promontory, Russia
Green River Shale, Wyoming, USA
Cretaceous amber, California and Alaska, USA, and Canada
Miocene amber, Mexico
Carboniferous shales, North American mid-continent
Coal Measures, UK

€ O S D C P Tr J K T

snakefly

Eurypterid, an extinct marine arthropod

Other Arthropods

In addition to familiar insects and trilobites, there are many other members of phylum Arthropoda, which includes all animals with jointed, chitinous, external skeletons. As well as living forms, such as spiders, scorpions, lobsters, shrimp, crabs, and horseshoe crabs, there are interesting extinct forms, such as the so-called sea scorpions or Eurypterids (broad, winglike animals). Eurypterids included the largest known arthropods and in their day were probably some of the most effective and lethal carnivores. Like their living relatives, they inhabited a variety of environments, from fresh and brackish water to shallow marine.

Eurypterids are often found in dark shales, probably representing brackish marine environments. A variety of early arthropods, including the Eurypterids, had pincers. Modern horseshoe crabs are good examples of primitive arthropods. Eurypterids are common in Silurian and Devonian rocks. They are often preserved as carbonized outlines in shales and limestones.

Look for animals with large pincers, segmented abdomens, and long, tapering tails called telsons. True crustaceans, such as crabs and shrimp, are found either as carbonized outlines or in three dimensions, especially those crabs with a thick shell. Arthropods with pincers range in size from mites less than 1⁄48 inch (0.5 mm) long, to giant Eurypterids more than 10 feet (3 m) in length.

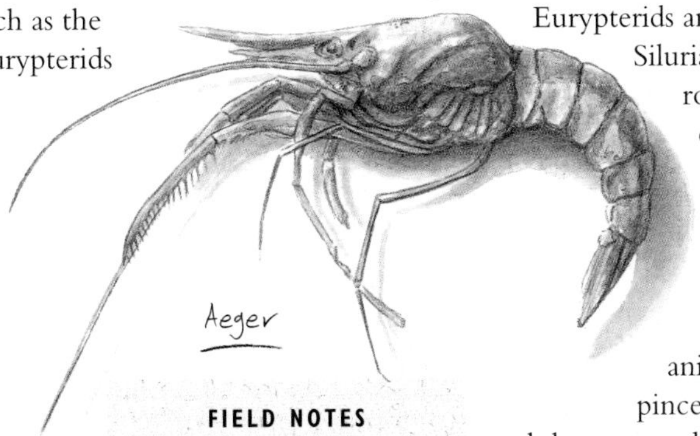

Aeger

Limulus

FIELD NOTES

Mazon Creek, Illinois, USA
Alum Shale, southern Sweden
Silurian rocks, Scotland, Norway and New York, USA
Cambrian rocks, Wisconsin and Missouri, USA
Devonian shales and limestones, New York, USA and throughout Europe

€ O S D C P Tr J K T

A selection of ostracods

Ostracods

Ostracods are small, bean-shaped crustacean arthropods, related to the lobsters, shrimp, and crabs. Like clams or brachiopods, they have two shells, joined by a dorsal hinge. The shell encloses a subtly segmented body with a full set of appendages. Ostracods are usually up to ⅙ inch (between 0.5–4 mm) in length, so use a hand lens or microscope for detailed study.

Ostracods are omnivorous and occur in both fresh and marine waters to a depth of 9,000 feet (2.8 km). They are usually associated with the bottom sediments (benthic), or are found living on the stems of plants. A few forms are planktonic and a few burrow in mud. They are one of the most numerous of all crustaceans and their fossils are very common throughout the geological record. Most species are widely distributed and have a short range in time, so they are a major biostratigraphic tool. They occupy a range of environments in terms of temperature and salinity.

The tiny bivalve shells are composed of chitin and calcium carbonate. They are joined along the upper edge and frequently have thin places that look like eyes. Many ostracod shells are highly decorated with nodes, swellings, ridges, and grooves, although some are completely smooth. Fossil shells are easiest to find in limestones, shales, and marls representing both marine and freshwater environments. The largest ostracods are up to ¾ inch (2 cm) in length. The first freshwater ostracods appeared in the coal-forming Carboniferous swamps.

palaeocopid ostracod

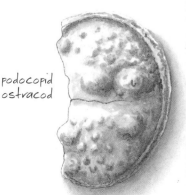

podocopid ostracod

FIELD NOTES

Permian rocks, Kansas, USA
Jurassic rocks, South Dakota, USA
Miocene rocks, Florida, USA
Carboniferous rocks, Illinois and Oklahoma, USA
Tertiary rocks, Europe
Mesozoic rocks, Australia
Cretaceous limestones, Texas, USA

Є O S D C P Tr J K T

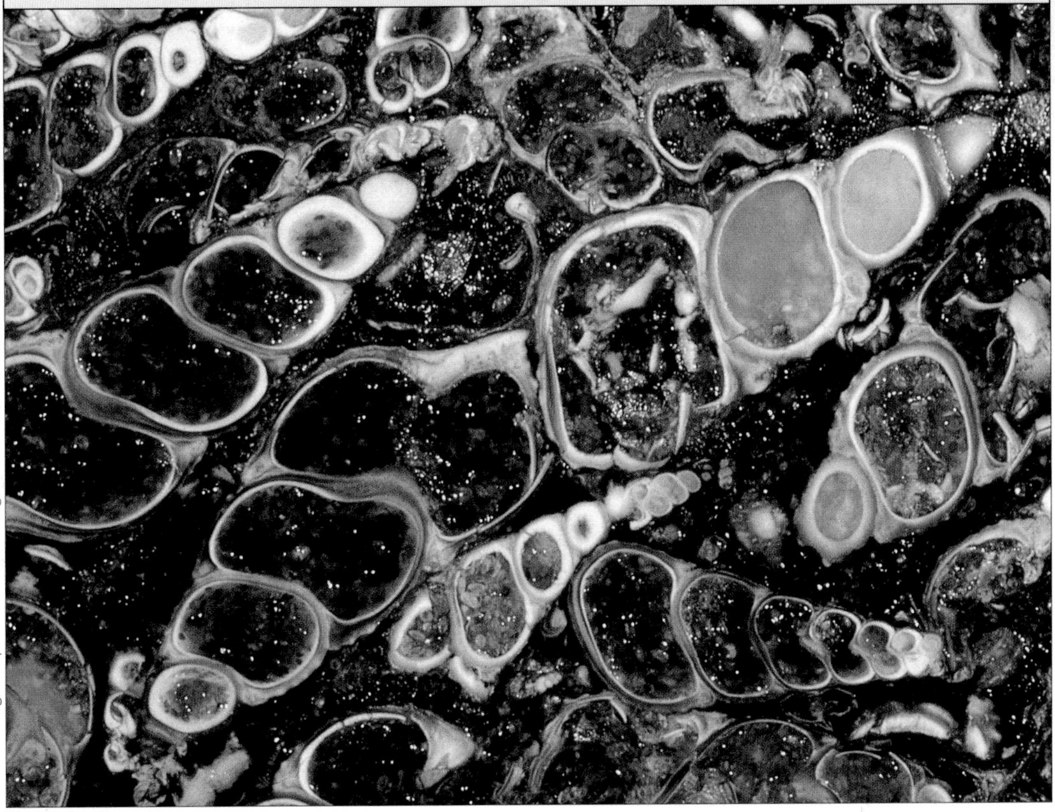

Turritella, gastropods in a slab of agate

Gastropods

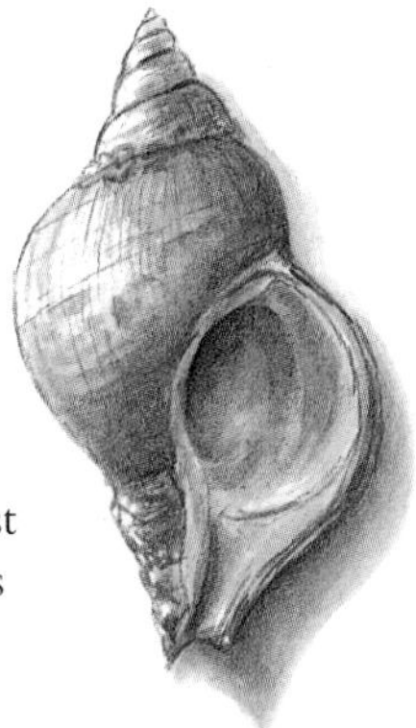

Gastropod means "stomach foot" and these mobile mollusks use their muscular "foot" to propel themselves. They have a mouth and true eyes. Most gastropods secrete a single, whorled, calcareous shell that fossilizes readily. Ecologically, they run the gamut from carnivores to herbivores. Many gastropods have a hardened radula, a type of toothed tongue, that they use to scrape away at food. Gastropods are the only group of mollusks that have successfully adapted to life on land, as snails and slugs. Most fossil and living species, however, are marine or freshwater. Marine varieties lived in many environments, most in shallow water, but some have been recovered from depths of more than 16,000 feet (5 km). The most ancient gastropod shells show no coiling, but subsequent forms began to coil on a single plane, similar to the shelled cephalopods. Later forms have a complex, three-dimensional growth that results in some form of spire. The most significant evolutionary changes within the group concern the twisting or torsion of the body organs, related to shell whorling, and the development of lungs in terrestrial forms.

Planorbis

Neptunia

Gastropods are usually preserved as entire shells and, to a lesser extent, as empty molds in certain limestones. Sometimes, even the original colors are preserved. A complete specimen may also include the flat calcareous operculum, or door, that covered the aperture, or main opening, during life. Some fossil gastropods are found joined to fossil crinoids, having once fed on the fecal debris produced by the crinoid.

FIELD NOTES

Early and middle Paleozoic limestones, Germany
Paleozoic limestones, Appalachian Mountains, USA
Carboniferous shales and limestones, Indiana and Illinois, USA
Cambrian limestones, Missouri, USA
Mississippian limestones, Belgium
Cretaceous limestones, mid USA

€ O S D C P Tr J K T

Turritella

Eurydesma, Permian

Bivalves

Also known as the pelecypods, bivalves include oysters, clams, and mussels, and have bodies encased between two calcareous shells, or valves, that are joined along a flat hinge plate by an elastic ligament. Superficially, bivalves bear a resemblance to brachiopods and the two are often confused by amateurs. In brachiopods, which also have two valves, one shell is usually larger than the other. Unlike brachiopods, bivalve mollusks have a plane of symmetry that passes between the shells, making the left and right valves, in most forms, mirror images of one another. Bivalves have a set of muscles for closing the shells, but they open automatically when the muscles relax. Where the shells join at the beak, or umbo, there are ridges, or teeth, of various types that help to hold the shells closed while the animal is alive.

Pitar

Bivalves are entirely aquatic and are found in fresh and ocean waters. Most forms are attached and immobile as adults, but some are capable of limited movement by "jet propulsion". They live in many environments, from the quiet, calm, deep ocean, to freshwater lakes and turbulent streams. They are abundant in shallow marine waters, where they live on hard or soft surfaces or burrow through soft sediment. They first appear in the Cambrian and are abundant in Mesozoic and Cenozoic rocks.

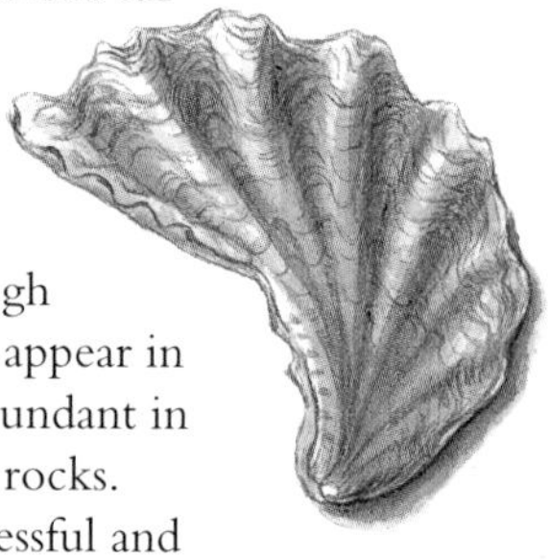

Ostrea

One of the most successful and best preserved groups is the oysters, which are particularly abundant in Cretaceous rocks around the world. Their thick, durable shells may even be preserved through ancient storms that ripped up the oysters and deposited them in beds that mimic oyster reefs.

Bivalve evolution has been well documented with many detailed studies on Cretaceous oysters, and bivalves are often used as important time markers.

FIELD NOTES

Cretaceous limestones, central Texas, USA
Jurassic limestones, UK
Triassic rocks, California, USA
Cretaceous limestones, Kansas, USA; France and UK
Early Paleozoic rocks, Ohio and New York, USA
Jurassic rocks, Idaho, USA

€ O S D C P Tr J K T

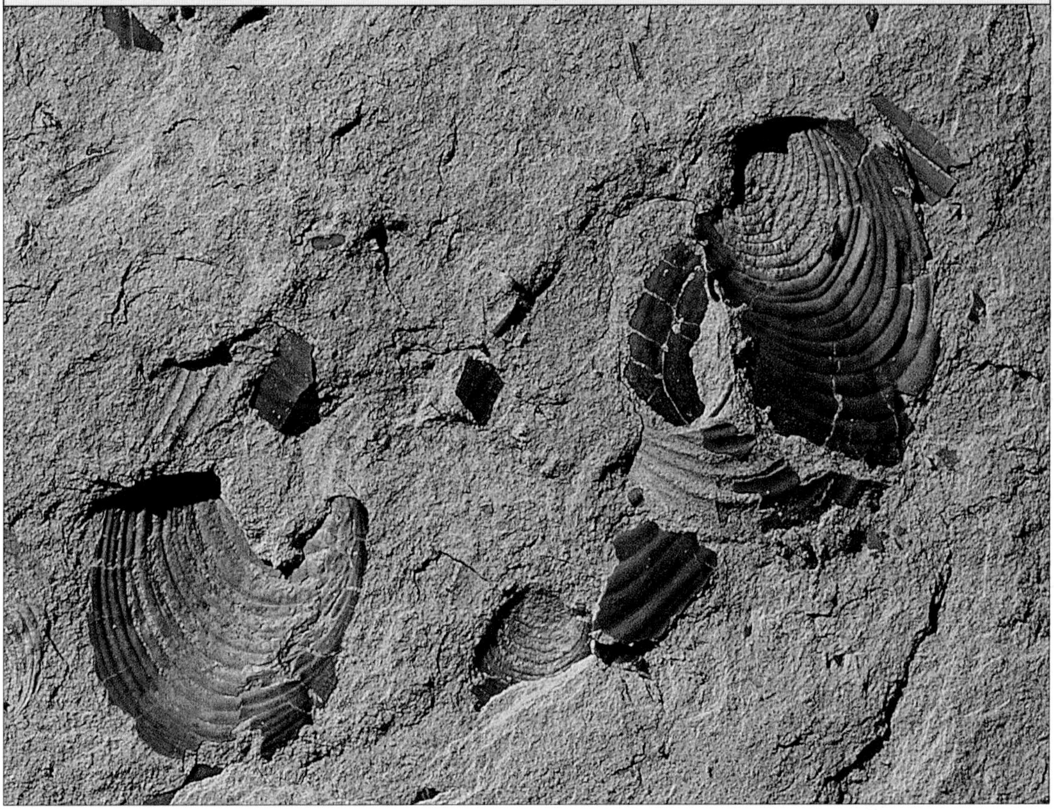

Buchia, Cretaceous

Lyropecten

Goniophora

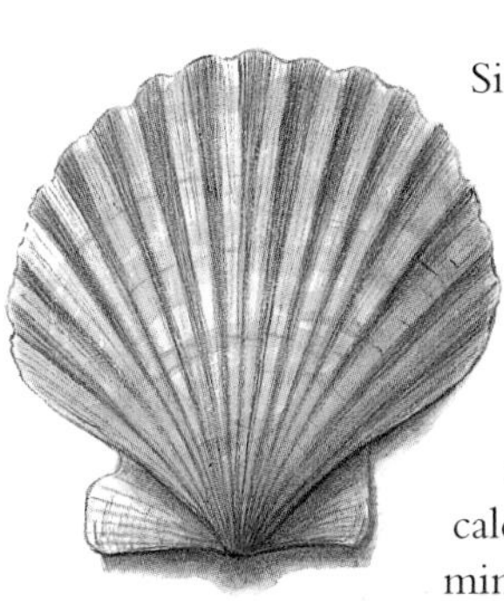

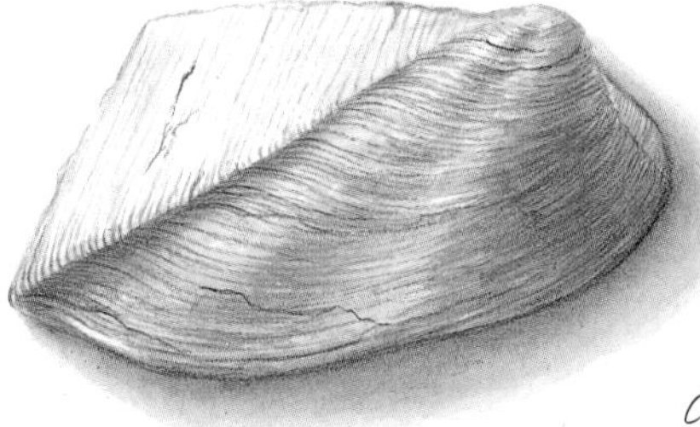

Since they live in many environments, preservation of their shells is likely in many different kinds of stone. The shells are made of multiple calcareous layers of the mineral aragonite, the mineral that gives the luster to mother-of-pearl. Over time, natural processes often convert the aragonite to calcite and destroy the luster.

Unlike brachiopods, many bivalves are found as isolated, single shells, because the shells open after death when the ligaments relax. Burrowing forms have a greater chance of being found with both valves intact. Some bivalves can even burrow into rock and hard wood.

Bivalve shells may be only a few millimeters in diameter, or more than a foot (30 cm) across. Look for entire shells especially among oysters and forms with thick valves. Also look for molds of shells—many shells are dissolved after being encased in rock and only the void is left behind. Many fossil bivalve shells have small holes drilled in them—these may have been made during attacks by either gastropods or sponges.

Bivalve Shells

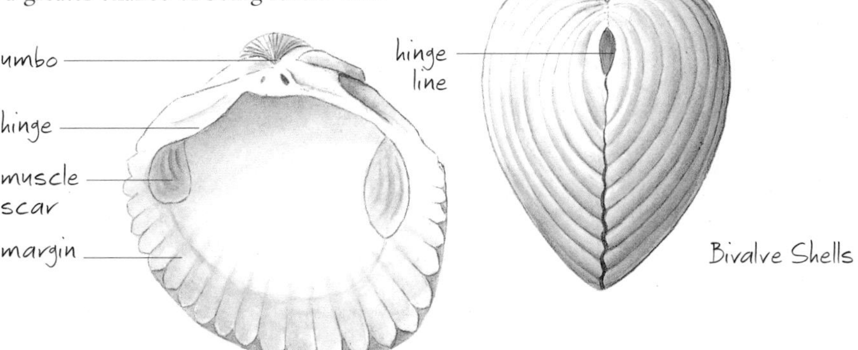

Dactylioceras (ammonite), Jurassic

Nautiloids and Ammonites

Cephalopods are highly active and very well developed marine carnivores. Although many forms had coiled or straight shells, some, such as octopus and squid, appear never to have developed major skeletons. These animals have complex eyes and brains, tentacles with sucker pads, and parrot-like beaks with which they consume their prey. Cephalopods use jet propulsion to move through water. The oldest known octopus is from the late Cretaceous of Palestine.

Ammonites, which are now extinct, and nautiloids were perhaps the most agile and intelligent invertebrate carnivores, living in ocean waters of all depths. Since these animals lived in the water column rather than on the sea floor, the rocks in which their shells are found may not actually reflect the environment in which they lived. Ammonites and nautiloids had straight or coiled chambered shells, in which the old living chambers were either empty or lined with mineral (cameral) deposits that acted as ballast.

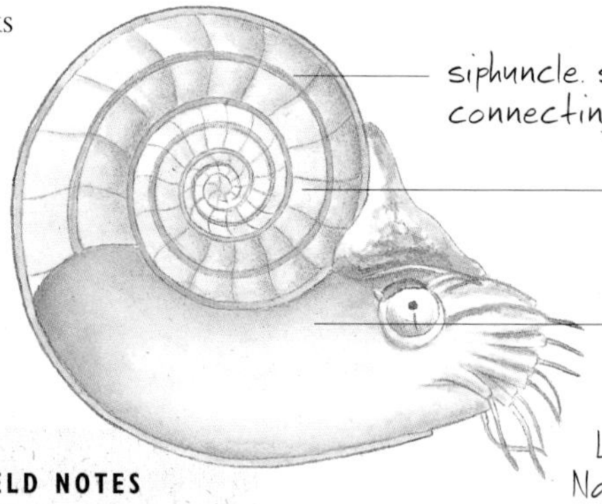

siphuncle. spiral tube connecting chambers

cameral chambers

living chamber

Living Nautiloid

Older chambers were separated by walls called septae that left distinct suture marks where they joined the outer shell. In nautiloids, these are seen as a simple sigmoid curve, but in ammonites, there are many types of complicated, jagged suture lines. Detailed lineages of ammonite evolution have been arrived at

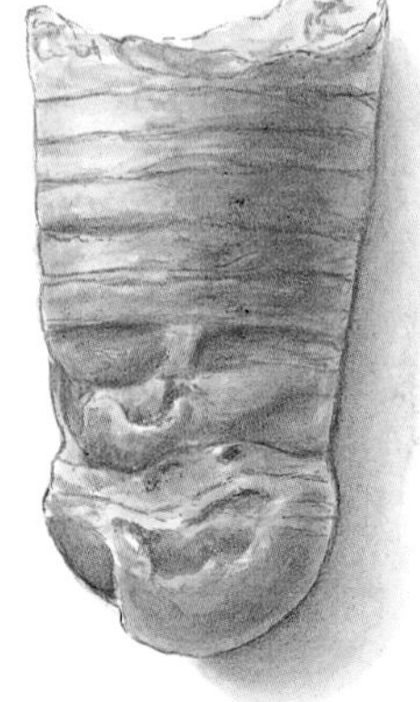
section of nautiloid

FIELD NOTES

Cretaceous limestones, Texas and Kansas, USA, and Europe

Triassic rocks, California, USA

Devonian rocks, New York, USA

Permian limestone, central Europe

Cretaceous lagoonal shales (dwarf forms)

Jurassic rocks of the Himalayas

Є O S D C P Tr J K T

Ammonites: Asteroceras (large shells) and Promicroceras (small shells)

by studying the suture patterns.

The abundance of ammonites and their presence in many types of sedimentary rock makes them one of the most useful biostratigraphic indicators of all. Coiled cephalopods are very common in the middle and late Paleozoic and the Mesozoic. Ammonites became extinct at the end of the Cretaceous. The small numbers of nautiloid surviving after this time make them relatively rare in Cenozoic rocks.

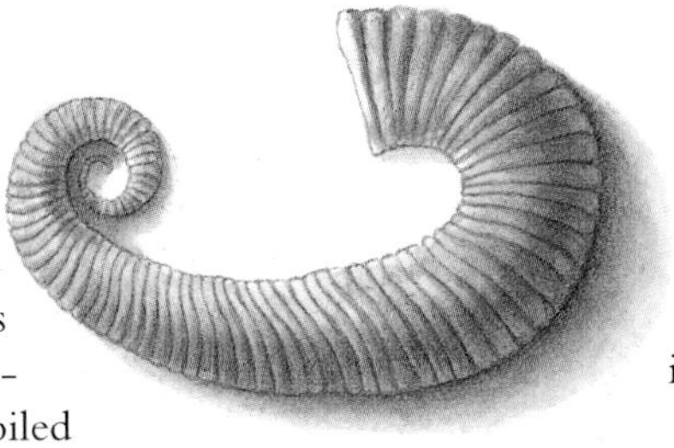

heteromorph ammonite

Nautiloids and ammonites were creatures of open tropical seas, and are abundant in limestones from tropical reefs. They may also be found in shales both from deeper water and from restricted lagoons. Some of the more interesting forms are dwarf species that presumably lived in the hyper-saline waters of such lagoons, but most are found in limestone.

Like the bivalve mollusks, the shells were made of the pearlescent calcareous mineral aragonite. In most fossil shells, the aragonite has been converted to calcite and the pearly luster has been lost, but in some forms, usually preserved in shales, the shells retain their beautiful luster. Unfortunately, the shells of most ammonites were so thin that they are rarely preserved as fossils. Usually, they are found as internal molds, preserving the delicate suture pattern so important in identification.

Groups of straight nautiloids, often hundreds at a time lying in the same orientation, are found in the Atlas Mountains of Morocco. These are often sold in nature stores and rock shops and are frequently polished so that any remaining shell is destroyed. The smallest ammonites occur in the dwarf faunas and adults may be only a fraction of an inch (3 or 4 mm) across, although some coiled ammonites more than 9 feet (3 meters) in diameter have been found.

ammonite Dactylioceras

ammonite Cophinoceras

■ *Glossopteris*, a fossil seed fern

Plant Fossils

Isolated spores in Ordovician rocks mark the first appearance of plant life in previously untenable terrestrial environments. Much hardier than their aquatic ancestors, land plants developed tough skeletons to stand against wind and the forces of gravity. It is those tough skeletal materials, which first appear in late Silurian and early Devonian rocks, that come to us today in the form of vast coal deposits and in a fragmented but plentiful fossil record. Over a long period, plants also evolved methods of reproduction that relied on water as well as on the wind, insects, mammals, and birds.

Land plants occupy almost every terrestrial environment, but the chances of fossilization are greater in low-lying swampy areas in temperate climates than they are in the tropics, where plants are more abundant. Tropical plants are rarely preserved because rapid bacterial action destroys the material before it can be buried.

Dicroidium

Rhacopteris

FIELD NOTES

Carboniferous coals, Appalachian Mtns, USA
Carboniferous rocks, Illinois, USA, and throughout UK
Petrified Forest NP, Arizona, USA
Cretaceous sandstones and shales, Argentina and Rocky Mountains, USA, Canada
Permian rocks, India

Є O S D C P Tr J K T

The earliest plants were simple photosynthetic stems with no branches, topped by spore-bearing structures that barely poked up into the sunlight. They still needed to be in water for some parts of their reproductive cycle. In a steady succession of advances through the Paleozoic and into the Mesozoic, land plants developed many innovations, including woody tissues that allowed them to grow tall, vascular tissues that allowed liquids to move up and down the stems, as well as leaves and branches that increased the area available for photosynthesis.

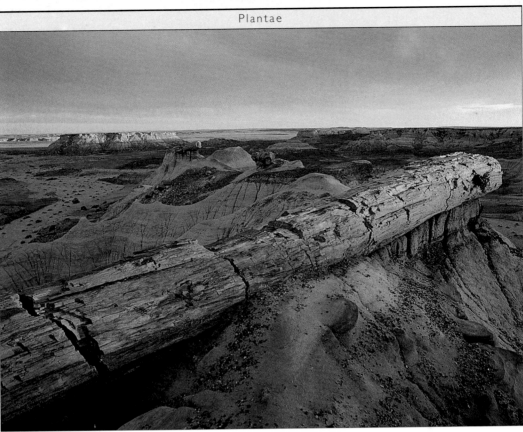

Petrified Forest National Park, Arizona, USA

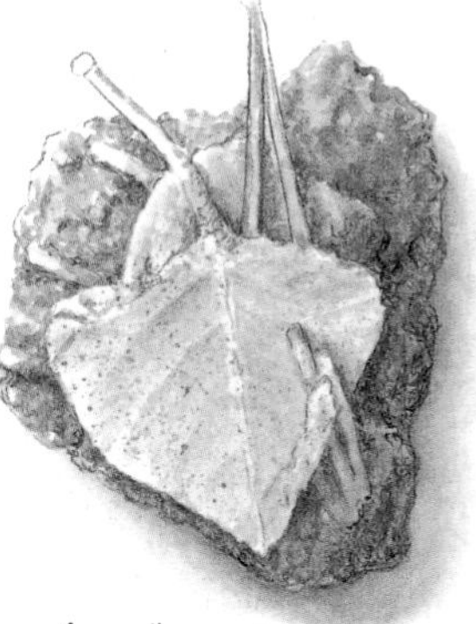

fossil
pine cone

The oldest plants reproduced asexually, their spores being blown through the air. Eventually, plants reproduced sexually, their seeds made fertile by pollen carried by wind or insects.

Plant fossils are very common, especially in Carboniferous and Tertiary rocks. The great coal swamps date from the Carboniferous, and much of the coal mined today comes from this period. Plant fossils can usually be found in deposits associated with coal beds. Many Tertiary rocks have poorer grades of coal where these fossils are better preserved. Plant fossils are common in any rocks that represent sediments deposited in the terrestrial environment. So look for fossil leaves and stem parts in sandstones representing rivers and streams and in shales that represent ancient lakes and river flood areas.

Unlike animals, plant fossils are usually found as separate portions, entire plants being a great rarity in the fossil record. To make matters worse, different plants may have nearly identical leaves, branches, or roots, so many plant fossils are given names based only on shape and not on real botanical relationships. Plants may be preserved as carbonized impressions in shales or nodules, or as three-dimensional casts of trunks. In some coals that are calcified, very old fossil plant material is preserved in such detail that it can be extracted and studied under a microscope. Unfortunately, some of the prettiest fossil plant material has the worst level of preservation, with no original detail remaining.

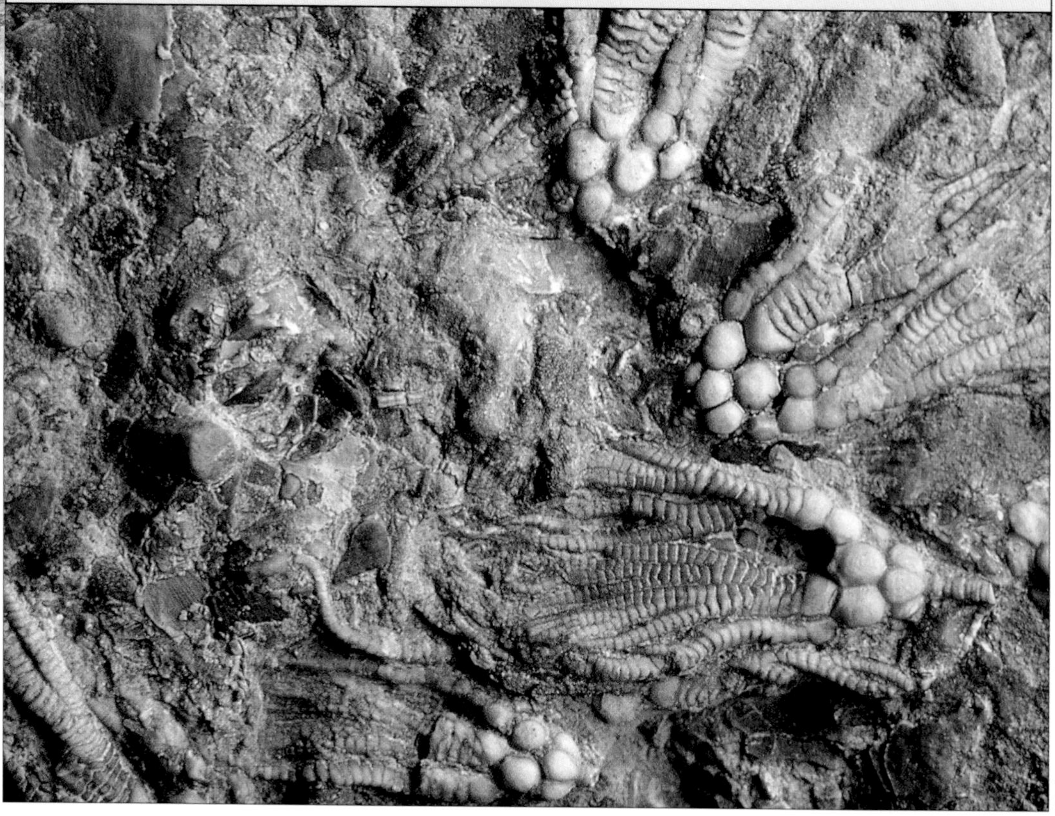

Protaxocrinus (crinoid), Silurian

Echinoderms

Echinoderms are attached or mobile organisms with an internal skeleton made of interlocking calcite plates (echinoderm means "spiny skin"). In some individuals there may be hundreds of separate calcite plates, each a single calcite crystal. They are unique among animals in having an internal hydraulic system, or water vascular system that pressurizes their thousands of inflatable tube feet. The attached forms are filter feeders, while the mobile forms, such as sea stars and sea urchins, may be carnivores, herbivores, or deposit feeders.

The oldest forms seem to have been attached and limited to ocean waters of normal salinity, living on hard bottoms, usually in vast numbers, or attached to floating pieces of wood.

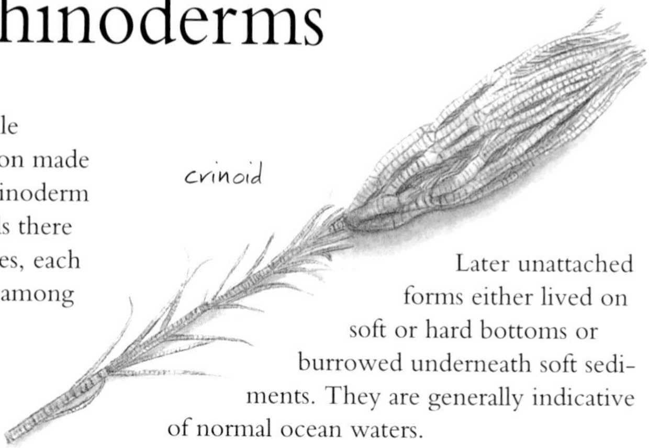

crinoid

Later unattached forms either lived on soft or hard bottoms or burrowed underneath soft sediments. They are generally indicative of normal ocean waters.

In the Cambrian and Ordovician, there was a wide variety of unusual attached and unattached echinoderms that became extinct. It has been suggested that one form, the mitrates, were ancestors of vertebrates, but this is highly unlikely. The crinoids, which survive today, typify the attached forms, with their root-like holdfasts (stems) made of poker-chip-like columnals, and head (calyx) formed of articulated arms joined to the cup-shaped body.

Look for echinoderm fossils in shales and especially limestones. Although it is not unusual to find entire sea urchin fossils, entire articulated stemmed forms are extremely

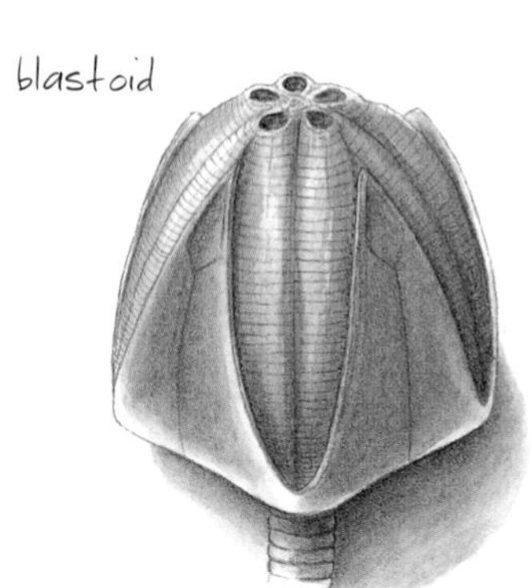

FIELD NOTES

Ordovician rocks, Sweden, Canada, and Ohio, USA
Silurian limestones, UK
Cretaceous limestones and shales, central Texas, USA
Silurian and Devonian rocks, New York and Illinois, USA
Carboniferous limestones Appalachian Mountains, USA

€ O S D C P Tr J K T

Plesiocidaris (echinoid), Jurassic

echinoid *Tylocidaris*

rare. Shortly after death, these fragile animals were usually disarticulated into many small polygonal calcite plates and columnals. Isolated columnals are among the most common fossils in many Paleozoic limestones, and in some cases, the entire limestone is made of countless columnals. In shales, articulated specimens are less likely to be found, but single calcite plates and entire calyces are more common and may be in good condition.

Expect to find more columnals than other body parts. The columnals themselves may have many different shapes, from the ubiquitous round to pentagonal or even star-like shapes. The central cavity (lumen), through which living tissue passed, may be very small, or most of the diameter of the columnal, or it may also be star-shaped.

Unfortunately, it is nearly impossible to identify a crinoid from the columnals alone. Sea stars are very rare as fossils. Sea urchins begin to show up in abundance in the late Paleozoic, and by the Cretaceous are very common fossils. Regular sea urchins have a symmetrical body, usually round, and represent forms that lived and moved about on the sea floor. It is not unusual to find isolated plates and spines of regular sea urchins in Carboniferous and Permian marine shales. Irregular sea urchins are bilaterally symmetrical and are streamlined for burrowing in soft sediments. Burrowing sea urchins are found in Cretaceous shales as entire body fossils. Their horizontal burrows may be wrongly identified as bones.

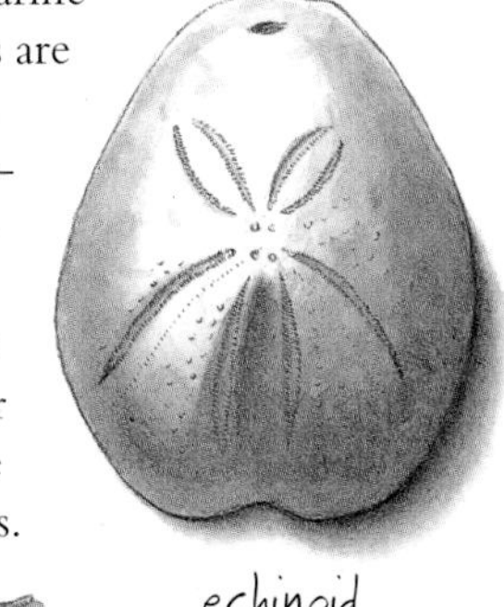
echinoid *Micraster*

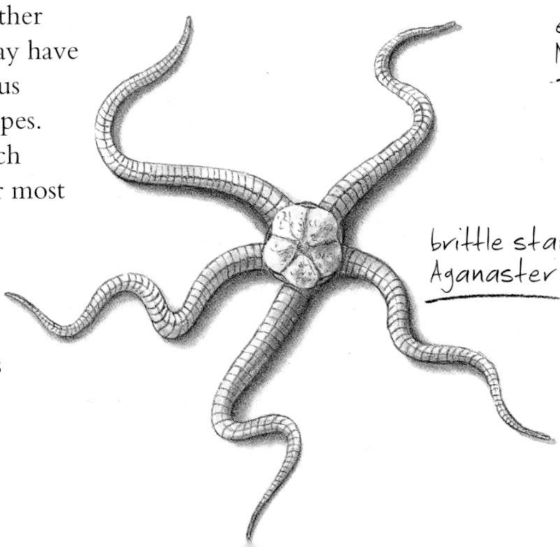

brittle star *Aganaster*

Knightia, Eocene fish

Vertebrates: Fish

Vertebrates are highly mobile, complex organisms with internal skeletons. The oldest vertebrates are fish, and the earliest fish are the jawless agnathans, which had heavy head skeletons, but only a cartilaginous internal skeleton. Closely related are the placoderms, the first jawed fish, also with heavy head shields, and the cartilaginous chondrichthyians, including modern sharks and rays. The most derived fish (osteichthyians) have bony internal skeletons and include all modern bony fish as well as the early ancestors of land-dwelling vertebrates.

The first fish appeared in freshwater environments but rapidly spread from lakes and rivers to lagoons and then to open seas. By the Silurian, fish occupied all water depths from lakes and streams to deepest oceans, so expect to find fish fossils in nearly any environment. With the exception of placoderms, all major groups of fish are still alive, although older groups are reduced in number and diversity. Lampreys and hagfish are all that remain of the agnathans, but these have lost their head shields. Sharks and rays have tooth-like bones in their scales and their numerous generations of teeth are common fossils.

Bony fish are now the most numerous. The lone surviving

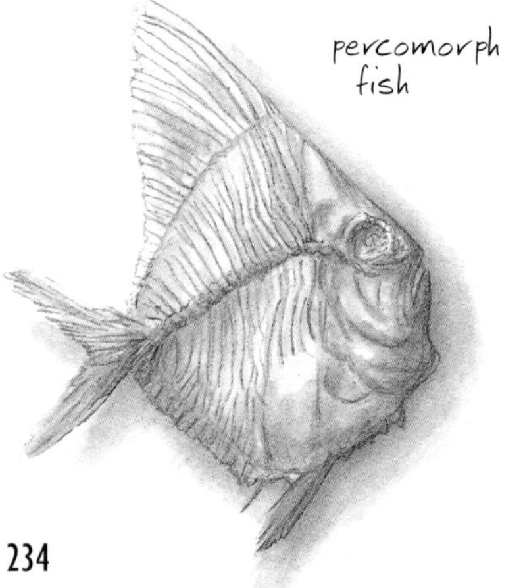

FIELD NOTES

Green River Shale, USA
Old Red Sandstone, UK
Niobrara Formation, central USA
Permian shales, central Texas, USA
Chalk Marl, Sussex, UK (coprolites)
Harding Sandstone, Utah, USA
Cretaceous limestones, central and southern USA
Holzmaden, Germany

€ O S D C P Tr J K T

Dapedium, Jurassic fish

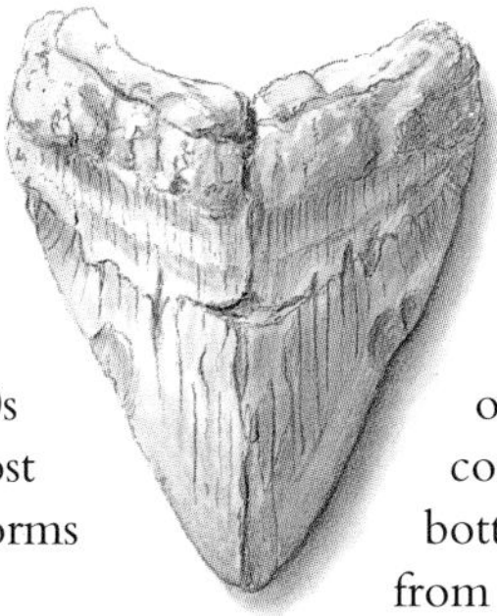

shark tooth

member of the group of bony fish that gave rise to land-dwelling animals, the coelocanth, was discovered in the 1930s in deep water off the African coast. Most bony fish respire with gills, but some forms have lungs and can breathe air.

As abundant as fish are, they are relatively rare as fossils. The bony teeth of sharks fossilize well and are common in some environments. Otherwise, fish are so easily broken up after death that they make poor candidates for fossilization. In some minor dark shales, fish bones can be very common. Since most fish are active swimmers, their fossils are likely to be spread through a variety of rocks. Look for them in shales, especially marine black shales or shales deposited in freshwater lakes. Fossil bones may be abundant in lake shales, where they are preserved on bedding planes. Fossil fish fecal material, coprolites, may be locally common in Permian shales of brackish or fresh water. Bones may also be concentrated in sediments on lake bottoms, where they have been protected from decay by lack of oxygen. Whole fish are more common in lake shales. In rare instances, in certain limestones, whole fish are preserved, uncrushed, in three dimensions. Many such fish from South America are available in rock stores. In such cases, the limestone can be delicately dissolved leaving an intact, articulated skeleton that looks almost modern.

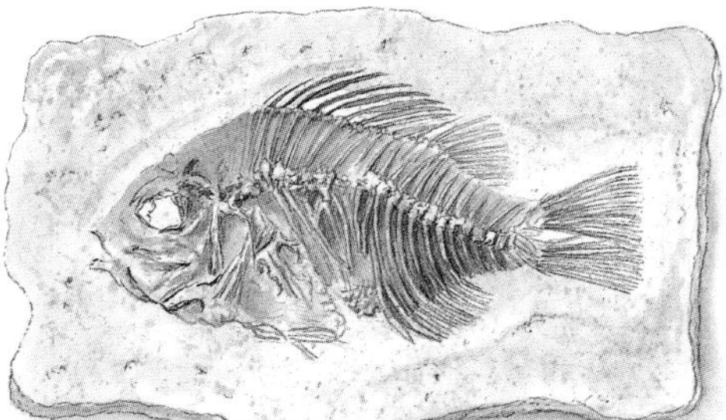

Priscacara

cast of Diplomystus

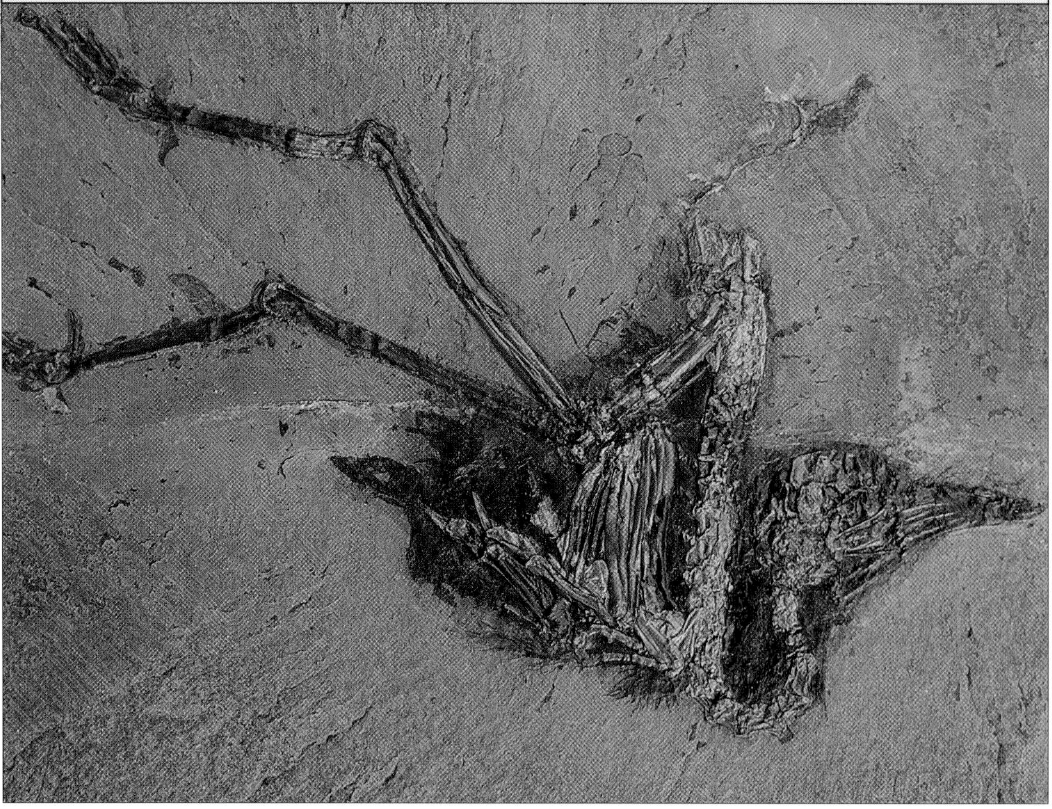

Fossil bird (rail)

Other Vertebrates

The higher vertebrates are among the largest and most active animals on Earth. They have a bony internal skeleton and are well adapted for living and moving around. Not only do they inhabit the land surface, but vertebrates fly, burrow, and some, such as seals, whales, ichthyosaurs, and plesiosaurs, have even moved back to living in the sea. In the Devonian, the first land-dwelling animals, the semi-aquatic amphibians, evolved from bony fish. In the Carboniferous, the first animals that could live on land without a larval stage or needing water for fertilization, appeared. These were the reptiles, and in the Triassic, one major branch gave rise to the mammals. The other branch, representing all living reptiles, went on to spawn the great lineages of giant reptiles, flying and aquatic reptiles, and the humble lizards and snakes. All major lineages of terrestrial vertebrates survive today, although not necessarily in their early forms or diversity.

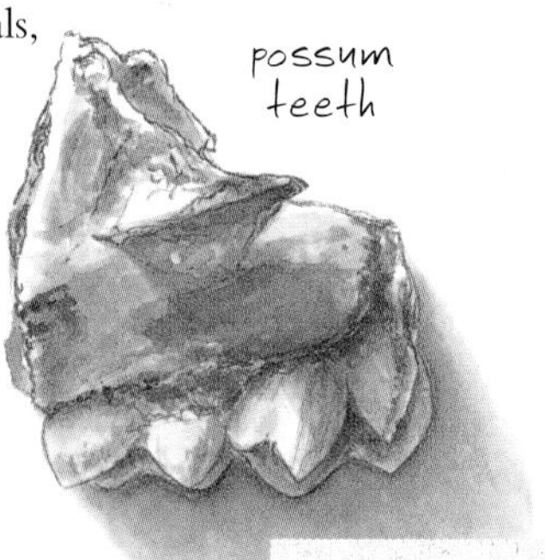

possum teeth

Higher vertebrates have occupied nearly all possible environments. Amphibians seem to have preferred tropical realms. Ancient reptiles, although mostly cold-blooded, are found in a variety of environments, reflecting their ability to adapt to extremes. Mammals have been able to move into very cold environments because of their ability to generate body heat.

In general, the fossil remains of higher vertebrates are uncommon for many reasons, the major one being that the land surface is not conducive to preservation. Although they may

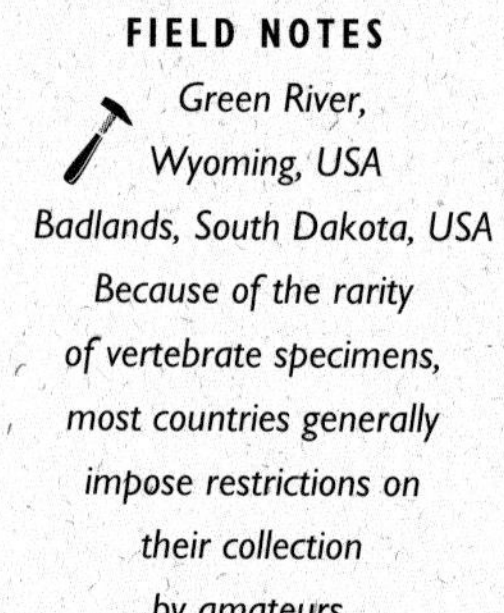
FIELD NOTES

Green River, Wyoming, USA

Badlands, South Dakota, USA

Because of the rarity of vertebrate specimens, most countries generally impose restrictions on their collection by amateurs.

€ O S D C P Tr J K T

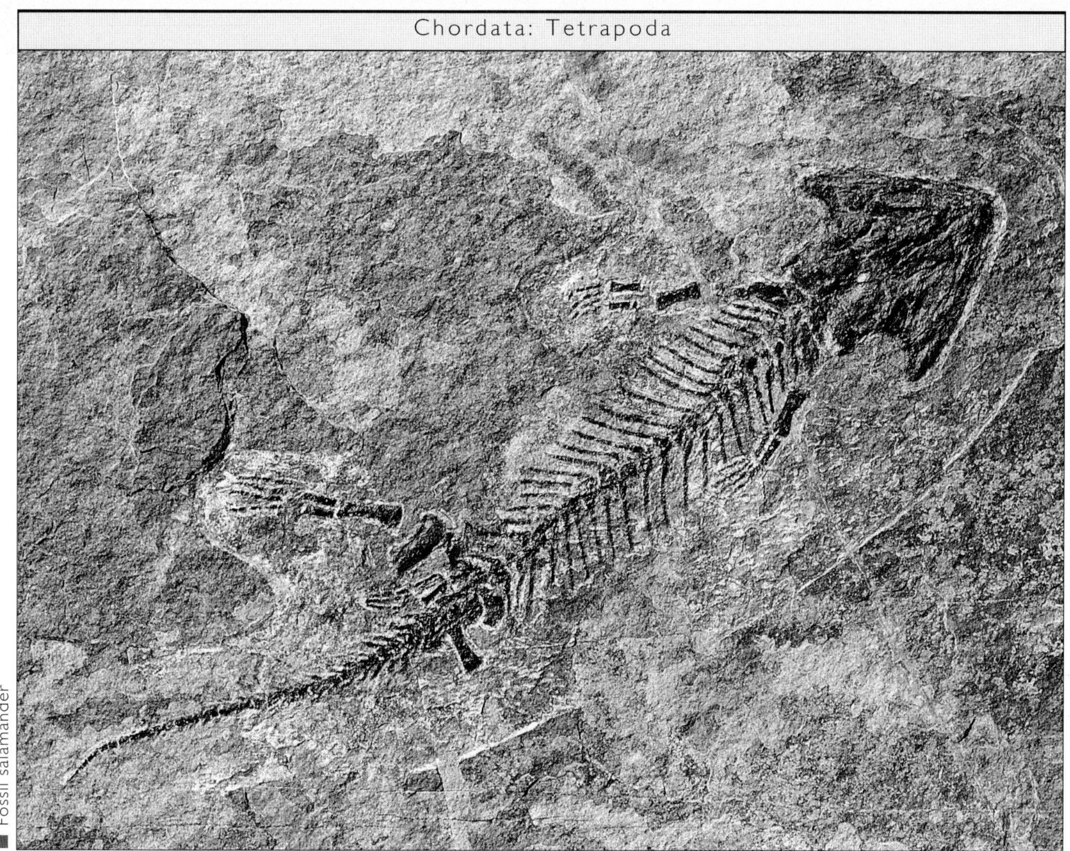
Fossil salamander

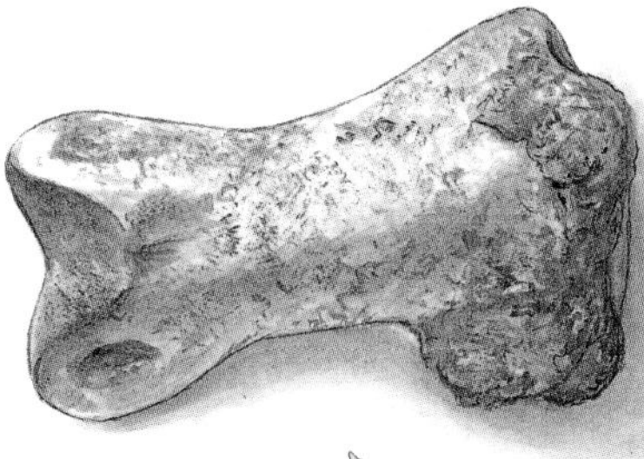

be abundant in certain places, they are usually difficult to find, difficult to remove correctly from rock, and are often fragmented. For these reasons, collecting fossil vertebrates requires technical skills and sophisticated preparation devices and storage facilities.

kangaroo toe-bone

Arguably, the best place to look at fossil vertebrates is in museum collections, where they have been carefully prepared for display. Vertebrate remains are likely to be found in any rocks that represent terrestrial or marine environments, but since vertebrates are rarely preserved, the search may be lengthy. When bone is found, it means there may be other bones in the vicinity.

rodent jaw (lower)

Look for higher vertebrates in association with shallow marine rocks, with lake deposits, and in the bases or sides of old river channels. Cave deposits may yield cave-dwelling vertebrates. Bones are rarely preserved in ancient soil deposits because chemical processes in soils tend to destroy them. Bone might seem hard, but it is readily destroyed. Fossil bone is very delicate and should be stabilized with preservative agents. Large vertebrates should be removed only by experienced people.

Mastodon

The vertebrate parts most likely to be fossilized are teeth, which are made of dense enamel. Ancient mammals are well known from their fossil teeth. Many reptilian and amphibian teeth are similar and, when found in isolation, may be difficult to identify. The most unusual vertebrate fossils are those found frozen in the permafrost that stretches across the northern hemisphere from Asia into North America.

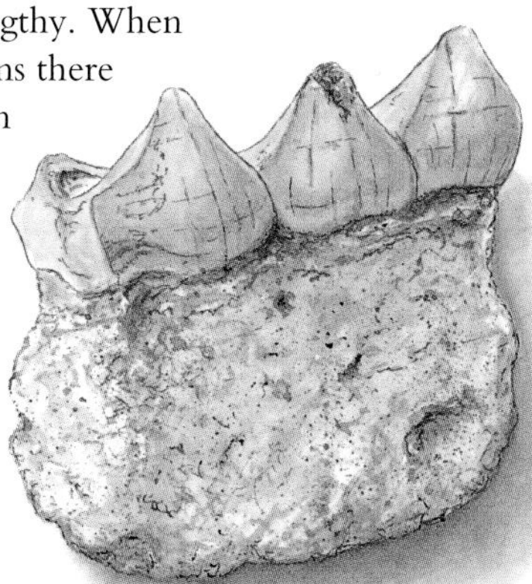

Fossil Sites

Ediacara Hills, South Australia

The Ediacara Fauna

There has been life on Earth for 3,600 million years, but early single-cell creatures left few fossils. The earliest fossils, 3,500 million years old, are bacteria from Australia and we know that more complex animals existed 1,000 million years ago because we have found their burrows.

The earliest fossils of multi-cellular organisms are between 700 million and 570 million years old. These soft-bodied animals take their collective name, the Ediacara Fauna, from the Ediacara Hills in South Australia, where they were found in astonishing abundance in 1947. Isolated specimens had been found earlier in England and Africa, and similar collections are now known from England, Wales, parts of Europe, southwest Africa, the USA, Canada, and Russia.

Ediacaran faunas are found in what were once shallow marine environments, possibly intertidal areas such as mudflats or tidal pools, where animals became stranded. Early work in Australia recognized them as members of modern groups. Altogether, about 30 species, thought to be animals such as segmented worms, jellyfish, and sea pens, were recognized, as well as a few seemingly unrelated oddballs.

Jellyfish comprise the largest group within the Ediacara Fauna with about 15 recognized species. *Mawsonia* is one of the larger and more common at about 5 inches (13 cm) in diameter. It is named after Sir Douglas Mawson, an Australian paleontologist who collected some of the first Ediacara specimens.

There are three species of segmented worm in the Ediacara Fauna, including *Dickinsonia*, about 3 inches (7.5 cm) long. Close inspection reveals what appears to be a central gut lobe, but no eyes or mouth. There are also three species of soft coral, called sea

Dickinsonia costata, a segmented worm.

FIELD NOTES

Ediacara Hills, South Australia
Charnwood Forest, UK
Mistaken Point, NFLD, Canada
South Australian Museum, Adelaide, South Australia
American Museum of Natural History, New York, and Smithsonian Institution, Washington, DC, USA
British Museum, London, UK

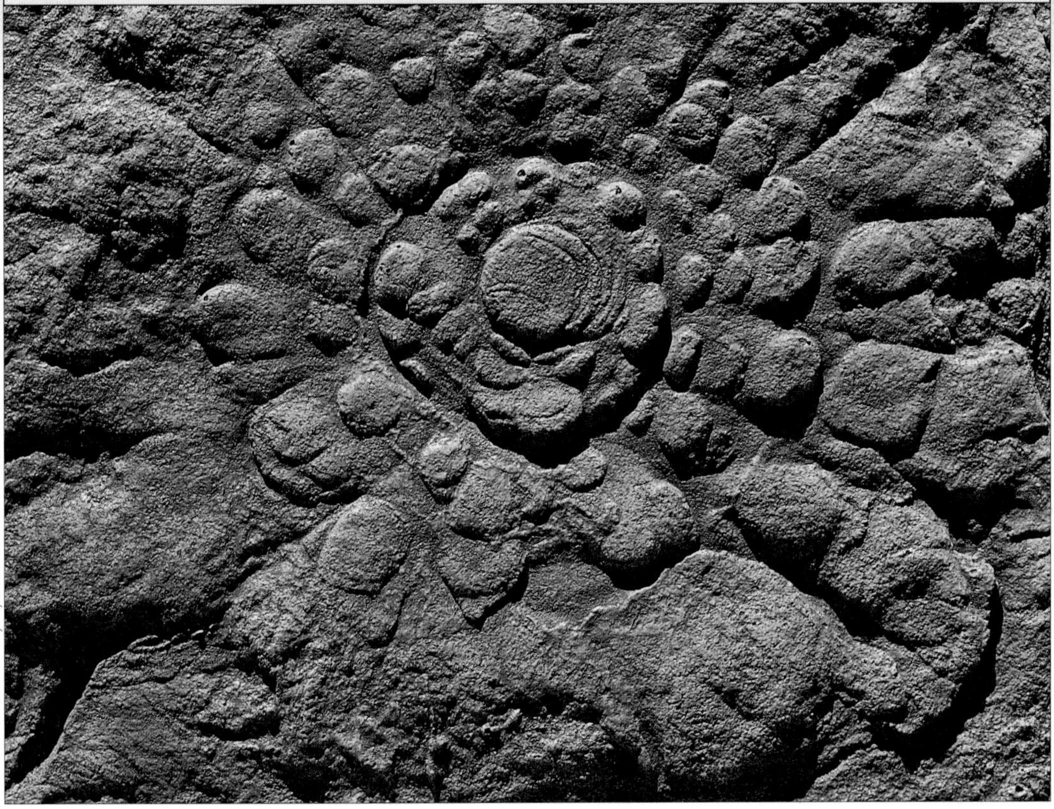

Mawsonia, a fossil jellyfish

pens, colonies of which look like feathers up to 24 inches (60 cm) long.

Among the "weirdos" in the Ediacara Fauna is *Tribrachidium*, a circular animal about 1 inch (2.5 cm) in diameter, with three radiating "arms" across its upper surface. Another oddity is *Parvancorina*, more ellipsoid in shape and up to 1 inch (2.5 cm) long with a broad, anchor-shaped ridge on the upper surface. It has been suggested that *Parvancorina* may be an early arthropod.

Attempts to understand Ediacaran organisms tried to fit them into previously known groups, but while some Ediacaran fossils resemble modern animals, closer inspection reveals structural details that exclude them. Furthermore, although differing in form, Ediacaran fossils follow a basic plan, which suggests that they may all be related in a single major group that has no modern descendants.

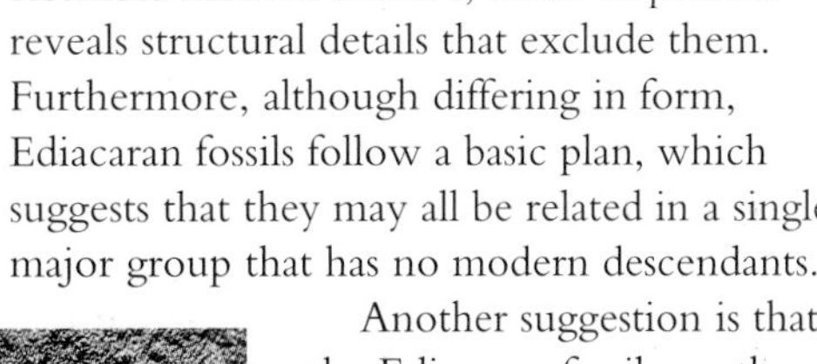

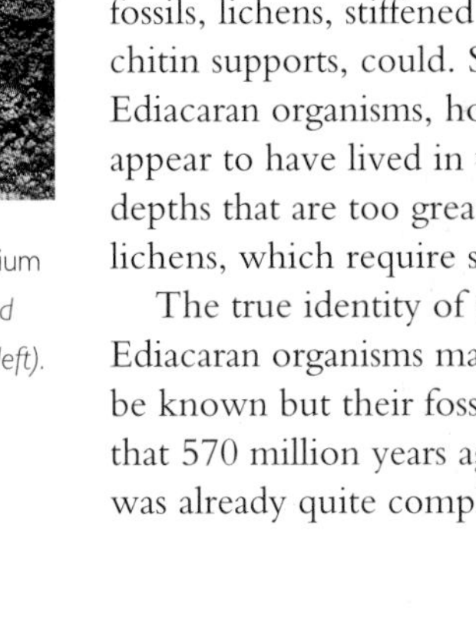

Tribrachidium (above) and a jellyfish (left).

Another suggestion is that the Ediacaran fossils are the remains of lichens, a composite of fungi and algae common in modern terrestrial environments. They are also known in fresh-water and shallow marine environments. While soft-bodied animals, such as jellyfish, are unlikely to leave the types of impression seen in Ediacaran fossils, lichens, stiffened by chitin supports, could. Some Ediacaran organisms, however, appear to have lived in water depths that are too great for lichens, which require sunlight.

The true identity of the Ediacaran organisms may never be known but their fossils tell us that 570 million years ago, life was already quite complex.

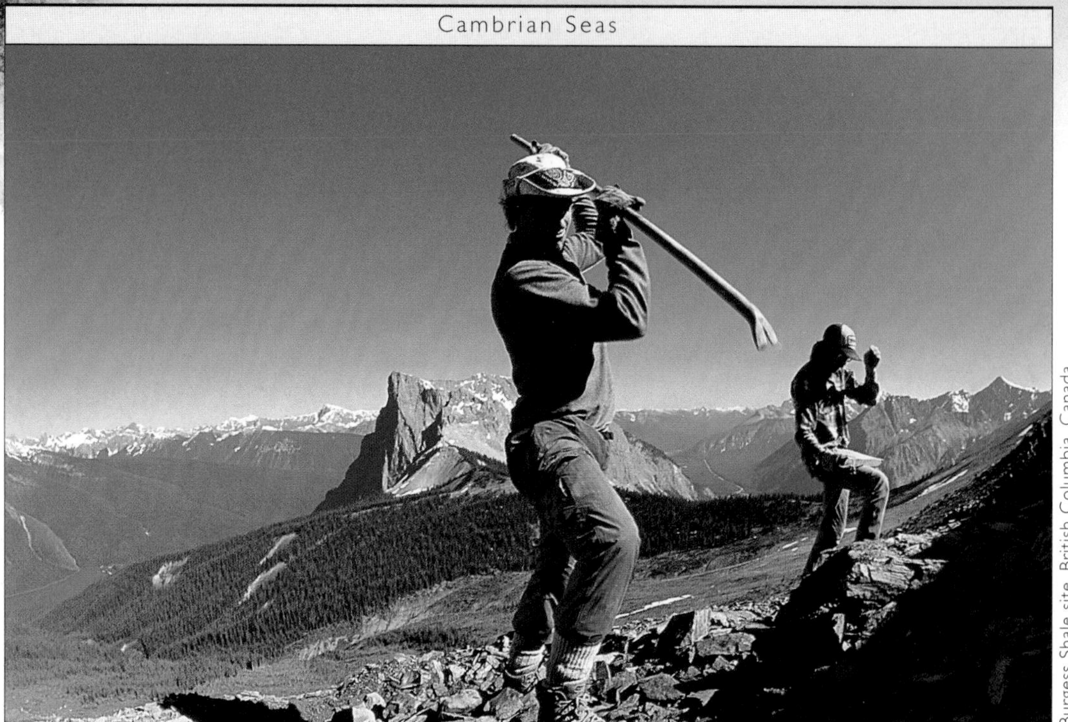

Working at Burgess Shale site, British Columbia, Canada

The Burgess Shale Fauna

During the Cambrian, life diversified rapidly from a few relatively uncomplicated forms to a vast number of quite complex organisms. Although rapid by geological standards, this process took almost 60 million years to develop fully. While the rudiments of this explosion of life can be understood from the fossils of hard-shelled organisms, it is the rare occasions where soft-bodied organisms are fossilized that better illustrate how dramatic it was. The best known soft-bodied fauna of the Cambrian comes from the Burgess Shale in British Columbia, Canada.

The Burgess Shale Fauna was discovered on the side of a mountain in 1909 by the famous American paleontologist Charles Doolittle Walcott (1850–1927), during a geological survey. The fauna dates to the middle of the Cambrian, about 540 million years ago, and contains more than 140 species of animal, most less than an inch (2.5 cm) or so in length. The reason so many different types of delicate organisms were preserved seems to be that the area was originally in very deep water just beyond the continental shelf.

FIELD NOTES

Burgess Shale, British Columbia, Canada

American Museum of Natural History, New York, USA

Smithsonian Institution, Washington, DC, USA

Є O S D C P Tr J K T

Animals living on that shelf would occasionally be swept over the edge and settle into the soft, fine-grained muds at the bottom where the water was too deep for scavenging animals to destroy their delicate carcasses.

One of the truly astonishing features of the Burgess Shale Fauna is that it contains many types of organism that probably became extinct by the end of the Cambrian. These were not simply different species of known types of organism, but whole body plans as different from each other as crabs are from spiders. While many of

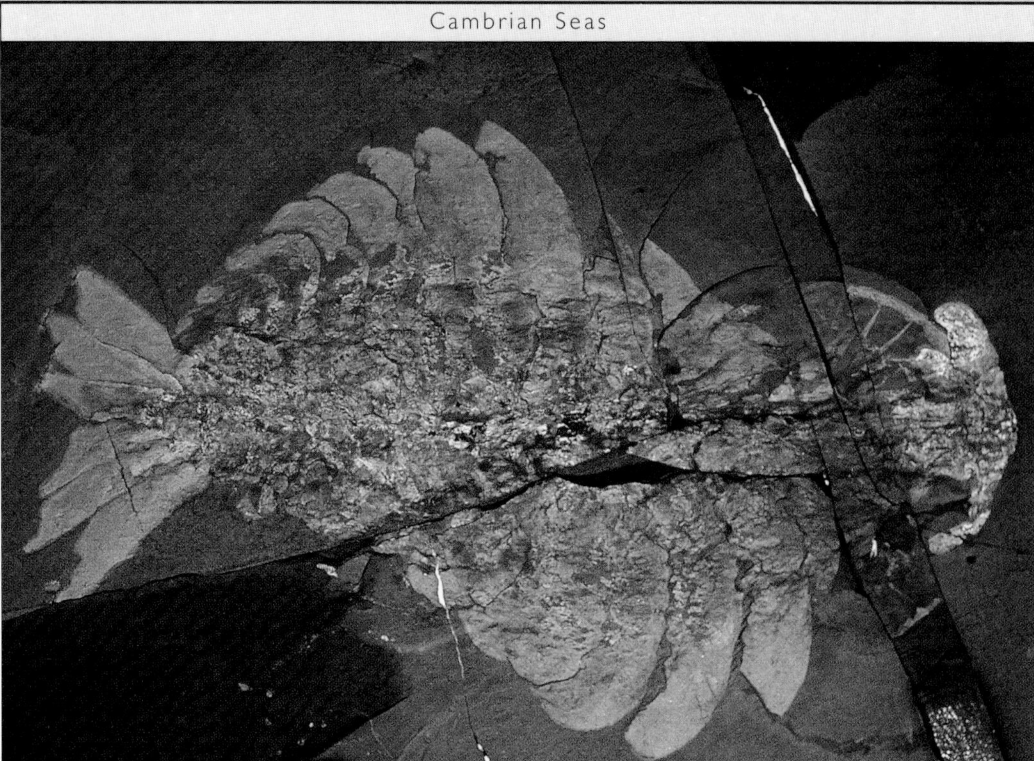

Anomalocaris from the Burgess Shale

The Burgess priapulid Ottoia *(left) in its burrow.*
The grasping organs of Anomalocaris *(below).*

these organisms appear to be types of arthropod, there is also a large number of organisms whose wider relationships are obscure. In fact, there are more basic types of organisms in the Burgess Shale Fauna than there are in the world today, a feature that challenges much of our understanding of the development of life.

Traditionally, we have accepted that life gradually gets more complex and diversified through time, but the Burgess Shale Fauna teaches us that the development of life has occurred as a selection process, with only a few types of organism surviving from earlier faunas that were once much more diverse. Although the Burgess Shale Fauna is the most widely known of the Cambrian soft-bodied faunas, there are other such faunas now recognized from around the world, including China, the USA, and Australia. These all tell the same strange story as the Burgess Shale—that by the mid Cambrian, life was amazingly diverse.

One of the most interesting animals from the Burgess Shale is *Anomalocaris,* a carnivore up to 2 feet (60 cm) long. It seems to have been composed mostly of soft tissue and, being large, it readily disintegrated after death. Initially, various pieces were described as different animals—the grasping organs were thought to be prawns. Recent finds, including complete specimens, reveal they all belonged to one animal.

Marrella splendens, *arthropod*

■ One of the sites where the Hunsrückschiefer Fauna is found

The Hunsrückschiefer Fauna

One of the most important fossil faunas from the middle Paleozoic era is the Hunsrückschiefer Fauna from Bundenbuch, Wissenbach, and Gemünden in the German Rhineland. These fossils come from the Hunsrück Shales, which are lower and middle Devonian in age. The Hunsrückschiefer Fauna includes sea stars, trilobites, other arthropods, and cephalopods.

What is amazing about the Hunsrückschiefer and Burgess Shale faunas is that, as well as the more typical hard elements, the soft parts of organisms are preserved along with whole organisms consisting entirely of soft parts. Fossil specimens from the Hunsrückschiefer Fauna are preserved as pyrite within the shale and reveal exquisite detail when X-rayed. Usually, only the hard parts of an organism are fossilized, so the soft anatomy, particularly of organisms that are now extinct, can never be properly understood from most fossil remains. From such rare cases of fossils of the soft anatomy, we gain valuable insights into the complete organism. For example, study of cephalopod specimens from the Hunsrückschiefer Fauna reveals that some types carried their shells on the outside of their bodies while others had a layer of soft tissue surrounding the shell. Such detail can rarely be deduced solely from the fossil remains of the animal's hard parts.

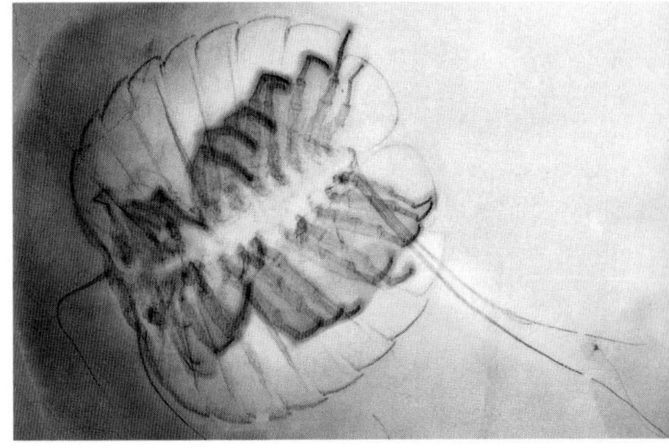

An X-ray of Cheloniellon calmani, *an unusual arthropod.*

FIELD NOTES

Bundenbuch, Wissenbach, and Gemünden, Germany

Senckenberg Museum, Frankfurt, Germany

Institut für Paläontologie, Bonn, Germany

€ O S D C P Tr J K T

In the Hunsrückschiefer Fauna, remains of some organisms not usually fossilized are preserved. *Mimetaster* is a bizarre arthropod that accounts for about half the known specimens of the Hunsrückschiefer Fauna. Curiously, despite its abundance

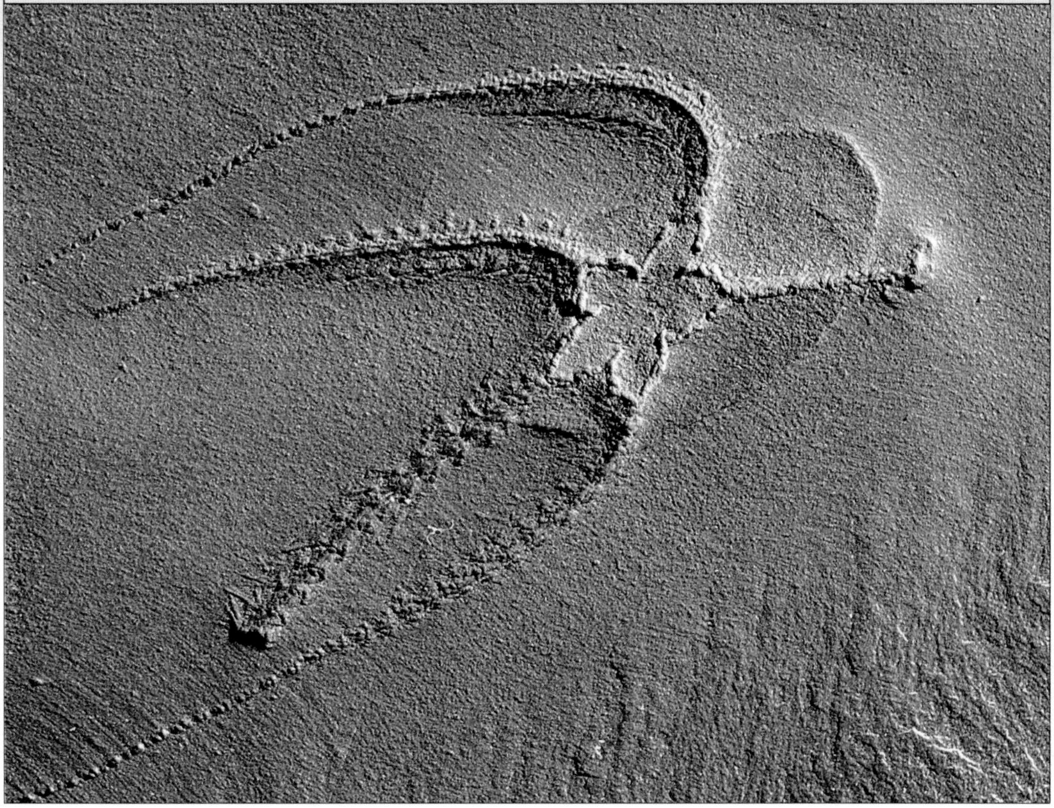

A brittle star from the Devonian, Germany

here, *Mimetaster* is known from nowhere else. It has eyes on stalks and two pairs of strong legs. Another strange arthropod is *Vasconisia*, which has a two-part shell covering much of its body. It does not appear to have eyes, but has three pairs of walking legs on its head. Two pairs of extra appendages near the mouth seem to have been used for manipulating food. *Mimetaster* was probably a deposit feeder, processing sediments for microsopic organisms, while *Vasconisia* was probably a carnivore or a scavenger.

While these two unusual animals seem to be related to one another, their wider affinities are obscure. They are possibly related to *Marrella* from the Burgess Shale Fauna and they seem to represent a whole group of arthropods seen only in faunas with exceptional preservation. The key point about deposits such as the Hunsrückschiefer and the Burgess Shale faunas is that they provide unique windows on a broad spectrum of life at a particular moment of time, including organisms without hard parts to leave fossils and the soft anatomy of others. This makes them so important to our understanding of the history of life that they have been given the name *Lagerstätten*, meaning "mother lodes".

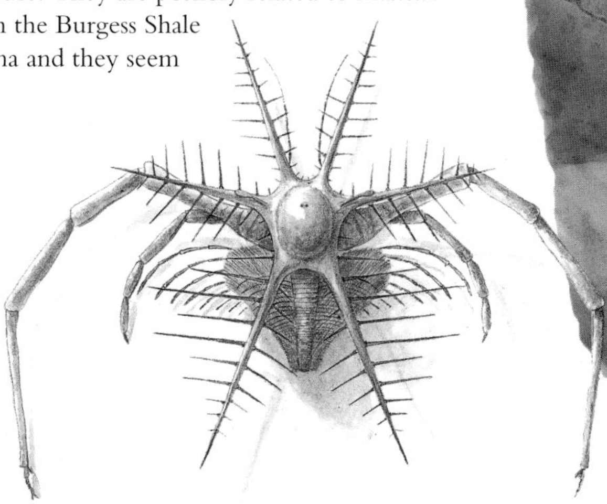

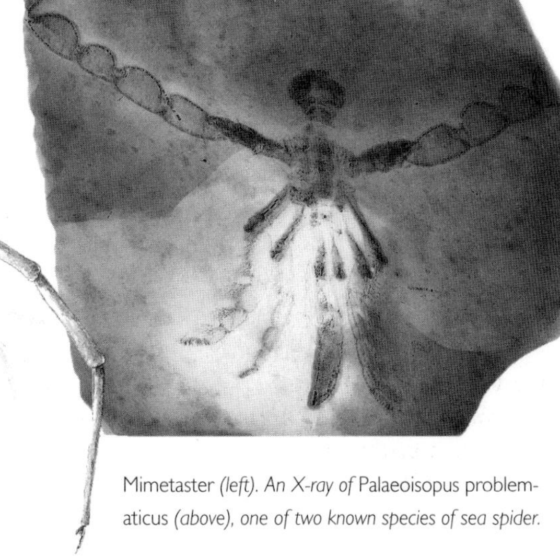

Mimetaster *(left). An X-ray of* Palaeoisopus problematicus *(above), one of two known species of sea spider.*

■ Mazon Creek fossil site

Mazon Creek

The Carboniferous, divided in the USA into the Mississippian and Pennsylvanian subperiods, saw the development of the first significant forests. Vegetation was so abundant that it produced the first major coal deposits, from which the Carboniferous takes its name. These forests were capable of supporting complex ecosystems.

Unfortunately, coal deposits are not particularly good locations for preserving fossils, but associated sediments often are. Such is the case in Illinois, in the area called Mazon Creek. Here, some 300 million years ago, a river delta ran through an extensive swamp. In many respects, Mazon Creek would have been like modern Louisiana, except that it would have been warmer, the whole location being only a few degrees from the equator at that time.

Mazon Creek is special because the fossils from the area are in a particularly fine state of preservation. These organisms were quickly buried in mud at the bottom of the river system, and bacterial action, while decomposing the organic material, also generated a hardened zone around the remains. These are now found as concretions, particularly hard balls of rock, which, when split open, reveal exquisitely preserved organisms. Such nodules or concretions are often found in soil dumps after strip-mining.

There are numerous exposures bearing concretions scattered over Grundy, Will, Kankakee, and Livingston counties in Illinois. It is certainly worth inspecting road and rail cuts, cliffs, quarries, and other excavations in these areas. Concretions are also found associated with the mines of LaSalle County. The abundance and exceptional preservation

FIELD NOTES

Mazon Creek, northeastern Illinois, USA

Illinois State Museum, Springfield, Illinois, USA
Field Museum of Natural History, Chicago, Illinois, USA
Burpee Museum of Natural History, Rockford, Illinois, USA

Є O S D C P Tr J K T

A Mazon Creek fern, Neuroptis *(above). A spider-like arachnid (right): rarely fossilized, 14 orders of spiders have been found here.*

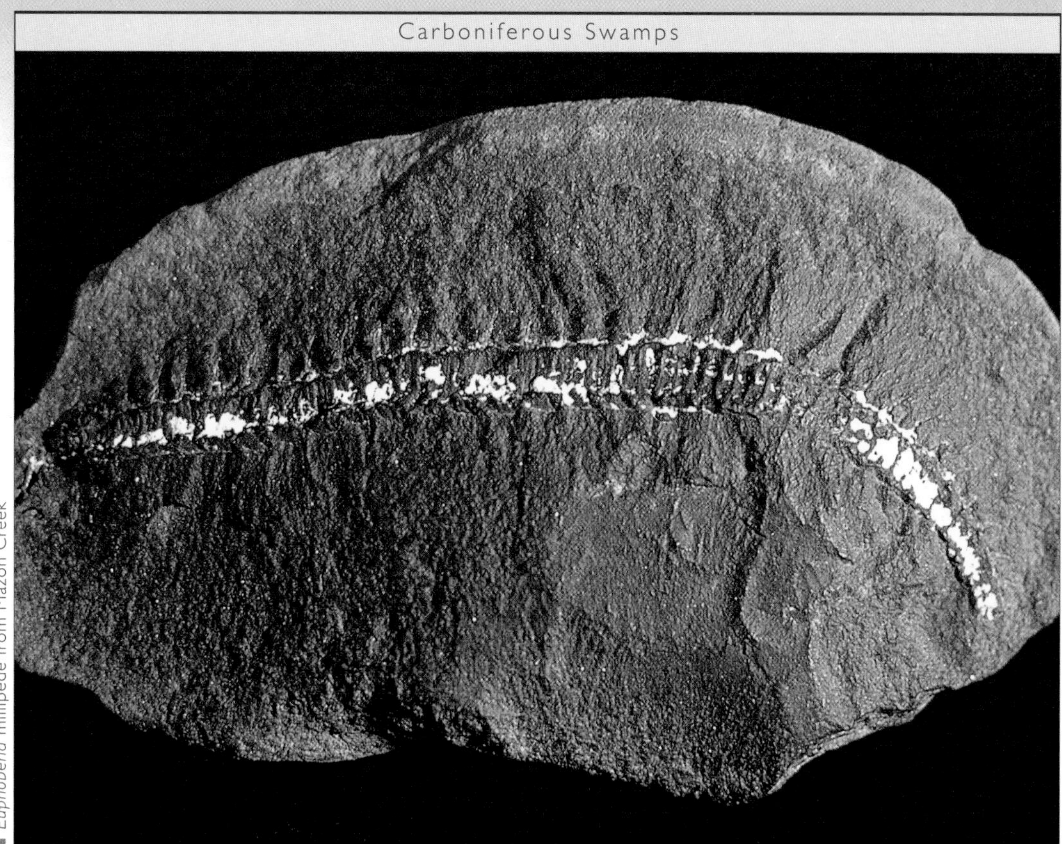

Euphoberia millipede from Mazon Creek

of fossils from Mazon Creek has made them favored specimens in collections.

The most bizarre and enigmatic organism to be recovered from Mazon Creek is a cigar-shaped, segmented animal with a spade-like tail and a shield-like snout with a toothed claw. Called *Tullimonstrum gregarium* after its finder, Francis Tully, these strange beasts still resist all attempts to place them within any known group. The curious *Tullimonstrum* has become the official state fossil for Illinois.

The Mazon Creek Flora contains more than 400 plant species from about 130 genera. Dominant among these plants are horsetails, ferns, and club mosses. The Mazon Creek Fauna, consisting of more than 320 species, can be divided into two parts, the Essex and Braidwood faunas. The Essex Fauna represents the animals that would once have lived in the shallow bays of the river system, while the Braidwood Fauna consists of terrestrial and freshwater creatures from the area. Animals recovered from the Essex Fauna include jellyfish, worms, clams, snails, shrimp, and fish, while the Braidwood Fauna includes insects, millipedes, centipedes, scorpions, spiders, ostracods, shrimp, horseshoe crabs, fish, and amphibians.

Coprolites are also found in Mazon Creek nodules and more exceptional specimens include young lampreys with the yolk sac still attached. The mixture of marine and freshwater organisms is thought to be the result of storm surges bringing marine creatures into the delta system.

Tullimonstrum *(left)*, Octomedusa pieckorum *jellyfish (below). The delicate tentacles are not usually preserved.*

The Karroo region of South Africa

The Karroo Beds

The Permian and Triassic witnessed one of the most important developments in the history of life: the evolution of mammals from reptilian ancestors. While obviously descended from reptiles, modern mammals have a significantly different structure. The links between the reptilian grade and the mammalian grade remained poorly understood until the significant fossil beds of the Karroo Basin in southern Africa were studied systematically from the late 1890s to the 1930s.

The Karroo Basin, a massive structure covering most of what is now South Africa, consists almost entirely of Permian and Triassic sediments. Earlier sediments in the basin reveal that there were glaciers to the north, but later sediments indicate a warmer, more seasonal environment. Throughout the Permian and Triassic, the Karroo Basin would have been a complex of rivers draining into a southern sea.

The animals living there evolved from reptiles to mammals and their fossilized remains are quite abundant through the basin.

Found in older sediments towards the bottom were remains of reptiles that are beginning to show mammalian affinities—animals such as the dicynodonts, with two dog-like fangs at the sides of their mouths, and the wolf-like *Lycaenops*. In common with mammals, these mammal-like reptiles had one major hole in the side of their heads to allow for the attachment of jaw

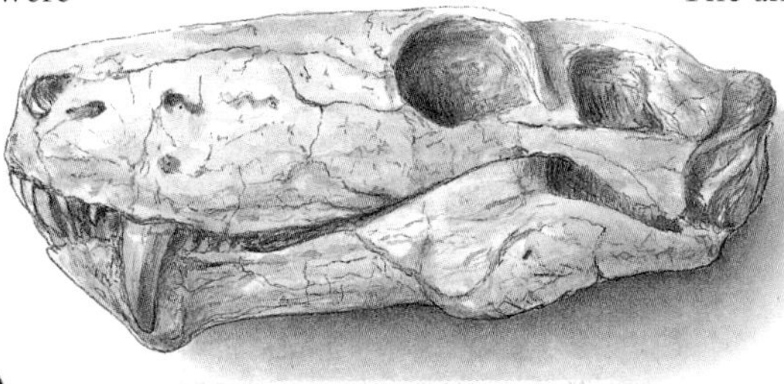

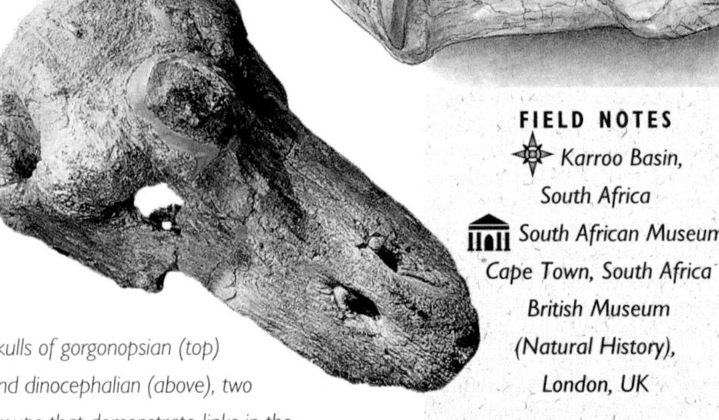

Skulls of gorgonopsian (top) and dinocephalian (above), two groups that demonstrate links in the transition between reptiles and mammals.

FIELD NOTES

Karroo Basin, South Africa

South African Museum, Cape Town, South Africa

British Museum (Natural History), London, UK

€ O S D C P Tr J K T

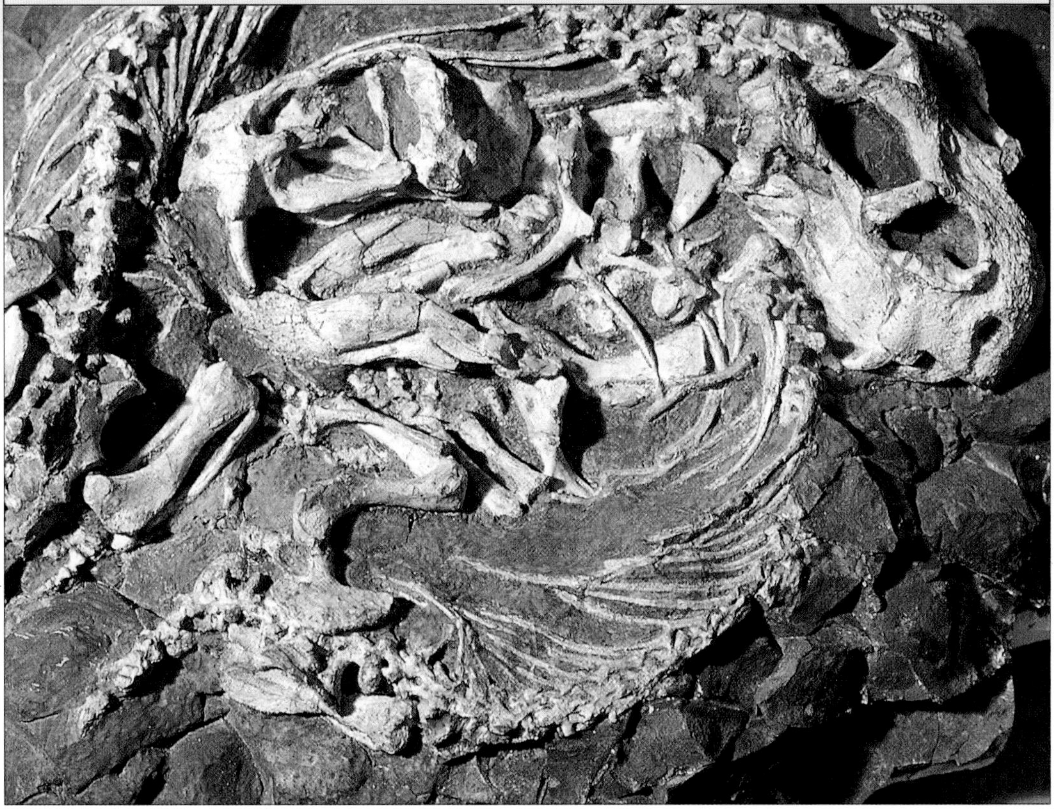

Two *Diictodon*, curled up in a sand-filled burrow

muscles but, in common with more primitive reptiles, they had a jaw made up of four major bones with a single bone in the ear.

Because an almost continuous record of animals living in the area over a period of about 50 million years is preserved in the sediments of the Karroo Basin, the evolutionary history of many groups can be tracked accurately from their fossils. By the time the latest sediments were formed, the primitive mammal-like reptiles of the Permian had been replaced by forms more closely resembling true mammals—in animals such as *Thrinaxodon*, the jaw comprised, principally, a single bone and the other reptilian jaw bones performed a new role as extra bones in the ear.

Oudenodon *skull (left), after preparation.* Large Bradysaurus *skeleton (below) being excavated.*

Mammal-like reptiles and their descendants, reptile-like mammals, dominated the Karroo Basin for most of the Permian and Triassic, but towards the end of the Triassic, a new group of animals emerged that would dominate terrestrial habitats for the next 160 million years. Early dinosaurs, such as the 40-foot (12-meter) long *Euskelosaurus*, are found in the youngest sediments of the Karroo Basin. It is not clear why, with such a promising start, mammals should have given way to dinosaurs in the late Triassic, but for the entire reign of the dinosaurs, mammals were rarely larger than a modern cat and were relegated to small, peripheral habitats.

One of the most famous paleontologists to work the Karroo Beds was Robert Broom, a Scot who later discovered the first fossils of the early hominid *Australopithecus*.

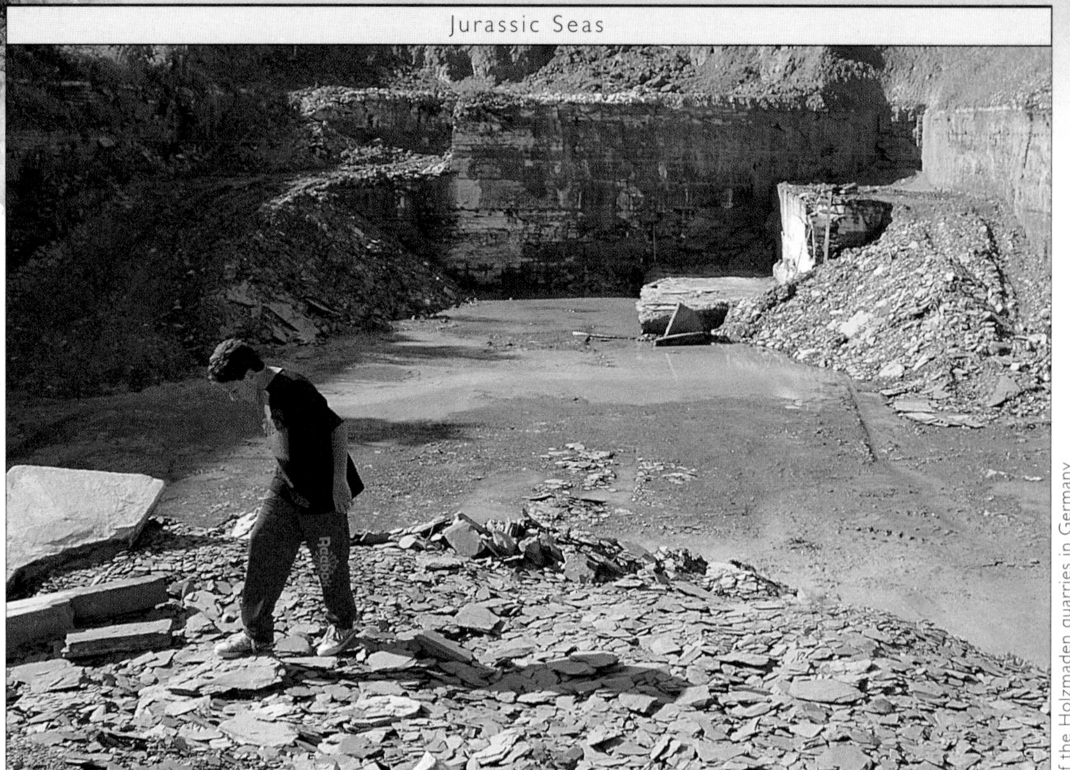

One of the Holzmaden quarries in Germany

Holzmaden and Solnhofen

For most of the Jurassic, large parts of western Europe were covered by warm, shallow seas, and associated with these seas are two most important fossil sites. Holzmaden, near Stuttgart, Germany, is early Jurassic, while Solnhofen, in Bavaria, dates from toward the end of that period. At both sites, the fossils are recovered during stone-quarrying operations.

Horseshoe crab from Solnhofen.

Fossils of marine creatures from the Jurassic, such as the dolphin-like ichthyosaurs, long-necked plesiosaurs, and short-necked pliosaurs, are found in many places, but rarely as complete skeletons. At Holzmaden, however, complete articulated specimens are abundant, usually with every detail of the bones faithfully preserved. Some specimens are so perfect that they have stomach contents and even unborn young inside females—at least one unfortunate ichthyosaur has been preserved in the act of giving birth. Occasionally, the skin and body tissues of the animals are also preserved, showing complete outlines and certain details of internal structure. Such rare preservations offer unique insights into the biology of these long-extinct creatures—we now know that ichthyosaurs had a dorsal fin and gave birth to live young at sea rather than coming ashore to lay eggs.

FIELD NOTES

Quarries near Holzmaden, Germany
Quarries around Solnhofen, Germany
Museum Hauff, Holzmaden, Germany
Eichstätt Museum, Bavaria, Germany
Senckenberg, Frankfurt, Germany

€ O S D C P Tr J K T

Many types of fish and invertebrates are also known from Holzmaden, all preserved in the same exquisite detail. One specimen, a petrified log some 30 feet (9 meters) long, is completely encrusted in crinoids up to 6 feet (1.8 meters) long. This breathtaking specimen

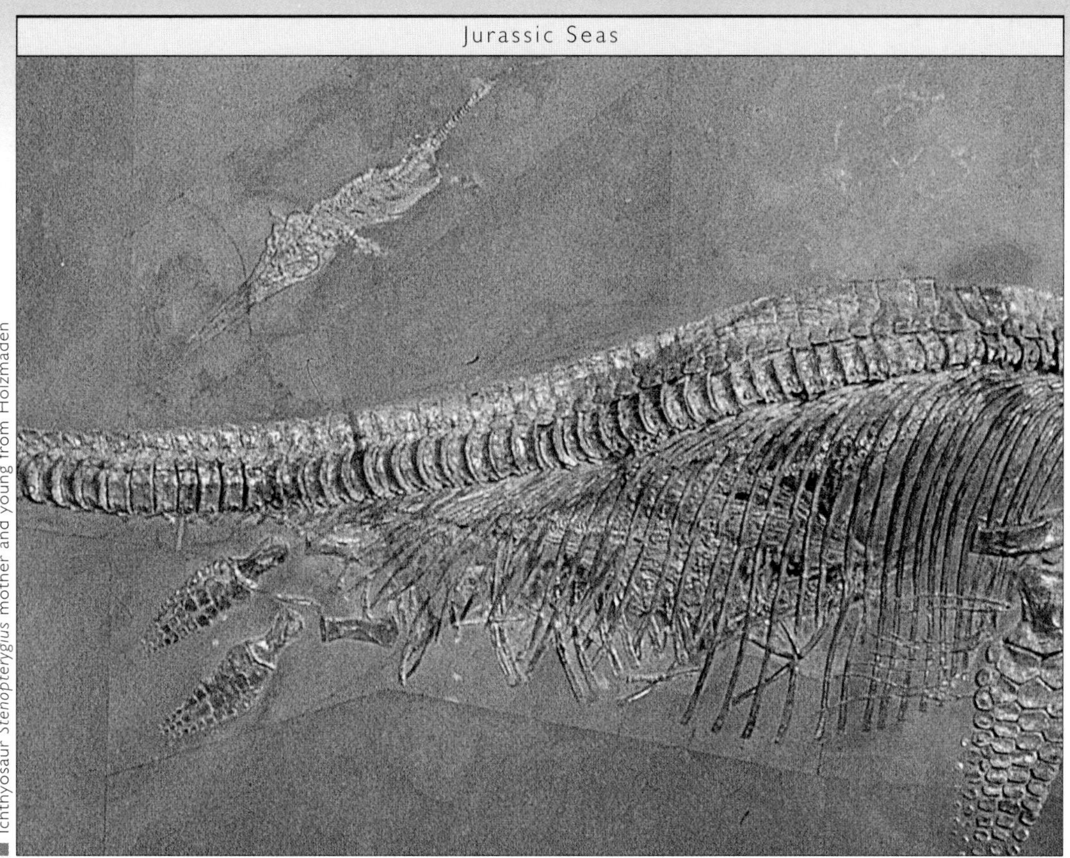

Ichthyosaur *Stenopterygius* mother and young from Holzmaden

covers an entire wall of the Museum Hauff, which is located near the deposits. Holzmaden specimens are relatively common and are on display in many natural history museums around the world.

Holzmaden represents the bottom of a large, shallow sea, while Solnhofen represents quiet lagoons behind extensive coral and sponge reefs. In this environment of soft, fine sediments, the remains of animals living in and around the lagoons are well preserved. The sea creatures from Solnhofen are very diverse, including jellyfish, worms, bivalves, ammonites, prawns, shrimp, lobsters, barnacles, crinoids, horseshoe crabs, echinoids, sea stars, and fish. Horseshoe crab fossils are often found by following tracks made as they crawled to their last resting place.

From the land come the fossils of insects, plants, small dinosaurs, lizards, crocodiles, and flying reptiles (pterosaurs). But perhaps the most important fossils from Solnhofen are the seven amazing specimens of *Archaeopteryx*. These show a small dinosaur with two wings, making it the first bird—clear impressions of feathers cover its body. If it were not for the exceptional preservation qualities of the Solnhofen limestone, it is unlikely that impressions of the feathers would exist, and the unique place *Archaeopteryx* holds between birds and dinosaurs would have been more difficult to understand.

The flying reptile Rhamphorynchus *(above) and a reconstruction of* Archaeopteryx *(right), both from Solnhofen.*

Opal mine at Lightning Ridge, NSW, Australia

Opal Fossils

Opals are among the most beautiful of gemstones, their deep images of flashing light and range of colors prized by people through the ages. Opal is a silicate, a mineral composed of silicon dioxide, the same molecule that forms glass. When water saturated with silica percolates through a host rock, the silica oxidizes to silicon dioxide and turns to opal. Solid silicon dioxide usually accumulates in cavities and veins in the host rock. These spaces can be bubbles and irregularities in the structure of the rock, but occasionally, the space is the natural mold made by an organism buried within the sediment. In such cases, when the silicon dioxide turns to opal, an opal fossil is formed.

Although found in many places, most of the world's gem-quality opal comes from early Cretaceous deposits in Australia. These sediments collected on the bottom of a huge inland sea and the inhabitants of that sea were often buried on the sea floor. In some cases, those remains have become opal fossils, the most abundant of which are shellfish, such as bivalves and gastropods. Opalized belemnites (a kind of straight-shelled nautiloid), opalized bones, and even whole skeletons have been found, but these are quite rare. Only about a dozen reasonably complete opalized skeletons are known.

Most of the bones and skeletons are of marine creatures, particularly plesiosaurs and pliosaurs, but some ichthyosaurs have

Polished opalized fossil belemnite (left) and gastropod shells (right).

FIELD NOTES

Coober Pedy and Andamooka, South Australia
White Cliffs and Lightning Ridge, New South Wales, Australia

The Australian Museum, Sydney, Australia
The South Australian Museum, Adelaide, Australia

€ O S D C P Tr J K T

Opalized marine fossils in polished slab

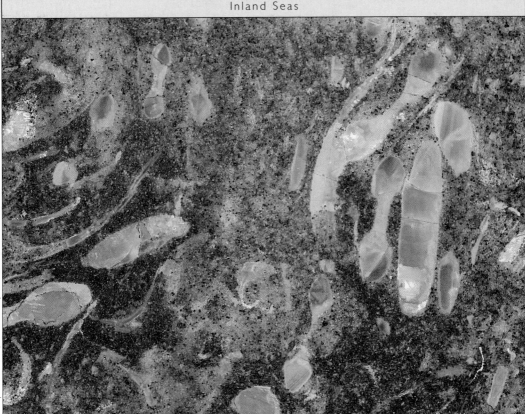

also been found. One particular specimen of a pliosaur became famous in Australia in 1993 when there was a possibility that it might be sold overseas. Within a few months, a public appeal raised almost $400,000, mostly through the efforts of primary-school children, to save Eric (the nickname the pliosaur acquired). Eric is now on display in the Australian Museum, Sydney.

Opalized jaw of Kollikodon, *an egg-laying mammal. Backlighting (below) reveals the internal structure.*

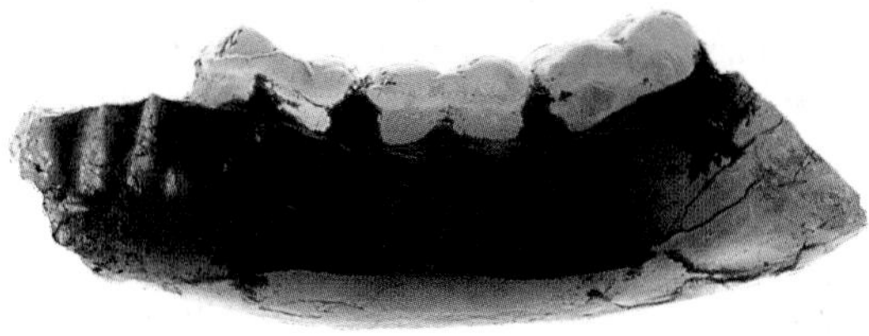

Terrestrial animals preserved as opal are extremely rare, but they have been recovered from Andamooka in South Australia and Lightning Ridge in New South Wales. Terrestrial animals, including dinosaurs and mammals, are known only from isolated elements that had been washed into the sea. The only opalized mammal fossils known from Australia from the Age of the Dinosaurs are two pieces of jaw from Lightning Ridge. Similarly, some Australian dinosaurs, such as *Walgettosuchus* and *Rapator*, are known from only a few, isolated, opalized bones.

Opal fossils are found by opal miners as they tunnel through the ground in search of gems. Usually, an opal fossil is formed from common opal with little of the characteristic flashes of color seen in gem-quality opal, but occasionally, an opal fossil composed of high-quality gem opal is found. These greatly sought-after specimens are prized pieces in collections worldwide.

The two small mammals known from opal fossils are *Steropodon* and *Kollikodon*, cat-size animals that are also monotremes, the group of egg-laying mammals that includes the modern echidna and platypus. The low, rounded teeth of *Kollikodon* indicate that it probably fed on prey that required crushing, such as clams and mussels. The sharper teeth of *Steropodon* indicate that it was a carnivore eating softer items.

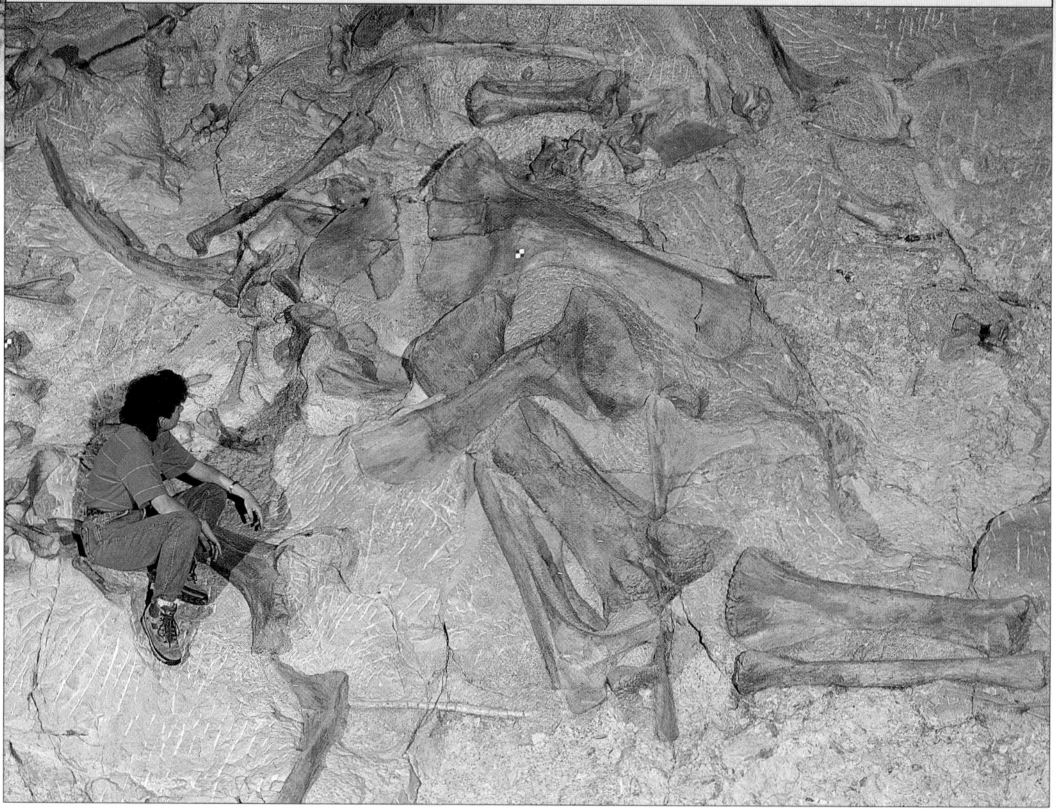

■ *Apatosaurus* and *Diplodocus* bones, Colorado, USA

Dinosaur Sites

For 160 million years, from the late Triassic to the end of the Cretaceous, dinosaurs ruled the Earth. Their fossilized remains have been found on every continent, including Antarctica, but their distribution is skewed to favor a few special regions. One of the most prolific is the western part of North America. Dinosaurs lived on land, usually a poor place for creating fossils, but during the Mesozoic Era, the Age of the Dinosaurs, much of western North America was covered by the flood plains of enormous river systems that drained into an inland sea. These flood plains were home to many different dinosaurs and other animals.

Centrosaurus *bone bed excavation.*

An area known as the Dinosaur Triangle, between Vernal and Price in Utah and Grand Junction in Colorado, is particularly rich in significant dinosaur deposits. Most of these are Jurassic in age along with some Triassic and Cretaceous deposits. Perhaps the most well-known site is Dinosaur National Monument, east of Vernal. In this one quarry the remains of nearly 100 individuals, representing 10 species, have been found. Most are types of long-necked dinosaur, such as *Diplodocus*, *Apatosaurus*, *Camarasaurus*, and *Barosaurus*, but there are other plant-eaters, such as *Stegosaurus*, *Dryosaurus*, and *Camptosaurus*, as well as meat-eaters, such as *Allosaurus* and *Ceratosaurus*. This unusual abundance is a result of bodies collecting in a sandbar as they were washed downstream in a huge river. Dinosaur National Monument is open to the public and many fine specimens can be

FIELD NOTES

Dinosaur National Monument, Utah, USA

Cleveland-Lloyd Quarry, Utah, USA

American Museum of Natural History, New York, USA

British Museum (Natural History), London, UK

Royal Tyrrell Museum, Drumheller, Alberta, Canada

Є O S D C P Tr J K T

Centrosaurus bone bed in Alberta, Canada

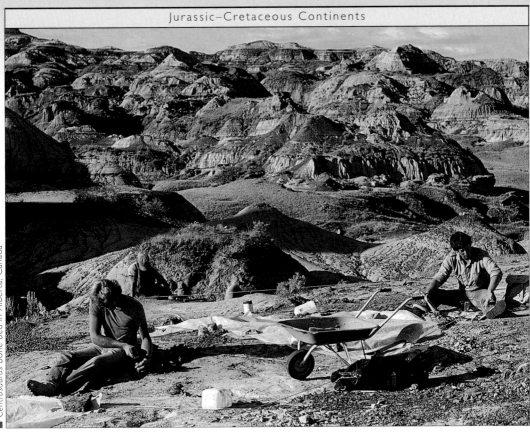

seen, still partly embedded in the quarry face. Other sites in the Dinosaur Triangle include the Cleveland-Lloyd Quarry south of Price.

Younger, Cretaceous-age deposits to the west and north of the Dinosaur Triangle contain a different set of dinosaurs. Dinosaur Provincial Park in Alberta, Canada, has been set aside to protect its vast fossil deposits. Coupled with the Royal Tyrrell Museum near Drumheller, this area is a mecca for dinosaur enthusiasts because of the beautiful specimens that are on display in the museum or have eroded out of the ground.

The Alberta specimens come from toward the end of the dinosaurs' reign and include a variety of strange duck-billed dinosaurs and various types of horned dinosaur.

In one bone bed, lie the remains of thousands of individuals of a horned dinosaur similar to *Centrosaurus*. This species appears to have lived in huge herds, perhaps comprising tens of thousands of individuals, that migrated across the continent in search of food. These migrations entailed dangerous river crossings during which many animals died, their remains swept away by the river. The carcasses collected where the flow of water slowed, producing the bone beds we find today.

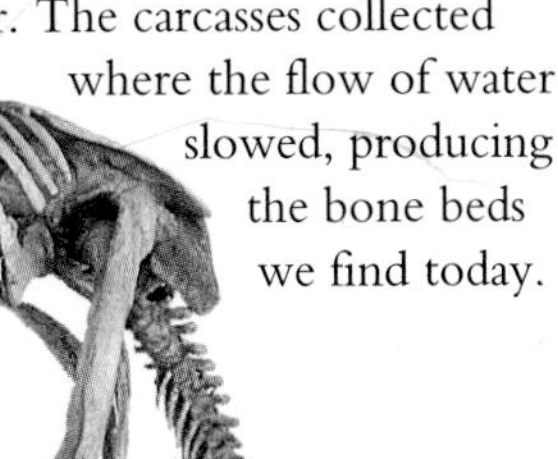

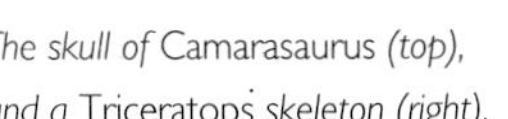

The skull of Camarasaurus *(top), and a* Triceratops *skeleton (right).*

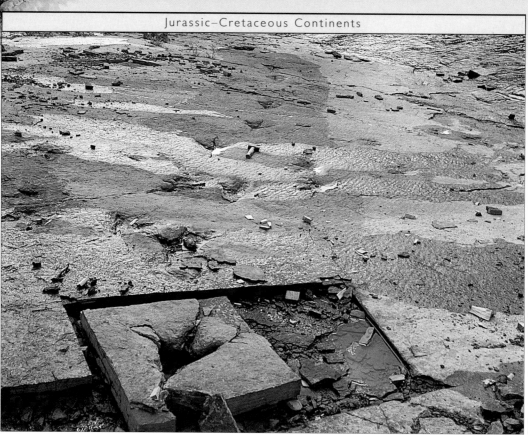

Trackway of meat-eating dinosaurs in Alberta, Canada

Dinosaur Eggs and Tracks

Fossilized tracks, eggs, and babies tell us more about dinosaur behavior than their fossilized bones. While fossilized tracks are reasonably abundant, fossilized eggs are rarer and are usually found only as fragments of shell.

Shell fragments were found in southern France as early as 1859, but the first find of whole dinosaur eggs, including complete nests and juveniles, was made in Mongolia in 1923 by a team from the American Museum of Natural History. Since then, dinosaur eggs have been reported from more than 200 sites in North and South America, Europe, Asia, and Africa. One of the best known sites is the appropriately named Egg Mountain in Montana, USA. Here, a whole dinosaur rookery has been discovered with evidence of many hundreds of nests.

Part of five parallel trackways in the Morrison Formation in Colorado, USA.

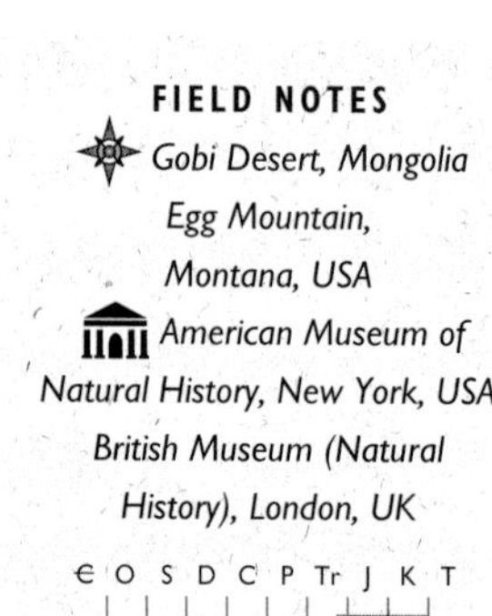

Dinosaur eggs are generally much smaller than most people expect—the largest found so far is 12 inches (30 cm) long and 10 inches (25 cm) in diameter, with a capacity of 5.8 pints (3.3 liters). This is not even twice the dimensions of an ostrich egg, yet it belongs to a 40-foot (12-meter) dinosaur.

From various finds of eggs and nests we know that at least some dinosaurs were similar to birds in the way they gathered in rookeries, built their nests, and looked after their young. At Egg Mountain, the baby dinosaurs are known to have stayed in the nest for an extended period after hatching because they have smashed the egg shells by tramping around in the nest, and because young dinosaurs that have advanced well beyond the hatchling stage have been found fossilized still inside the nests. Further, the teeth of these baby dinosaurs show wear from feeding, so the parents must have brought food back to the nests for the young. These and

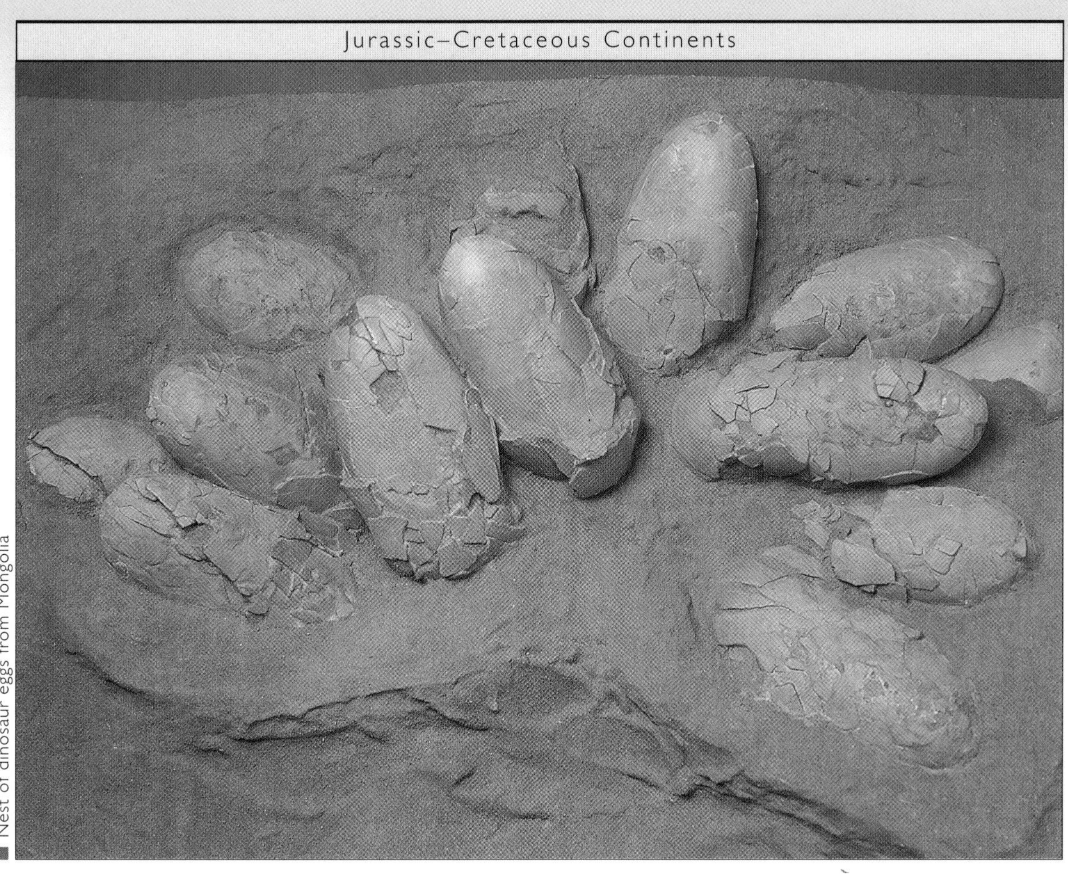

Nest of dinosaur eggs from Mongolia

many other surprising details of dinosaur behavior were gleaned from careful research at the Egg Mountain site. The dinosaur that nested there was named *Maiasaura* ("good mother lizard") in honor of its superior parenting abilities.

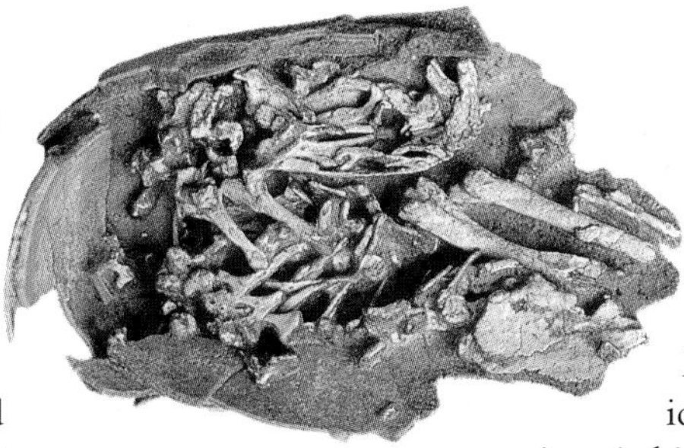

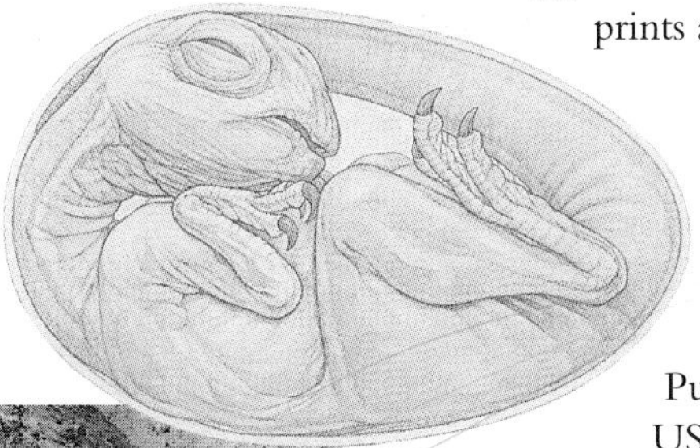

Dr Roy Chapman Andrews at a nest of dinosaur eggs (below) in Mongolia, 1926. Developing dinosaur embryo (right) and a skeleton of an unhatched dinosaur embryo (above).

Dinosaur footprints and trackways have been recorded from all over the world. Early finds of dinosaur footprints from Massachusetts were wrongly identified as belonging to gigantic birds. The longest known footprints are more than 3 feet (90 cm).

We know from various trackways that many dinosaurs lived and moved in groups. Further, these groups were often highly structured. Trackways at Purgatoire River, Colorado, USA, show a number of large, long-necked dinosaurs walking with the juveniles in the middle. This same pattern is seen in social animals today and is a way of protecting the young from attack by predators.

The identity of dinosaur eggs and tracks is usually difficult to establish. In one case, a Mongolian nest was thought to belong to a small, plant-eating dinosaur *Protoceratops* and the skeleton of a meat-eater found nearby was thought to be an egg thief *Oviraptor*. The eggs have since been shown to belong to *Oviraptor*.

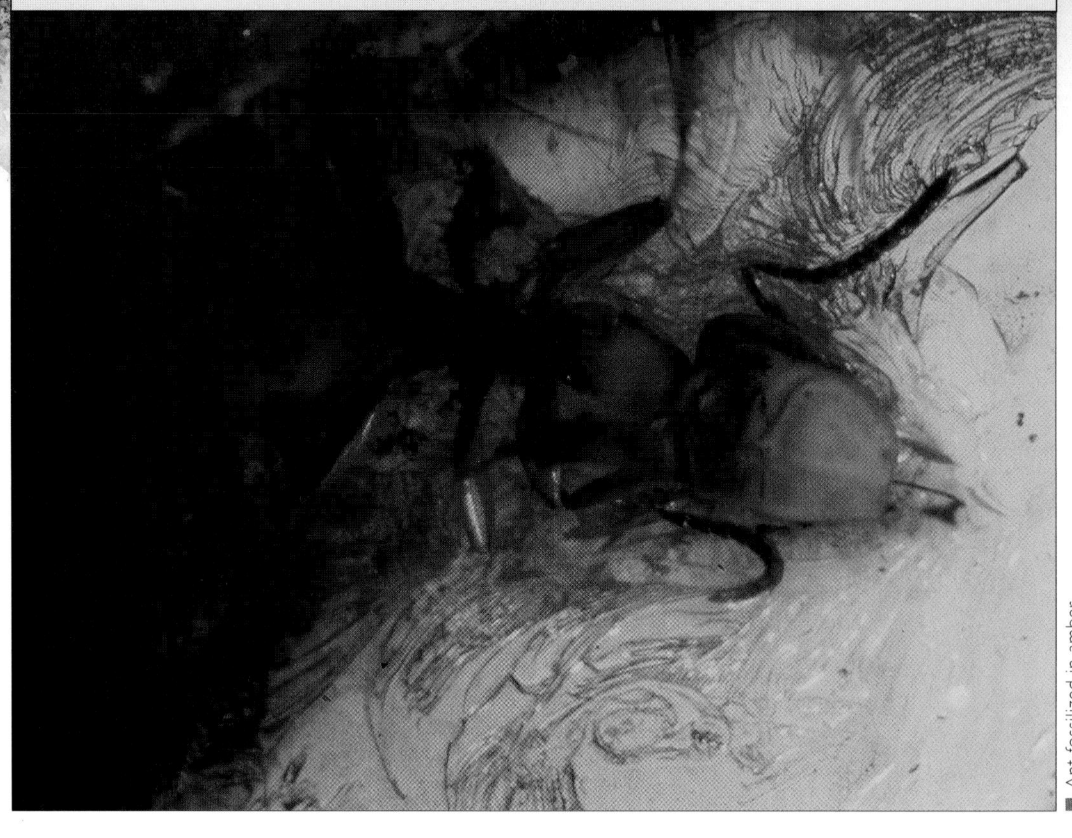

■ Ant fossilized in amber

Amber Fossils

Chrysopilus, *a fossil fly in amber.*

Amber is resinous tree sap that has hardened. The trees that originally produced the sap were usually conifers, but some angiosperms also exuded sap that was converted to amber. Technically, amber is not a mineral, but in many cultures, it is classed with precious stones for use in jewelry and ornaments. Last century, amber was melted down to produce high-quality lacquers and varnishes. People have long been attracted by its beauty. A wedge-shaped piece was found in a 10,000-year-old burial site in Denmark. Most amber is less than 70 million years old, but some types are 100 million years old.

As old, hardened tree sap, amber can be thought of as a fossil, but the most widely recognized fossils associated with amber are the insects and other small animals that occasionally became trapped in the sap while it was still viscous. The vast majority of amber fossils are insects, but other organisms, such as spiders, centipedes, and even small lizards, are found preserved in this way.

Many aspects of amber fossils are unique, and their importance cannot be overstated. An animal trapped in amber is preserved in exquisite detail, unparalleled by other forms of fossil preservation. Usually, even microscopic hairs on the most minute of insects are faithfully preserved without any crushing or distortion. Amber fossils also provide a detailed record for groups of organisms that are very delicate and rarely preserved in other ways. Insects and spiders are very poorly represented in the fossil record and what fossils are known of representatives of these groups are usually badly

FIELD NOTES

Cliffwood, New Jersey, USA
The Baltic Coast
The Dominican Republic
American Museum of Natural History, New York, USA
British Museum (Natural History), London, UK
Smithsonian Institution, Washington, DC, USA

€ O S D C P Tr J K T

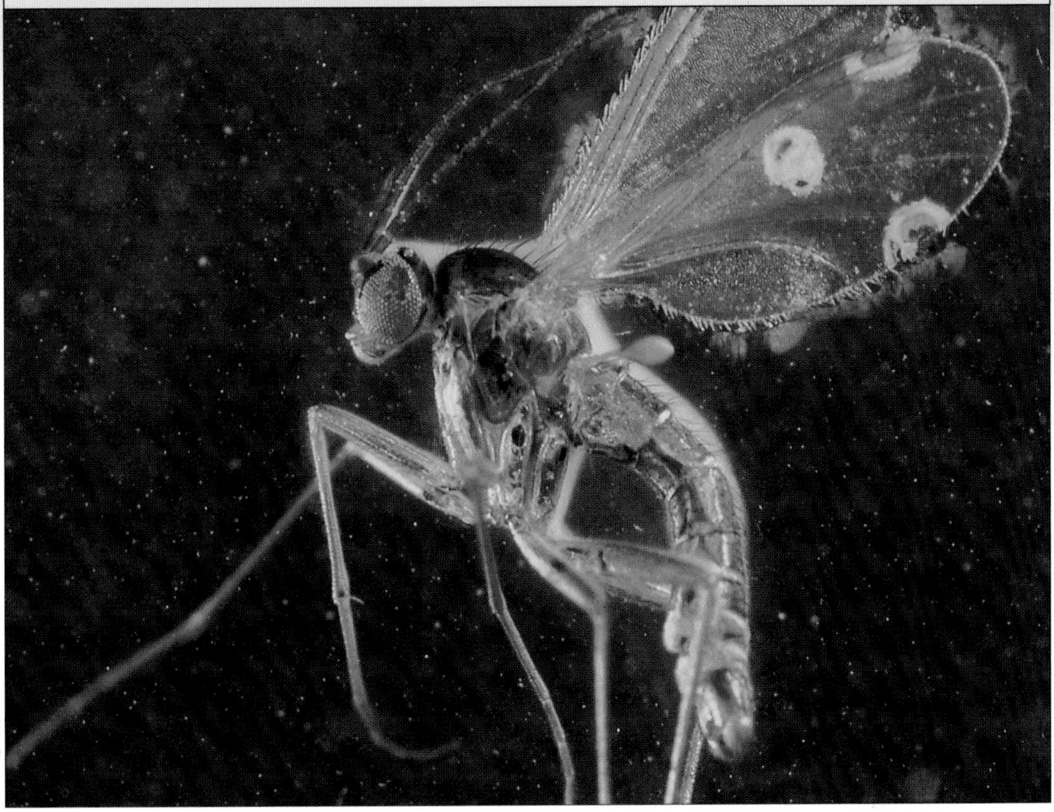

Fungus gnat in amber, Eocene

crushed and distorted. Without the amber fossils of insects and spiders, little would be known about the evolutionary history of either group.

While amber fossils represent one of the most ideal forms of fossil preservation, it is not perfect. Once trapped in the tree sap, the organism decays and dehydrates leaving a carbonized mass inside a perfect mold of the original organism. In some cases, insects in amber have been dissected, and carbonized films representing the remains of major muscle groups and internal structures have been identified. However, despite the fantastic stories in the film *Jurassic Park*, we have not been able to retrieve DNA from the food of ancient insects. Small fragments of DNA from some amber insects have been identified, making it possible to compare their ancient DNA with that of contemporary relatives and calibrate the rate of change. Paleontologists have recently been able to reactivate bacterial spores from inside 40-million-year-old bees trapped in amber.

Amber with a 35–40-million-year-old moth.

Many fine specimens of insects in amber are available for purchase through rock and museum shops. The buyer should, however, be wary, as amber fossils are easily faked using living insects and synthetic amber-like resins.

Although there are tests available to distinguish real and synthetic amber, the best advice for the amateur is to deal only with reputable outlets and to avoid unusually cheap specimens.

While very little amber contains insects, such fossils are known from many sites around the world, including the coast of the Baltic Sea, the Isle of Wight, and New Zealand. Amber is usually associated with coal and shale deposits.

Ants in amber.

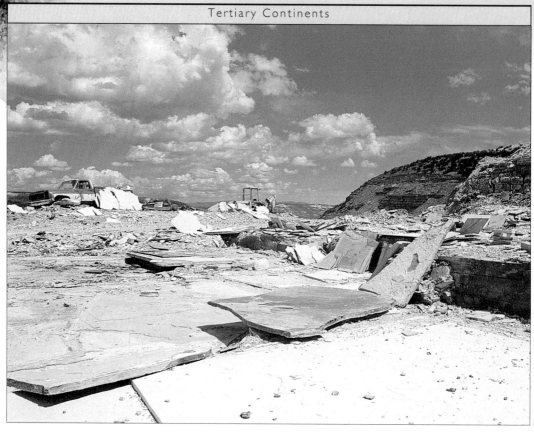

■ A quarry at Fossil Butte National Monument, Wyoming, USA

After the Dinosaurs

After the extinction of the dinosaurs some 65 million years ago, a new group, the mammals, assumed dominance of the major terrestrial environments and later made significant forays into both the air and sea. Mammals actually arose at about the same time as the first dinosaurs, in the late Triassic, but during the reign of the dinosaurs, they remained small and fairly inconspicuous in the fossil record. It is argued that if dinosaurs had not become extinct and left the field free of large vertebrates, mammals may not have become the dominant group they now are.

By the early Tertiary, the mammals had diversified into an astonishing variety of forms and early representatives of most of the major groups had made their first appearance. At this time, the continents of the Northern Hemisphere were still close enough to each other to allow exchange among them of some groups of animals. Primates, for example, the group that today includes monkeys, apes, and humans, were reasonably abundant in North America, a continent from which they later disappeared. Similarly, more primitive types of mammal, such as marsupials, were quite common in the northern hemisphere, but by the middle of the Tertiary, they were restricted to South America and Australia.

The early to mid Tertiary was a relatively warm, moist time, and forests dominated much of the planet. Animals living in forests tend to be smaller than those in more open habitats and generally more diverse. North America was

Fossil stingray from the Green River Formation, Wyoming, USA.

FIELD NOTES

Fossil Butte National Monument, Wyoming, USA

The Big Horn, Montana, USA

American Museum of Natural History, New York, USA

United States Museum of Natural History, Smithsonian Institution, Washington, DC, USA

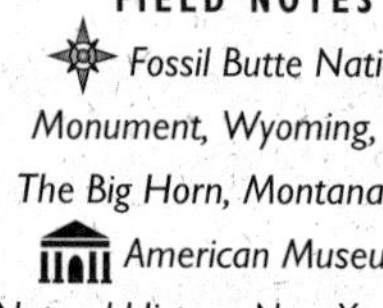

Fossil fish from the Green River quarry, Wyoming, USA

Crocodile skull (right) and a primate skull, Plesiadapsis *(below).*

typical for this period and a number of very important sites record the evolution of mammals at this time. At Fossil Butte National Monument in Wyoming, numerous fossils from the early Paleocene through to the mid Eocene have been discovered. While most widely known for the thousands of beautifully preserved fossil fish from the Green River Formation, other animals, such as turtles, crocodiles, birds, and mammals, have also been recovered there. Farther to the north, in Montana, is the Big Horn Basin, which covers a similar range of ages as those of the Fossil Butte.

In early Tertiary deposits of western North America, some types of mammal fossils are so common that they are used as index fossils to correlate distant rock units. Mammals are useful as index fossils because their teeth are hard (more likely to become fossilized) and distinctive in shape for each different species, so they can be readily identified. For example, the teeth of the small mammal *Haplomylus* are commonly used as index fossils in this area to identify rock of very early Eocene age.

While the period after the demise of the dinosaurs is often referred to as the Age of Mammals, it is important to realize that many other groups of animals have also had evolutionary success during this time. Birds have diversified to a degree that parallels the success of the mammals. Lizards, snakes, crocodiles, and turtles have also diversified, but not to the same extent as mammals or birds. In the seas, the advanced bony fish (the teleosts) rapidly diversified into thousands of species representing many different forms.

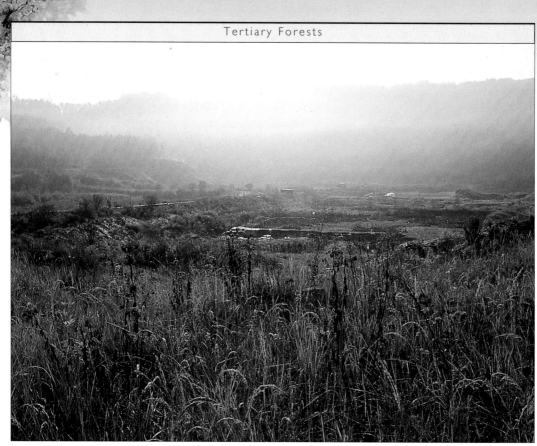

Grube Messel, Darmstadt, Germany

The Age of Mammals

Fossil deposits representing the age of mammals are usually more common than those from the age of dinosaurs, even though dinosaurs were around for more than twice as long. Younger fossils are better represented than older ones, mainly because older fossil deposits are more likely to have eroded or to have been covered by younger sediments.

Worldwide, there are many fine and unusual deposits of mammal fossils. One of the most fascinating is the Eocene site, Grube Messel, near Darmstadt in Germany. Messel is for vertebrates what the Burgess Shale and the Hunsrückschiefer faunas are for invertebrates—one of those extremely rare situations where the soft parts of organisms were preserved along with hard skeletons. About 150 years ago, quarrying work began at Messel for brown coal and oil shales from which kerosene and other petroleum products were extracted.

Small wallaby skull from Riversleigh.

Fossils were found during these activities and removed to various museums. It was found that when one side of these fossils was carefully prepared and embedded in resin, delicate removal of the remaining rock from the other side could reveal not only the articulated skeleton of the animal, but sometimes also the skin, fur, feathers, internal organs, and gut contents. On occasion, beetles are found with the brilliant, iridescent colors of their wing covers preserved intact.

Messel represents what was once a rain forest area with a diversity of plant and animal

FIELD NOTES

Grube Messel, near Darmstadt, Germany

Riversleigh, Queensland, Australia

Senckenberg Museum, Frankfurt, Germany

Queensland Museum, Brisbane, Australia

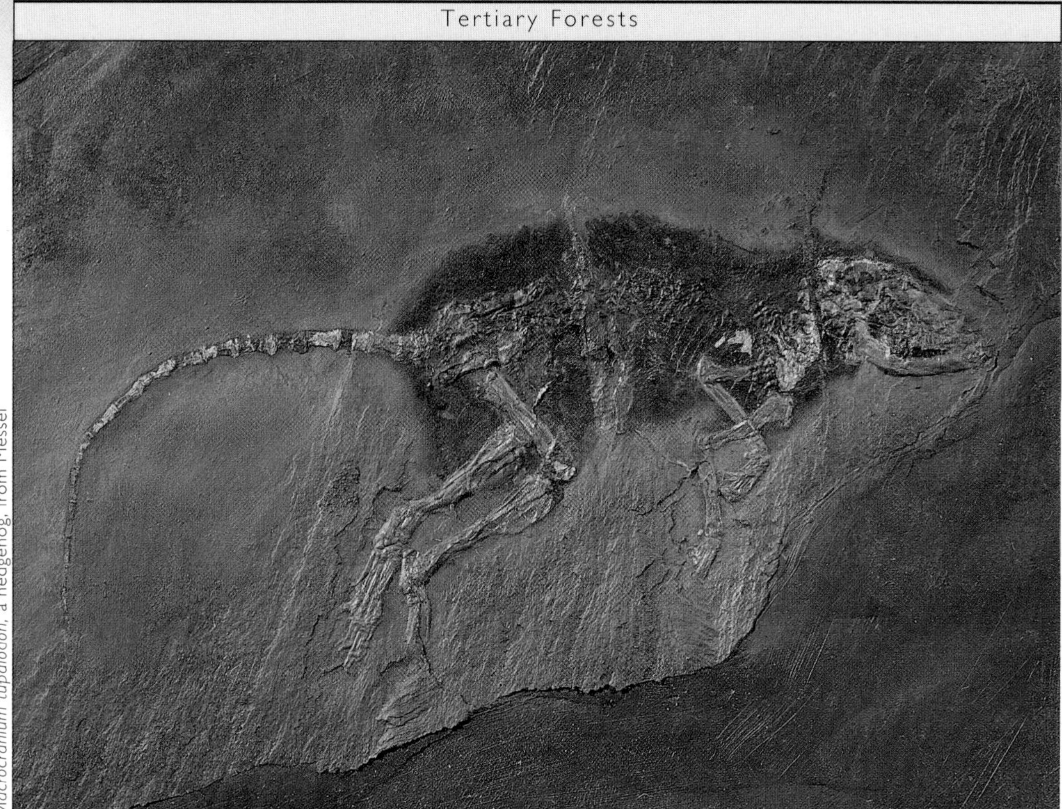

Macrocranium tupaiodon, a hedgehog, from Messel

life. Apart from the many species of plant, this site has produced fossil insects, fish, frogs, turtles, reptiles, birds, and a huge variety of mammals.

On the other side of the world, Riversleigh, in Queensland, Australia, provides a unique and exceptional window on the development of vertebrate life in Australia. It is remarkable because of the abundance of fossils found there and their near-perfect preservation, although animal soft tissue is not preserved in these specimens.

Riversleigh (above), North Queensland, Australia, has produced an abundance of new fossils. Bandicoot skull and jaw (left).

The ancient platypus Obduradon.

In the early Miocene, 25 million years ago, Riversleigh was a rain forest growing on an extensive area of limestone and teeming with animals. As the limestone dissolved in the water on the forest floor, animals or skeletons that fell into pools and ponds gradually became encased in limestone, an excellent preserving medium. They can be released by dissolving the surrounding rock in mild acid, leaving behind skulls, teeth, and bones as perfectly preserved as the moment the animal died. Fossil insects, crustaceans, snails, fish, lungfish, frogs, crocodiles, snakes, turtles, lizards, birds, and a variety of mammals, including egg-laying mammals, marsupials, and bats, have all been recovered from Riversleigh.

Tar pits in the Los Angeles Basin, California, USA

Tar Pits

In the heart of metropolitan Los Angeles is a unique and remarkable fossil deposit, the Rancho La Brea Tar Pits. These disclose a relatively complete record of literally thousands of animals and plants that lived in the area during the past 40,000 years.

The Los Angeles Basin contains significant reserves of oil, to the extent that there are working oil derricks in the heart of downtown Los Angeles. The area is also tectonically active and the energy from this activity has cooked portions of the oil deposits into a thick, gooey soup and assisted in the transport of this material through the rock units. At Rancho La Brea, these natural asphalts periodically break through to the surface where they form pools. Surface pools of asphalt act like adhesive flypaper and are literally capable of holding an elephant. Once stuck in the tar, animals and plants are quickly buried, and the petroleum-rich sediments are an excellent preservative for the hapless creatures.

The variety of life recovered from Rancho La Brea is remarkably diverse, including many types of plant, insect, ostracod, bivalve, gastropod, and many different vertebrates. Famed for its beautifully preserved large mammals, including bison, horses, sabertooth cats, wolves, coyotes, giant sloths, camels, antelopes, and bears, the site has also yielded other vertebrates, including small birds, reptiles, and amphibians. More rarely, elephantine mammoths and mastodons were found at Rancho La Brea as well as deer, a tapir, a jaguar, and a peccary.

While the site has produced an amazing diversity of animals, there is an unusual twist about

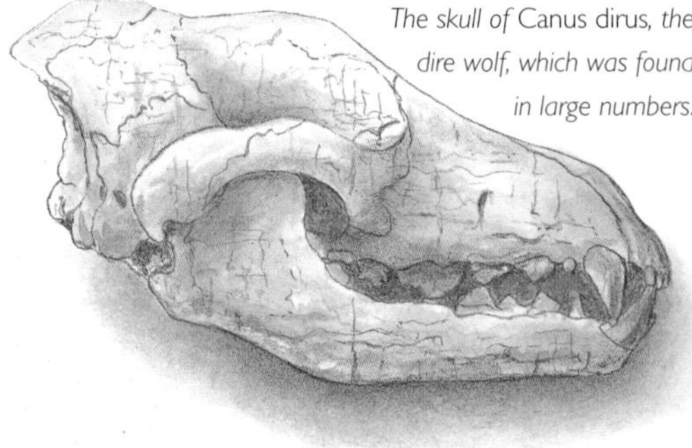
The skull of Canus dirus, *the dire wolf, which was found in large numbers.*

FIELD NOTES

Rancho La Brea Tar Pits, Los Angeles, California, USA

Los Angeles County Museum of Natural History, California, USA

George C. Page Museum, Los Angeles, California, USA

€ O S D C P Tr J K T

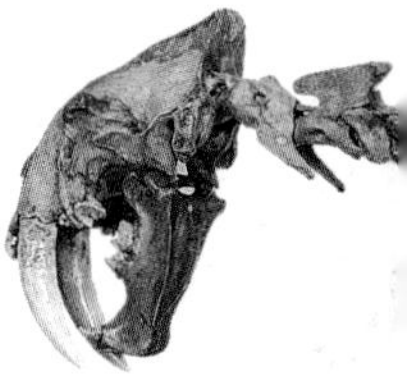

Excavation at Rancho La Brea site c. 1910

which ones are prominent. Predators, such as dire wolves (larger than, but similar to, the living gray wolf), saber-tooth and other large cats, together outnumber the large prey animals by more than 6:1. Usually, the ratio of predator to prey would be the other way around with prey outnumbering predators by about 20:1. It appears that Rancho La Brea was a death-trap for predators. One can imagine that a large prey species, such as a bison,

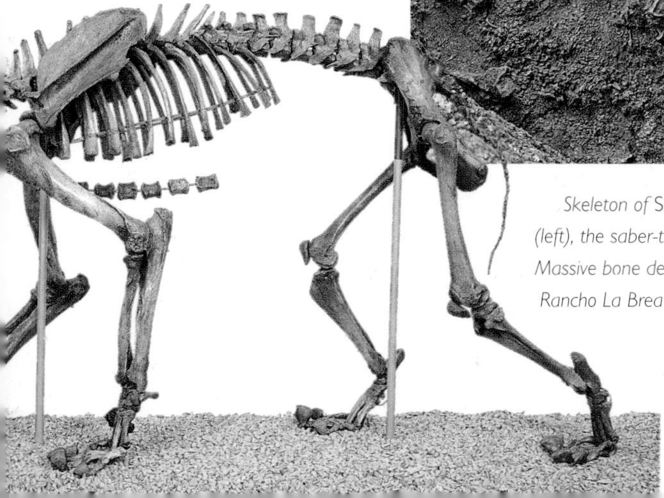

Skeleton of Smilodon *(left), the saber-tooth cat. Massive bone deposits at Rancho La Brea (above).*

hopelessly mired in the tar would be an irresistible attractant for numerous and varied predators who, in their attempts to gain a quick and easy meal, also became trapped in the sticky grave.

By about 11,000 years ago, the larger animals, including saber-tooth cats, native horses, and bison, had disappeared from the area. A short time later, in paleontological terms, a new animal appeared in the deposit. While only one specimen of this animal, a human, has been recovered, a quantity of human artifacts has also been found at Rancho La Brea.

By about 9,000 years ago, when this luckless individual was living in what is now the Los Angeles area, the animals and plants were essentially the same as those that we would recognize from the area today.

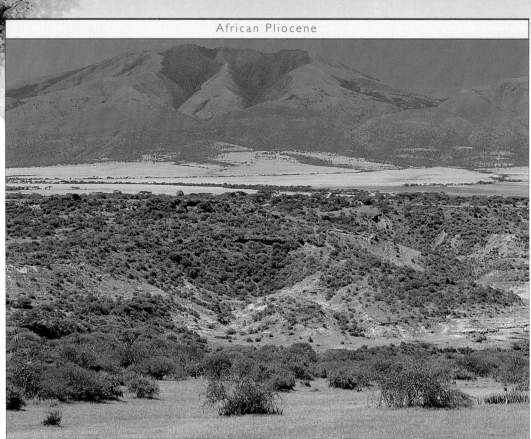

Olduvai Gorge, Tanzania

Early Fossil Humans

Of all the organisms preserved in the fossil record, no group fascinates us more than those that lead directly to ourselves. The fossil record of our ancestors, while not abundant, is complete enough to show clearly our common ancestry with apes and our evolution.

It was Charles Darwin who correctly suggested that Africa was the cradle of humanity. Fortunately, the eastern half of Africa has been rifting apart for the past few million years, thus providing the right kinds of rockbed to preserve human fossils at the time when apes, including humans, were evolving away from the other primates. In the early Miocene, about 18 million years ago, a baboon-sized primate called *Proconsul* appeared in Africa.

Proconsul has features that clearly indicate that it was more ape-like than other primates and it appears to represent a common ancestor for both humans and other apes. The human (hominid) and the African ape branches of the family tree seem to have split about seven or eight million years ago, but there are few fossils from this period. The earliest recognizable hominid, *Australopithecus ramus*, was found in Eritrea in 1994. This fossil is 4.5 million years old, and for the next 2.5 million years, other species of *Australopithecus* roamed Africa. Mostly shorter than 5 feet (150 cm), *Australopithecus* had a brain larger than that of a chimpanzee but smaller than a modern human. The faces of these creatures were more human-like than those of any other apes. *Australopithecus* has been found in a number of eastern and southern African sites, including Olduvai Gorge (Tanzania), Hadar (Eritrea), Koobi Fora (Kenya), and Swartkrans (South Africa). Footprints 3.5 million years old from Laetoli (Tanzania) indicate that *Australopithecus* walked fully upright, as do modern humans.

A second type of hominid, *Homo habilis*, appears in some of the sites where *Australopithecus*

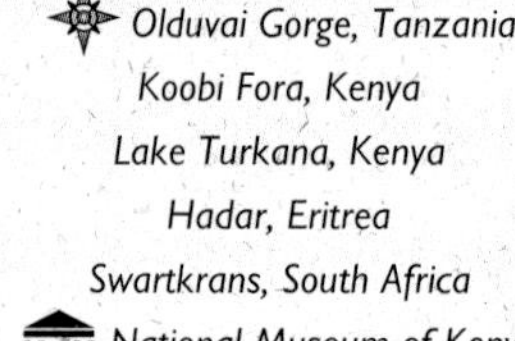

FIELD NOTES

Olduvai Gorge, Tanzania
Koobi Fora, Kenya
Lake Turkana, Kenya
Hadar, Eritrea
Swartkrans, South Africa
National Museum of Kenya, Nairobi, Kenya
National Museum of Ethiopia, Addis Ababa, Eritrea

€ O S D C P Tr J K T

Site at East Turkana, Kenya

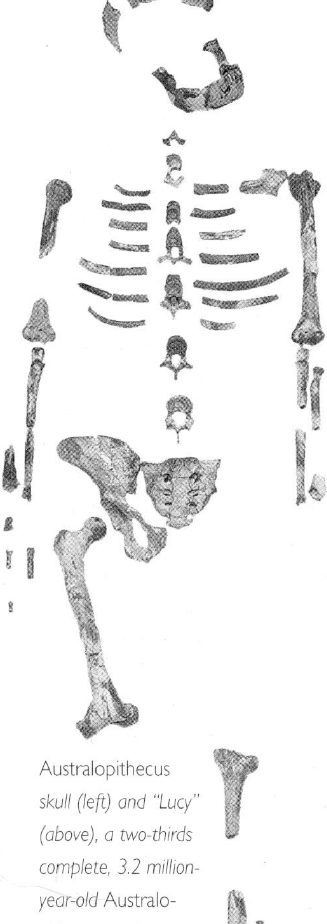

Australopithecus skull (left) and "Lucy" (above), a two-thirds complete, 3.2 million-year-old Australopithecus skeleton, found in Eritrea.

Hominid footprint trail (below), left in volcanic ash in Laetoli, Tanzania.

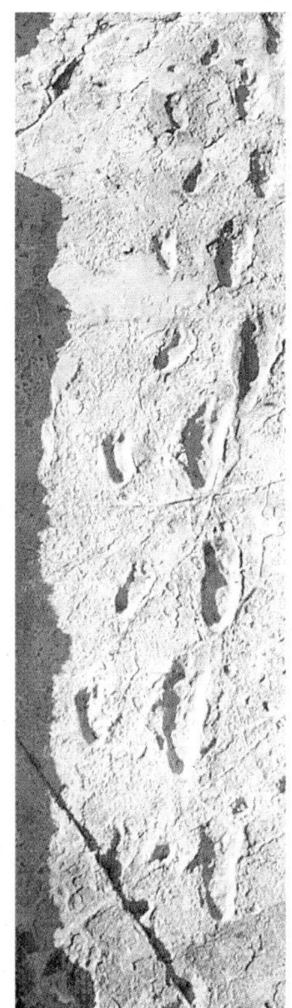

has been found, particularly Olduvai Gorge and Koobi Fora. This is the earliest member of the genus to which we belong.

Homo habilis was generally much larger than *Australopithecus,* with a larger brain and a more vertical face. It was also the first hominid to leave Africa—fossils have been collected from sites in the Middle East—but it was *H. erectus,* the next species of *Homo* to appear, that really began to explore the world.

The oldest specimens of *H. erectus*, found in Africa, are about 1.5 million years old; more modern specimens have been found in Asia and Europe. *Homo erectus* seems to have been the first user of fire, with hearths reported in China, Hungary, and France. *Homo erectus*, including the more popularly known Peking Man, Heidelberg Man, and Java Man, disappears from the record about 300,000 years ago. The only surviving hominid, *Homo sapiens*, went on to dominate the world.

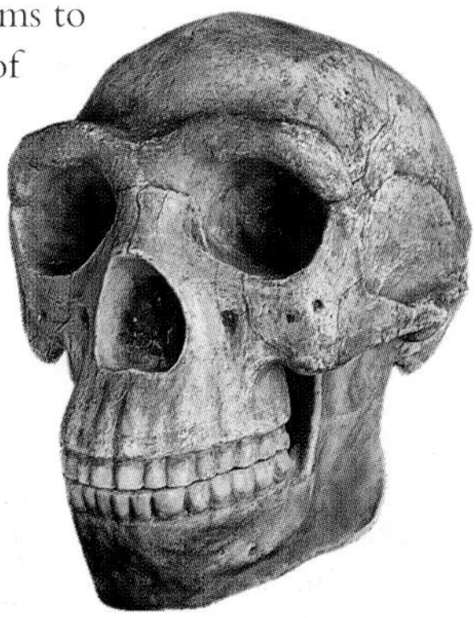

A skull of Homo erectus.

Partial skeleton of a Neandertal

Recent Fossil Humans

The origin of our species, *Homo sapiens*, is somewhat confused by an abundance, rather than a lack, of fossil evidence. Our species is quite varied in its nature as we gradually evolved from our ancestor, *H. erectus*, so the problem lies in the exact definition of when a hominid stops being *H. erectus* and becomes truly *H. sapiens*.

By 700,000 years ago, *H. erectus* had spread as far east as Java, and by 500,000 years ago, there were populations living in Europe. But the transition between *H. erectus* and *H. sapiens* appears to have occurred in Africa about 400,000 years ago. The earliest members of our species are termed archaic forms because they still retain traits of primitive members of our genus, such as thickened skulls and prominent brow ridges.

One group of archaic humans is the Neandertals, found throughout Europe and western Asia from 200,000 years ago until about 30,000 years ago. Retaining some archaic features, the Neandertals, nevertheless, had an average brain size slightly larger than our own. Despite this, their stone tools show few signs of development or improvement through time and their cultural expressions, such as burials, appear to be very limited. Other *H. sapiens* spread across Asia, and by 60,000 years ago, they had entered Australia. By about 11,000 years ago, they had crossed the Bering Straits from Russia to enter the Americas. The dispersal of *H. sapiens* into Australia and the Americas was assisted by the sea level being lower, leaving land bridges exposed—huge amounts of water were locked up in the glaciers during the last ice age.

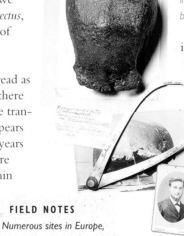

Eugene Dubois (below), who found the skull of Java man (left), later renamed Homo erectus, *in sedimentary deposits at a bend on the Solo River, Java.*

FIELD NOTES

Numerous sites in Europe, Africa, and Asia

Most national museums in Europe, Africa, and Asia
The American Museum of Natural History, New York, USA
The Smithsonian Institution, Washington, DC, USA

€ O S D C P Tr J K T

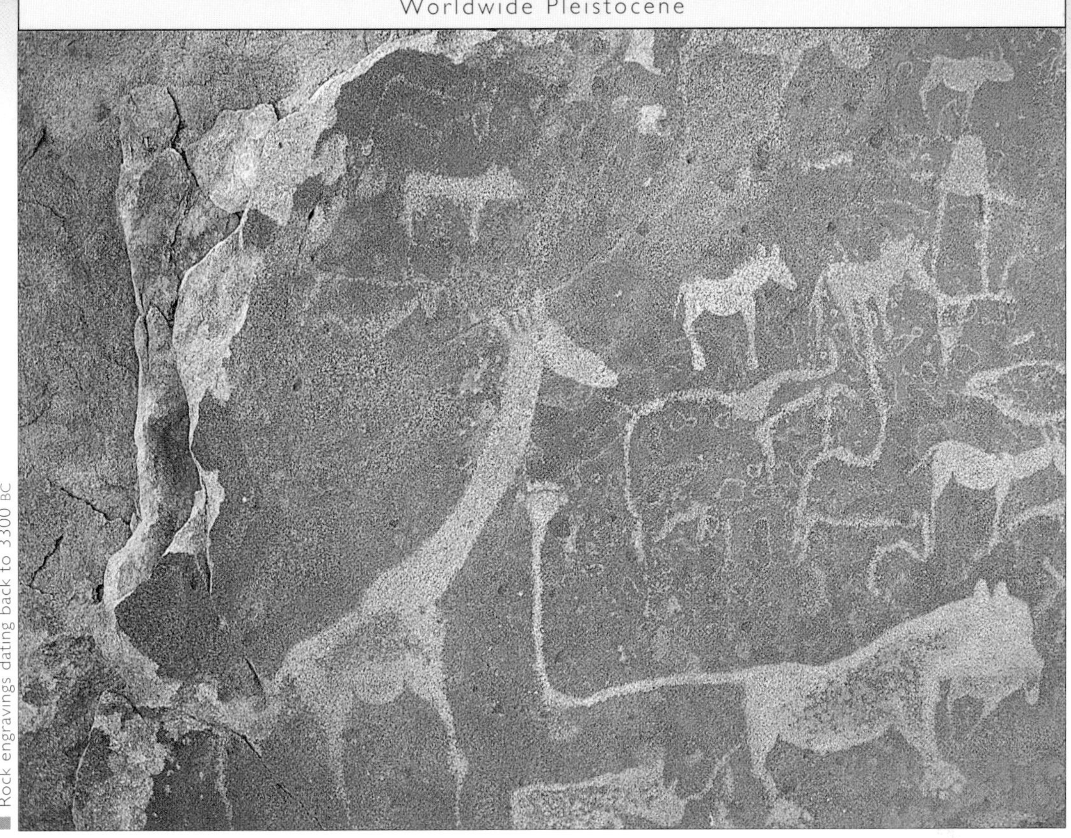

Rock engravings dating back to 3300 BC

Despite this assistance, the first Australians would still have had to cross significant expanses of sea, indicating that they may have been the first people to build seaworthy vessels.

About 35,000 years ago, a more modern form of *H. sapiens*, often called the Cro-Magnons, appeared in Europe and began to displace the Neandertals. The Cro-Magnons had a more developed culture than the Neandertals as well as more complex and intricate stone tools.

Skull and bones of a Neandertal (below) and the skull and mandible of a Cro-Magnon male (above right).

Set against the 3,500-million-year history of life on Earth, the development of our species in the past few tens of thousands of years has been dramatic, perhaps even frightening. By about 10,000 years ago, the basic principles of agriculture had been acquired, and by 5,000 years later, a few types of animals had been domesticated. The first walled settlements were established some 8,000 years ago, and the principles of working copper, lead, and gold were mastered 6,000 years ago.

By 3,500 years ago, some societies were complex enough to form large empires by conquering their neighbors. Warfare is a dominant theme in the recent history of *H. sapiens*, culminating in the development of weapons capable of destroying all life.

Populations of *H. sapiens* have reached numbers that threaten their continued existence and that of many of their fellow organisms. After evolving for 3,500 million years, it would be folly to squander the rich heritage of life.

Expedition to recover frozen mammoth, Siberia, 1901

Frozen Fossils

The ivory of mammoths from Siberia has been traded extensively in both China and Europe for perhaps the past thousand years. In 1707, Evert Ysbrand Ides, a Dutchman involved in this trade, reported that the frozen head and foot of a mammoth had been recovered from the icy tundra. Since then, the frozen remains of as many as 24 mammoths have been retrieved from all over northern Siberia, as well as frozen woolly rhinoceros, bison, and an array of other animals.

These animals date back to the last ice age, which ended 10,000 years ago. It was once thought that larger animals became trapped in crevasses in the glaciers, which then collapsed in on the hapless victims. More recent research shows that these animals tended to live where there was sandy ground that became boggy when thawed. Perhaps, during an exceptional summer, the ground thawed and the animals became mired, only to be deep frozen the following winter. Being essentially packed in ice, the carcasses of these animals are remarkably well preserved—claims that the meat is still fit for human consumption are probably exaggerated (although there are reliable reports of dogs eating it and of carcasses being

Woolly mammoths (above) were common in northern latitudes during the Pleistocene. A frozen mammoth leg (left).

FIELD NOTES

Numerous sites in Siberia
The Italian Alps
Paleontological Institute, Moscow, Russia
University of Innsbruck, Austria
British Museum (Natural History), London, UK

€ O S D C P Tr J K T

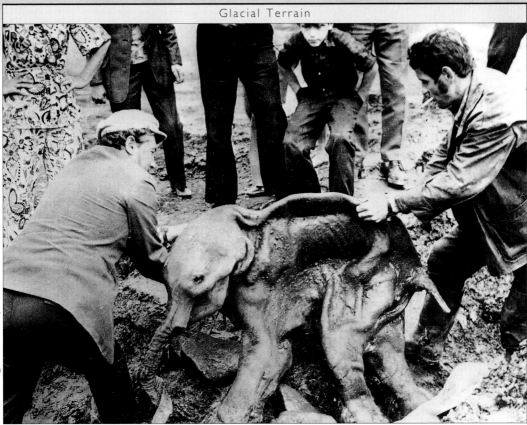

Recovering a frozen baby mammoth, Siberia, 1977

scavenged by wolves and foxes). The fur (or "wool") of these animals has also been preserved, along with the internal organs and stomach contents—one frozen mammoth was found to have been eating wild thyme, poppies, and buttercups just before its death.

Frozen animals are often found on river banks where the action of the river has eroded the tundra to expose the buried carcasses. This led to early suggestions that the mammoth was a burrowing creature that would die if it broke through the surface and was exposed to light and air. Natives of northern Siberia sought out these specimens, not only to collect their tusks for trade, but also to retrieve their bones, which they fashioned into bowls, cups, and small ornaments.

Frozen fossils are usually associated with what is now permanently frozen ground, such as the tundra in northern Siberia. However, a dramatic find was made in 1991 in a glacier high in the Alps between Italy and Austria. The remains of a frozen man, nicknamed Ötzi, had become exposed during the spring thaw. Ötzi was beautifully preserved along with his belongings, which included leather leggings, a loin cloth, a fur cap, a grass raincoat, and a jacket made of alternate strips of brown and golden fur. This 5,250-year-old traveler, with three broken ribs, appears to have been caught in a blizzard. He seemed poorly equipped for the environment in which he died and may have been fleeing some catastrophe.

This 5,250-year-old man (above) was found frozen in the Italian Alps in 1991. He was found complete with his axe (above) and other possessions of extraordinary workmanship.

The height of heaven, the breadth of earth, the abyss, and wisdom—who can search them out?

Apocrypha,
ECCLESIASTICUS 1:3

RESOURCES DIRECTORY

Further Reading

Rocks and Minerals

A Field Guide in Color to Minerals, Rocks and Precious Stones, by Jaroslav Bauer (Chartwell Books Inc, 1992). A very useful field guide to minerals, gemstones, and rocks, in which the minerals are grouped according to their color, and then in increasing hardness within each color group.

A Field Guide to Rocks and Minerals, by Frederick H. Pough (Constable, London). A guide to identification of specimens in the field, with more than 250 photographs.

Evolution of the Earth, by Robert H. Dott (McGraw Hill, 1994). A review of the geological history of the USA, period by period. Also covers the rest of the world, but in much less detail. Deals with the development of the theory of plate tectonics. Well illustrated with many maps, diagrams, and fossil photos and drawings.

Gems and Crystals from the American Museum of Natural History, by A. Sofianides and G. Harlo (Houghton Mifflin, 1990). The world's greatest collection, housed in the American Museum of Natural History, New York, NY, USA.

Gemstones of the Southern Continents, by Lin Sutherland, (Reed Books Pty. Ltd., 1991). A description of the natural gemstones, how they form and what controls their characteristic properties.

Geology of National Parks, by Ann G. Harris and Ester Tuttle (Kendal/ Hunt, 1990). Reviews the geology of 49 National Parks in the USA. Although its focus is quite specific, this is also a good, general geology text.

Geology of the Great Basin, by Bill Fiero (University of Nevada Press, 1986). Although dealing mostly with Nevada and Utah in the USA, this high-quality book explains many phenomena that have worldwide significance. Clear descriptions and many good diagrams.

Handbook of Gem Identification, (12th Edition), by Richard Liddicoat Jr (Gemological Institute of America, 1990). The standard gem identification reference, containing step-by-step procedures for simple and conclusive tests.

Minerals of the World: An Illustrated Encyclopedia of the World's Rocks and Minerals, by Rudolf Duda and Lubos Rejl (Arch Cape Press, 1990). A comprehensive guide to more than 600 of the world's minerals, arranged in order of increasing hardness, each photographed in the form in which it occurs naturally.

Minerals, Rocks and Fossils, by W.R. Hamilton, A.R. Woolley, A.C. Bishop (Hamlyn Publishing Group, 1992). A pocket-sized field guide with more than 600 superb color illustrations and photographs.

New Views on an Old Planet—A History of Global Change, by T.H. Van Andel (Cambridge University Press, 1994). The most recent book to explain in simple language the interrelationships within the Earth sciences, using the plate tectonics overview.

Prospecting for Gemstones and Minerals, by John Sinkankas (Van Nostrand Reinhold, 1970). A comprehensive introduction to starting a mineral collection.

Rocks and Minerals, by Chris Pellant (Dorling Kindersley, 1992). An excellent, full-color field guide to over 500 rocks and minerals. Includes a useful introductory chapter on mineral and rock formation and mineral identification.

The Crust of Our Earth, an Armchair Guide to the New Geology, by Chet Raymo (Simon & Schuster, 1986). An inexpensive guide to the fundamentals of plate tectonics, explained in clear and simple language with many helpful illustrations.

The Macdonald Encyclopedia of Rocks and Minerals, by A. Mottana, R. Crespi, and G. Liborio (Macdonald & Co Ltd, 1987). A dictionary of more than 350 minerals and rocks describing their appearance, physical properties, the environment in which they form, their occurrence and uses.

Stones, by R.V. Dietrich (Geoscience Press, 1989). A guide to the collection and identification of worn stones in the field.

The Audubon Society Field Guide to North American Rocks and Minerals, by Charles W. Chesterman (Alfred A. Knopf, 1990). A comprehensive, pocket-sized mineral identification guide with almost 800 photographs in full color.

The Field Guide to Geology, by David Lambert and the Diagram Group (Facts On File, 1988). Deals with a broad range of Earth-science topics in a clear and interesting way. Excellent diagrams to simplify even complex concepts.

The Volcanic Earth: Volcanoes and Plate Tectonics Past, Present & Future, by Lin Sutherland (UNSW Press, Sydney, 1995). The Earth's volcanoes and their geological settings as determined by plate tectonics. Up to the minute and well explained, with hundreds of excellent color photographs.

The Voyage of the Great Southern Ark, by Reg and Maggie Morrison (Ure Smith Press, 1991). A fascinating and highly readable account of the 4,000-million-year journey of the Australian continent.

Volcanoes & Earthquakes, by Linsay Knight (Time-Life Books, 1995). Part of the Nature Company Discoveries series for children, this is a well-illustrated and clearly explained introduction to the fundamentals of plate tectonics.

Fossils

An Illustrated Guide to Fossil Collecting, by R. Casanova, R.P. Ratkevich, (Naturegraph, 1981). A simple, easy to follow guide to fossil collecting.

Atlas of Invertebrate Macrofossils, by J.W. Murray (Longman, 1985). In spite of being rather technical, this book is still accessible to the amateur. It offers an extensive series of comparative pictures to help with the identification of fossil specimens.

Dinosaur Safari Guide, by V. Costa (Voyageur Press, 1994). A must for anyone who wishes to explore the numerous dinosaur sites of North America, including trackways, bone deposits, and museums.

Drawing and Understanding Fossils, by E. W. Nield (Pergamon Press, 1987). A practical and theoretical guide for beginners.

Field Guide Fossils, by Richard Moody (Chancellor Press, 1986). With more than 300 illustrations, this is a particularly useful book for identifying the common fossils of England and Europe.

Fossil, by Dr Paul D. Taylor (Dorling Kindersley, 1990). Part of the Eyewitness Guides for children, this is a great introduction to fossils with many color photographs and illustrations.

Fossils, by Roy A. Gallant (F. Watts, 1985). Aimed at young people, this book describes the processes of fossilization and evolution. It also demonstrates how fossils can tell us about the Earth's history.

Fossil Collectors Handbook, by Gerhard Lichter (Sterling, 1993). Translated from German, this book is a practical guide to finding, identifying, preparing, and displaying fossils.

Fossil Detective, by Christopher Forsey and Joyce Pope (Troll Associates, 1994). This children's book discusses the nature and origin of fossils, their collection and study, and what they can tell us about the past.

Fossils for Amateurs: A Handbook for Collectors, by R.P. MacFall, and J.C. Wollin (Van Nostrand Reinhold Company, 1972). Although a little old, this book provides a thorough coverage of techniques and procedures for the amateur fossil collector. It includes tips on collecting, preparing, preserving, storing, and cataloging fossils.

Fossils of the World, by V. Forek, J. Marek, J. Benes, Ed. Julian Brown (Arch Cape Press, NY, 1988). Lavish color hardback with photographs, line drawings and descriptions of hundreds of common fossils from around the world.

Hunting Dinosaurs: On the Trail of Prehistoric Monsters, by Louie Psihoyos (Capricorn Link, 1994). Produced in association with *National Geographic*, this is an entertaining and readable account of the history of dinosaur research with brilliant photographs.

Invertebrate Palaeontology and Evolution, by E.N.K. Clarkson (Allen & Unwin, 1979). An introduction, at first-year university level, to the various groups of common invertebrate fossils. It provides particularly good explanations of how once-living animals relate to their fossil remains.

Life in Amber, by George O. Poinar (Stanford University Press, 1992). This book surveys all the life forms that have been found preserved in amber, as well as describing the formation of amber and the location of major deposits. Also provides practical information on how to spot fakes.

Life of the Past, by N. Gary Lane (MacMillan Publishing Company, 1992). Excellent paperback, with photographs and line drawings, on the evolution of life on Earth.

Riversleigh, by Michael Archer, Suzanne Hand, and H. Godthelp (Reed, 1991). By focusing on the extraordinarily rich fossil site of Riversleigh in northern Australia, this book tells the story of the ancient rainforests of inland Australia.

The Audubon Society Field Guide to North American Fossils, by Ida Thompson (Alfred A. Knopf, 1992). A handy, pocket-size, comprehensive guide to North American fossils, with color photographs and diagrams.

The Illustrated Encyclopedia of Fossils, by Giovanni Pinna (Facts On File, 1990). Originally in Italian, this is a beautifully illustrated English translation.

The Practical Paleontologist, by Steve Parker and Raymond Louis Bernor (Simon and Schuster, 1990). An amateur's manual to the preservation and collection of fossils.

The Story of Fossils: In Search of Vanished Worlds, by Yvette Gayrard-Valy (Published in USA by Harry N. Abrams, 1994). Originally published in French in 1987, this is a lively and accessible introduction to paleontology.

World Palaeontological Collections, by R.J. Cleevely (Mansell, 1983). This is a handy guide to the best fossil collections throughout the world.

The Bone Hunters: the Heroic Age of Paleontology in the American West, by Url Lanham (Dover Publications, 1991). An entertaining history of paleontology in the USA.

ORGANIZATIONS

Clubs & Societies

USA

American Federation of Mineralogical Societies, PO Box 26523, Oklahoma City, OK 73126-0523, tel. (405) 525-2692

Central Texas Paleontological Society, Meets at the Little Walnut Creek Library, 835 W. Rundberg Lane, Austin, TX 78758. Contact: Don O'Neill, 16420 Edgemere, Pflugerville, TX 78660, tel. (512) 251-2848

Convair Rockhounds, Meets at CRA Auditorium, General Dynamics, Kearny Mesa, San Diego, CA. Contact: Bob Bailey, tel. (619) 466-5119

Dallas Paleontological Society, PO Box 710265, Dallas, TX 75371. Contact: Rocky Manning, tel. (214) 466-0483

Delaware Valley Earth Science Society, Contact: Nathan Schachtman, 408 Third Ave, Haddon Heights, NJ,

Delaware Valley Paleontological Society (DVPS), PO Box 686, Plymouth, PA 19462

Eastern Federation of Mineralogical and Lapidary Societies, E.D. Reynolds, Box F, Amenia, NY 12501

The Fossil Society, Cleveland Museum of Natural History, 1 Wade Oval, University Circle, Cleveland, OH 44106

Gem, Lapidary and Mineral Society of Washington, DC, 2131 No. Monroe St, 201 Arlington, VA 22207

Cobb County Gem & Mineral Society, P.O. Box 309, Marietta, GA 30061

Keene Mineral Club, 12 Partridgeberry Lane, West Swanzey, NH 03469 tel. (603) 352-5932

Kentucky Paleontological Society, 365 Cromwell Way, Lexington, KY 40503 Contact: Dan Phelps tel. (606) 277-3148 e-mail jfcost00@ukcc.uky.edu

Midwest Mineralogical and Lapidary Society of Dearborn, Michigan, PO Box 368, Dearborn, MI 48121

Orange County Mineral Society, Contact: Paul G. Kosara, Treasurer, 608 Stage Rd, Monroe, NY 10950 e-mail paulkosara@aol.com

Oregon Agate and Mineral Society, Meets at 8138 E. Stark St, Portland. Contact: Delora Compton, 7803 SE Morrison St, Portland OR 97215, tel. (503) 252-1736

Philadelphia Mineralogical Society, PO Box 33, Penllyn, PA 19422 Meets at the Schuylkill Center for Environmental Education, 8480 Hagy's Mill Rd, Philadelphia, PA 19128

San Diego Lapidary Society, Meets at 5654 Mildred St, Linda Vista, San Diego, CA Contact: Bob & Cathy Davis, tel. (619) 278-9987

San Diego Mineral and Gem Society, Meets at Spanish Village, Balboa Park, San Diego, CA Contact: Bill Defur, tel. (619) 276-3885

Society of Vertebrate Paleontology, 401 N. Michigan Ave, Chicago, Il, tel. (606) 11-4267

The Southern California Paleontological Society, Meets at the George C. Page Museum, 5801 Wilshire Blvd., Los Angeles, California 90036 Contact: John Tabor, 2608 El Dorado St, Torrance, CA 90503

Canada

Calgary Rock and Lapidary Club, 110 Lissington Drive, Calgary SW T2M 4L6

Club de Minéralogie de Montreal, Case Postale 305, St Michel H2A 3M1

The Vancouver Paleontological Society, Centerpoint Post Office, PO Box 19653, Vancouver, BC V5T 4E7

Walker Mineralogical Club, 100 Queens Park, Toronto M5S 2C6

United Kingdom

Amateur Geological Society, 25 Village Rd, Finchley, London NW3 ITL

Palaeontological Association, Dr Mike Barker, Department of Geology, University of Portsmouth, Portsmouth PO1 3QL, England. e-mail: BarkerMJ@cv.port.ac.uk

Scottish Mineral and Lapidary Club, 22b St Giles St, Edinburgh EH1 1PT

Sidcup Lapidary and Mineral Society, 16 Preston Drive, Bexley Heath DA7 4VQ

Wessex Lapidary & Mineralogical Society, New Alresford, Winchester, Hampshire

Australia

Lapidary Club of New South Wales, PO Box 425, Darlinghurst, NSW 2010

Mineralogical Club of Victoria, PO Box 146, Collins Sreet, Melbourne, Vic 3001

Qld Lapidary and Allied Craft Clubs Association Inc., GPO Box 1761, Qld 4001

The Fossil Club of New South Wales, Contact: Robert Murphy, 52 Lorenzo Cres, Rosemeadow, NSW 2560

The Riversleigh Society Inc.,
PO Box 876,
Kensington, NSW 2033

BELGIUM
Cercle de Minéralogie et Paléontologie de Belgique,
Delhasse St, 36,
1060 Brussels

ITALY
Gruppo Mineralogico Di Ivrea,
Contact: Carlo Alciati,
Viale Biella 17,
10015 Ivrea (To)

University and Museum Collections

USA
Academy of Natural Sciences, Philadelphia, Pennsylvania
Alaska Historical Library and Museum, Juneau, Alaska
American Museum of Natural History, New York, New York
California Academy of Sciences, Golden Gate Park, San Francisco
Carnegie Museum of Natural History, Pittsburgh, Pennsylvania
Denver Museum of Natural History, Denver, Colorado
Dinosaur Quarry Visitor Center, Dinosaur National Monument, Jensen, Utah
Earth Sciences Museum, Provo, Utah
Field Museum of Natural History, Chicago, Illinois
Great Lakes Area Paleontological Museum, Traverse City, Michigan
Los Angeles County Museum, Los Angeles, California
Museum of Northern Arizona, Flagstaff, Arizona
Museum of the Rockies, Bozeman, Montana
National Museum of Natural History (Smithsonian Institution), Washington, DC
Natural History Museum, Carlsbad Caverns National Park, Carlsbad, New Mexico
New York State Museum, Albany, New York
Peabody Museum, Yale University, New Haven, Connecticut
Rainbow Forest Museum, Petrified Forest National Park, Arizona
University of California Museum of Paleontology, Berkeley, California
University of Michigan Museum of Paleontology, Ann Arbor, Michigan
University of Wyoming Geological Museum, Laramie, Wyoming

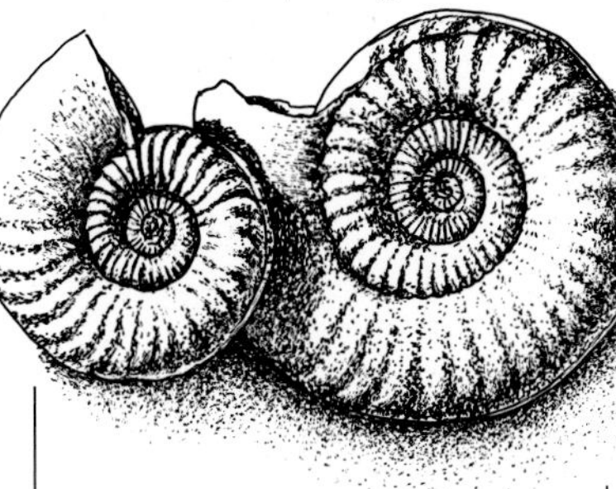

Utah Museum of Natural History, Salt Lake City, Utah

AUSTRALIA
National Museum of Victoria, Melbourne
Queensland Museum, Brisbane
South Australian Museum, Adelaide
The Age of Fishes Museum, Canowindra, NSW
The Australian Museum, Sydney
The Gold Museum, Ballarat
West Australian Museum, Perth

CANADA
Dinosaur Provincial Park, Patricia, Alberta
Fundy Geological Museum, Nova Scotia
Geological Museum, University of British Columbia, Vancouver
Musée de géologie de l'Université Laval, Quebec City, Quebec
Musée d'Histoire Naturelle, Miguasha, Quebec
National Museum of Natural Sciences, Ottawa, Ontario
Paleontological Museum and Research Institute, Drumheller, Alberta
Redpath Museum, McGill University Montreal, Quebec
Royal Ontario Museum, Toronto, Ontario
Royal Tyrrell Museum of Paleontology, Alberta, Canada

FRANCE
Harmas de J.H. Fabre, Séfignan-du-Comtat
Minerology Museum of the Paris School of Mines, Paris
National Museum of Natural History, Paris
Sorbonne Mineral Collection, Paris

GERMANY
Eichstätt Museum, Bavaria
Freiberg University of Mining and Technology, Freiberg
Institut für Paläontolgie, Bonn
Museum für Naturkunde der Humboldt-Universität, Berlin
Museum Hauff, Holzmaden
Mineralogische Staatssammlung München, Munich
Mineralogy Collection, Bayern
Senckenberg Museum, Frankfurt
State Museum for Mineralology and Geology, Dresden

ITALY
Museum of Geology and Paleontology, Florence
Museum of the Institute of Mineralology, Rome
Museum of Minerology, Bologna

NEW ZEALAND
Geology Museum of the University of Otago, Dunedin

RUSSIA
A.E. Fersman Mineralogical Museum, Moscow
Central Geological and Prospecting Museum, Leningrad
Museum of Earth Science, Moscow
Paleontological Institute, Moscow

SOUTH AFRICA
South African Museum, Cape Town

UNITED KINGDOM
British Museum (The Natural History Museum), South Kensington, London
The Geological Museum, London
The Hunterian Museum, Glasgow
The Royal Edinburgh Museum, Edinburgh

INDEX *and* GLOSSARY

In this combined index and glossary, bold page numbers indicate the main reference, and italics indicate illustrations and photographs.

C

G

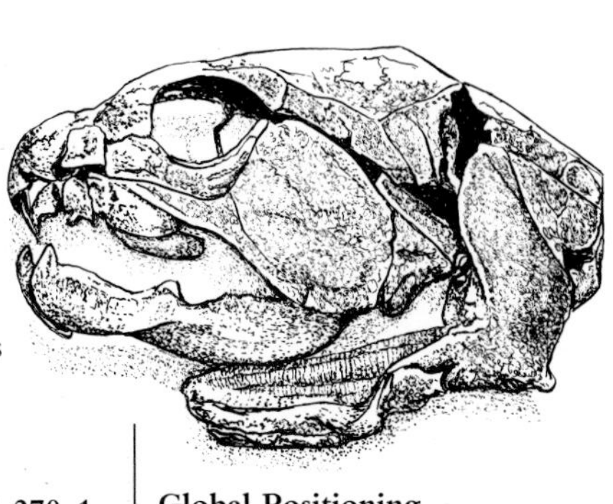

H

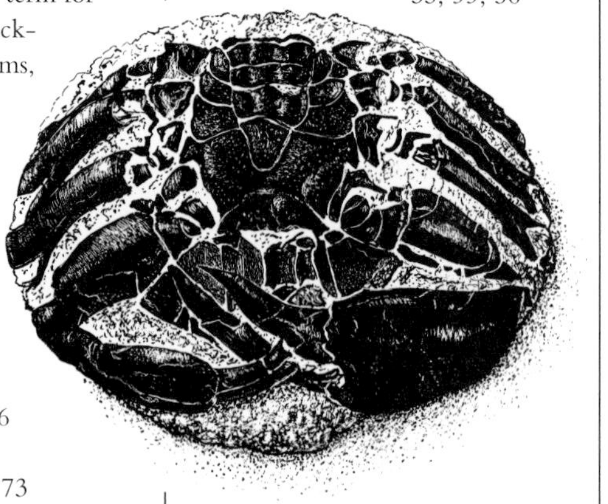

CONTRIBUTORS

Arthur B. Busbey III, B.S. and M.A. Geology, and Ph.D. Anatomy, is an Associate Professor of Geology at Texas Christian University, Austin, USA. He has published papers on fossil reptiles, especially crocodilians, and on satellite remote sensing and geographic information systems. He enjoys family camping trips and visiting museums.

Robert R. Coenraads, BA(Hons) Geology and Geophysics, M.Sc. Geophysics, Ph.D. Geology, works as a geological consultant, specializing in gemstone, gold, and base metal exploration. He lives in Australia and designs and leads geological-archeological field trips in the Americas and the Pacific region.

David Roots, B.Sc.(Hons) Geology, Ph.D. Marine Geophysics, is a research consultant at Macquarie University, Sydney, Australia. He specializes in structural geology, marine geophysics, and plate tectonic motor modeling. He leads groups on tours during which he explains the geological basis for various features of the scenery.

Paul M. A. Willis, BSc Geology/Zoology, PhD Vertebrate Palaeontology, is a consultant specializing in fossils and dinosaurs for primary school education. His thesis was on Australian fossil crocodiles. British-born and educated in Australia, he has traveled extensively throughout the world in the course of his studies.

CAPTIONS

Page 1: Ammonite fossil with preserved mother-of-pearl luster.
Page 2: Traces of malachite on Horseshoe Island, Antarctica.
Page 3: Detail of fossilized bone from Texas, USA.
Pages 4–5: Buildings carved into rock at Petra, in Jordan.
Pages 6–7: Fossil ferns from the Triassic, Karnische Alps, Germany.
Pages 8–9: Fossil starfish *Ophiperma egertoni.*
Pages 10–11: Precipitated limestone flows at Mammoth Hot Springs, Yellowstone National Park, Wyoming, USA.
Pages 12–13: Columnar basalt formations at Giant's Causeway, Northern Ireland, UK.
Pages 28–9: The Glasshouse Mountains in Queensland, Australia, 25 million-year-old plugs from a hot-spot volcano.
Pages 58–9: Buda limestone formation, Big Bend National Park, Texas, USA.
Pages 88–9: Agatized dinosaur bone from the Morrison Formation, Colorado.
Pages 114–15: A graptolite fossil from the lower Ordovician, Mount Hunneberg, Sweden.
Page 131 (inset bottom): Polished agate.
Pages 132–3: The opening to Leviathan Cave, a giant limestone cave in Nevada, USA.
Page 136–7: A polished slab of rhodochrosite.
Page 136 (inset top): Salt formations at Scammon's Lagoon, Baja, Mexico.
Page 136 (inset bottom): Obsidian detail.
Pages 184–5: Eroded limestone spires, Bryce Canyon, Utah, USA.
Page 185 (inset top): Eruption of Kimanura, Virunga National Park, Zaire.
Pages 202–3: Fossil site at Riversleigh, far north Queensland, Australia.
Pages 206–7: Kauri pine fossil from NSW, Australia.
Page 207 (inset top): Coral fossil.
Page 207 (inset bottom): Ammonite fossil, *Placenticeras* species, from Morocco.
Pages 238–239: Dinosaur footprints from Lower Jurassic, La Sal Mountains, Utah, USA.
Page 183 (inset top): Insect preserved in amber.
Page 183 (inset bottom): Cell structure in fossilized dinosaur bone.
Pages 272–3: Fossilized wood in Petrified Forest National Park, Arizona, USA.

ACKNOWLEDGEMENTS

The publishers wish to thank the following people for their assistance in the production of this book: Roger Bohn, Christopher Buykx, Melanie Corfield, Selena Hand, Robert Jones (Australian Museum, Paleontology), Margaret McPhee, Ross Pogson (Australian Museum, Mineralogy), and Dianne Regtop.

PICTURE AND ILLUSTRATION CREDITS

t = top, b = bottom, l = left, r = right, c = centre, i = inset
A = Auscape International; AA/ES = Animals Animals/Earth Scenes; AA&A = Ancient Art and Architecture Collection, London; AAP = Australian Associated Press Photo Library; AM = Australian Museum (photos by Maurice Ortega unless otherwise indicated); APL = Australian Picture Library; BCL = Bruce Coleman Limited, UK; Bettman = Bettman Archive; Bridgeman = Bridgeman Art Library, London; CSIRO = Commonwealth Scientific and Industrial Research Organisation, Australia; FLPA = Frank Lane Picture Agency; Granger = The Granger Collection, New York; IPPS = Icelandic Photo and Press Service; MCP = The Mazon Creek Project, Northeastern Illinois Univeristy; ME = Mary Evans Picture Library; MW = Mantis Wildlife; NGS = National Geographic Society; NHM = Natural History Museum, London; OSF = Oxford Scientific Films; PE = Planet Earth Pictures; PL = The Photo Library, Sydney; RGS = Royal Geographic Society, London; PR = Photo Researchers; SLNSW = State Library of New South Wales, Sydney SPL = Science Photo Library; TS = Tom Stack and Associates; VU = Visuals Unlimited; Werner = Werner Forman Archive

1 John Cancalosi/A 2 Ben Osbourne/OSF 3 GI Bernard/OSF 4-5 Nicola Sherriff/RGS 6-7 John Cancalosi/BCL 8-9 John Cancalosi/BCL 10-11 Jeff Foott/TS 12-13 Tom Till/A 14tl NHM; c Reg Morrison; b Jocelyn Burt/BCL 15tl Rich Buzzelli/TS; tr and b NHM 16tl Bob Buduwal, Peter Cook Collection/Australian Institute of Aboriginal and Torres Strait Islander Studies; tr Kristinn Ben/Mats Wibe Lund/Icelandic Photo; cr Geolinea; b ME 17t Sinclair Stammers/SPL/PL; cl Geolinea; cr John Cancalosi/A 18tl ME; tr Krafft/A; b Hulton Deutsch/PL 19t Francois Gohier/A; c ME/Explorer/A; b Bettman/APL 20t ME; bl Geoff Doré/BCL; br Explorer-Archives/A 21tl Paul Harris/RGS; tr ME; b Reg Morrison 22 Jay M Pasachoff/VU 23 Reg Morrison 24tl and background Geolinea; tr and b John Reader/SPL/PL 25t Tashi Tenzing; c Jane Burton/BCL; b NHM 26t NHM; bl J Fields/AM; br Rich Buzzelli/TS 27tl Kev Deacon/A; tr NHM; cr MR Phicton/BCL; b Jeff Foott/BCL 28-29 Reg Morrison 30l Reg Morrison; r George J Wilder/VU 31tl Granger; tc Alain Compost/BCL; tr Image Library, SLNSW, Sydney; bl Peter Ryan/SPL/PL; br NASA/GSFC/TS 32tl Dennis Harding/A; tr Scott Berner/VU; b Robin Smith/PL 33c Albert J Copley/VU; bc Gary Lewis/PL; br Fritz Prenzel/BCL 34 Dieter and Mary Plage/BCL 35t Keith Gunnar/BCL; c David Parker SPL/PL 36t NASA/Tsado/TS; b David Foster/VU 37tl Ferrero/Labat/A; tr Mats Wibe Lund/ IPPS; cl Bruce Davidson/OSF 38t NASA/ Tsado/Tom Stack/TS; bl Science VU/WHOI, D. Foster/VU 39tl Ferrero/Labat/A; l Bruce Davidson/OSF; r Mats Wibe Lund/IPPS 40 Restec, Japan/SPL/PL 41t David Matherly/VU; b David Bertsch/VU 42tl David Hardy/SPL/PL; c Werner Stoy/BCL; bl Mats Wibe Lund/ Bragi Gudmunds/IPPS; br David Roots 43tl Dieter and Mary Plage/BCL; tr Jay M Pasachoff/VU; c Mats Wibe Lund/Gardar Pálsson; b Glenn Oliver/VU 44t ME; c Austral; b AAP 45tl AAP; tr stf/str/AAP; c AAP; b NHM 46t Jane Burton/ BCL; bl Tony Waltham/Trent University; br Reg Morrison 47t Tony Waltham/Trent University; bc Jens Rydell/BCL; br Museum of Victoria, Mineralogy 48tl AJ Copley/VU; tc David B Fleetham/TS; tr Kevin Schafer/TS; b Fritz Prenzel/BCL 49t and c M Krafft/A; br Brian P Foss/VU 50tr Gary Milburn/TS; cl AJ Copley/VU; cr TSA/TS; b Rod Planck/TS 51t Jules Cowan/ BCL; b Oliver Benn/PL 52t NHM; c Jean-Paul Ferrero/A; b Reg Morrison/A 53t Hans Reinhard/BCL; c Jane Burton/BCL; b Reg Morrison 54t Andrew Syred/PL; c Jeff Foott Productions/BCL

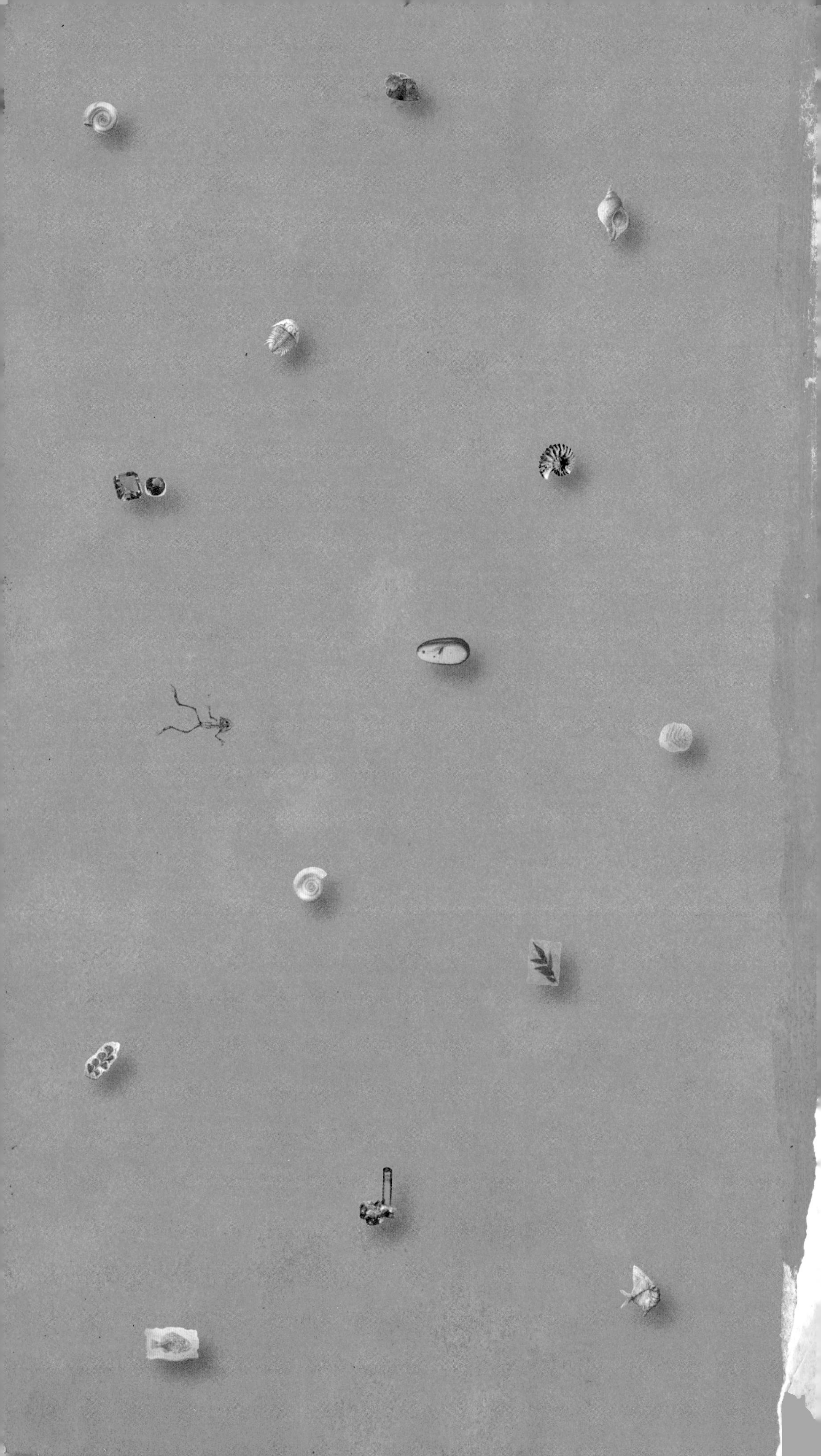